Methods in Molecular Biology

Series Editor
John M. Walker
School of Life and Medical Sciences
University of Hertfordshire
Hatfield, Hertfordshire, UK

For further volumes:
http://www.springer.com/series/7651

For over 35 years, biological scientists have come to rely on the research protocols and methodologies in the critically acclaimed Methods in Molecular Biology series. The series was the first to introduce the step-by-step protocols approach that has become the standard in all biomedical protocol publishing. Each protocol is provided in readily-reproducible step-by step fashion, opening with an introductory overview, a list of the materials and reagents needed to complete the experiment, and followed by a detailed procedure that is supported with a helpful notes section offering tips and tricks of the trade as well as troubleshooting advice. These hallmark features were introduced by series editor Dr. John Walker and constitute the key ingredient in each and every volume of the Methods in Molecular Biology series. Tested and trusted, comprehensive and reliable, all protocols from the series are indexed in PubMed.

Cancer Cell Biology

Methods and Protocols

Edited by

Sherri L. Christian

Department of Biochemistry, Memorial University of Newfoundland, St. John's, NL, Canada

Humana Press

Editor
Sherri L. Christian
Department of Biochemistry
Memorial University of Newfoundland
St. John's, NL, Canada

ISSN 1064-3745 ISSN 1940-6029 (electronic)
Methods in Molecular Biology
ISBN 978-1-0716-2375-6 ISBN 978-1-0716-2376-3 (eBook)
https://doi.org/10.1007/978-1-0716-2376-3

This Humana imprint is published by the registered company Springer Science+Business Media, LLC, part of Springer Nature.
The registered company address is: 1 New York Plaza, New York, NY 10004, U.S.A.

Preface

This volume provides detailed methods for the dissection of mechanisms underlying cancer cell biology. This book begins with fundamental techniques and considerations for culturing cancer cell lines and then more advanced techniques to study patient-derived tissue using xenografts and cryopreservation of tumor cells. The book then describes methods to analyze the co-culture of breast cancer cells with adipocytes in two-dimensional and three-dimensional systems. Next are a series of chapters that allows the researcher to analyze cancer cell physiology including cell proliferation, protein secretion by ELISA, flow cytometry-based detection of protein levels by conventional fluorescence and mass spectrometry, and multi-parametric immunofluorescence analysis of tumor tissue. Also included are analysis of protein expression by western blot, analysis of surface protein levels and recycling by biotinylation assay, and proteomics analysis by liquid chromatography mass spectrometry. This is followed by chapters describing a sophisticated analysis of cellular metabolism using the Seahorse assay, specific analysis of cancer cell mutations by CRISPR/Cas9, single cell RNA sequencing, along with a chapter on validation by reverse transcriptase quantitative PCR, and then bioinformatics analysis of RNA-seq data. Lastly, chapters describing the isolation and characterization of extracellular vesicles follow. This volume follows the highly successful *Methods in Molecular Biology* format, with each chapter including an introduction to their respective techniques; lists of the necessary materials and reagents; step-by-step, readily reproducible laboratory protocols; and tips on troubleshooting and avoiding known pitfalls.

Cancer Cell Biology aims to provide both the burgeoning and experienced cancer researcher with a comprehensive set of tools for the analysis of cancer cell biology in the lab.

St. John's, NL, Canada ***Sherri L. Christian***

Preface

Contents

Contributors

MARYA AHMED • *Department of Chemistry, University of Prince Edward Island, Charlottetown, PE, Canada; Faculty of Sustainable Design Engineering, University of Prince Edward Island, Charlottetown, PE, Canada*

EMMANUEL C. ASANTE • *Department of Biochemistry, Memorial University of Newfoundland, St. John's, NL, Canada*

SOOFIYAH P. AYAANI • *Warwick Medical School, The University of Warwick, Coventry, UK*

DAVID A. BARNES • *Department of Biochemistry, Memorial University of Newfoundland, St. John's, NL, Canada*

MAUNISH BARVALIA • *Life Science Institute, Department of Microbiology and Immunology, University of British Columbia, Vancouver, BC, Canada*

ERIN M. BELL • *Department of Cellular and Physiological Sciences, University of British Columbia, Vancouver, BC, Canada*

MARK D. BERRY • *Department of Biochemistry, Memorial University of Newfoundland, St. John's, NL, Canada*

JEANETTE E. BOUDREAU • *Department of Microbiology and Immunology, Dalhousie University, Halifax, NS, Canada; Department of Pathology, Dalhousie University, Halifax, NS, Canada*

FRANK BUCHHOLZ • *Medical Faculty and University Hospital Carl Gustav Carus, UCC Section Medical Systems Biology, TU Dresden, Dresden, Germany; National Center for Tumor Diseases (NCT) Dresden and German Cancer Research Center (DKFZ), Dresden, Germany; German Cancer Consortium (DKTK), Dresden, Germany*

ALBERT CABALLERO-SOLARES • *Department of Ocean Sciences, Memorial University of Newfoundland, St. John's, NL, Canada*

JENNA K. CAINES • *Department of Biochemistry, Memorial University of Newfoundland, St. John's, NL, Canada*

PIERRE CAUCHY • *Universitätskilinkum Freiburg, Freiburg, Germany; Max Planck Institute of Immunobiology and Epigenetics, Freiburg, Germany*

JENNIFER A. CHAN • *Department of Pathology & Laboratory Medicine, University of Calgary, Calgary, AB, Canada; Arnie Charbonneau Cancer Institute, University of Calgary, Calgary, AB, Canada*

MARK CHRISTIAN • *School of Science and Technology, Nottingham Trent University, Nottingham, UK*

SHERRI L. CHRISTIAN • *Department of Biochemistry, Memorial University of Newfoundland, St. John's, NL, Canada*

KRYSTA M. COYLE • *Department of Molecular Biology and Biochemistry, Faculty of Sciences, Simon Fraser University, Vancouver, BC, Canada*

PAMELA M. DEAN • *Department of Cellular and Physiological Sciences, University of British Columbia, Vancouver, BC, Canada*

SHEILA DROVER • *Immunology and Infectious Diseases Group, Division of BioMedical Sciences, Memorial University of Newfoundland and Labrador, St. John's, NL, Canada*

WASUNDARA FERNANDO • *Department of Pathology, Faculty of Medicine, Dalhousie University, Halifax, NS, Canada*

SCOTT D. FINDLAY • *Department of Biology, Massachusetts Institute of Technology, Cambridge, MA, USA; Department of Oncology, University of Alberta, Edmonton, AB, Canada*
TATE O. GOODMAN • *Department of Cellular and Physiological Sciences, University of British Columbia, Vancouver, BC, Canada*
MARCIA L. GRAVES • *Department of Microbiology and Immunology, University of British Columbia, Vancouver, BC, Canada*
JENNIFER R. HALL • *Aquatic Research Cluster, CREAIT Network, Ocean Sciences Centre, Memorial University of Newfoundland, St. John's, NL, Canada*
KENNETH W. HARDER • *Life Science Institute, Department of Microbiology and Immunology, University of British Columbia, Vancouver, BC, Canada*
JINHONG KIM • *Department of Biochemistry and Molecular Biology, Faculty of Medicine, Dalhousie University, Halifax, NS, Canada*
MARINA B. LEMIESZEK • *Department of Obstetrics and Gynecology, University of Alberta, Edmonton, AB, Canada*
STEPHEN M. LEWIS • *Atlantic Cancer Research Institute, Hôtel-Dieu Pavilion, Moncton, NB, Canada; Beatrice Hunter Cancer Research Institute, Halifax, NS, Canada; Department of Chemistry and Biochemistry, Université de Moncton, Moncton, NB, Canada*
MODELINE N. LONGJOHN • *Department of Biochemistry, Memorial University of Newfoundland, St. John's, NL, Canada*
PAOLA MARCATO • *Department of Pathology, Faculty of Medicine, Dalhousie University, Halifax, NS, Canada; Department of Microbiology and Immunology, Faculty of Medicine, Dalhousie University, Halifax, NS, Canada*
PAOLA A. MARIGNANI • *Department of Biochemistry and Molecular Biology, Faculty of Medicine, Dalhousie University, Halifax, NS, Canada; Department of Pathology, Faculty of Medicine, Dalhousie University, Halifax, NS, Canada*
JOVAN MIRCETIC • *Medical Faculty and University Hospital Carl Gustav Carus, UCC Section Medical Systems Biology, TU Dresden, Dresden, Germany*
LOIS M. MULLIGAN • *Division of Cancer Biology and Genetics, Cancer Research Institute, Queen's University, Kingston, ON, Canada*
J. PATRICK MURPHY • *Department of Biology, University of Prince Edward Island, Charlottetown, PE, Canada*
SARAH NERSESIAN • *Department of Microbiology and Immunology, Dalhousie University, Halifax, NS, Canada*
VANESSA NOEL • *Department of Biochemistry, Memorial University of Newfoundland, St. John's, NL, Canada*
RODNEY J. OUELLETTE • *Atlantic Cancer Research Institute, Hôtel-Dieu Pavilion, Moncton, NB, Canada; Beatrice Hunter Cancer Research Institute, Halifax, NS, Canada; Department of Chemistry and Biochemistry, Université de Moncton, Moncton, NB, Canada; Centre hospitalier universitaire Dr. GL Dumont, Réseau de santé Vitalité, Moncton, NB, Canada*
NIKITHA K. PALLEGAR • *Department of Biochemistry, Memorial University of Newfoundland, St. John's, NL, Canada*
POLINA V. PAVLOVICH • *Moscow Institute of Physics and Technology, Dolgoprudny, Russia*
HONG-DIEN PHAN • *Department of Biochemistry, Memorial University of Newfoundland, St. John's, NL, Canada*
DEVANAND M. PINTO • *National Research Council, Halifax, NS, Canada*

EDUARDO REYES-ALVAREZ • *Division of Cancer Biology and Genetics, Cancer Research Institute, Queen's University, Kingston, ON, Canada*
MATTHEW L. RISE • *Department of Ocean Sciences, Memorial University of Newfoundland, St. John's, NL, Canada*
CALVIN D. ROSKELLEY • *Department of Cellular and Physiological Sciences, University of British Columbia, Vancouver, BC, Canada*
JEREMY W. ROY • *Atlantic Cancer Research Institute, Hôtel-Dieu Pavilion, Moncton, NB, Canada; Beatrice Hunter Cancer Research Institute, Halifax, NS, Canada*
SHADY SAYED • *Medical Faculty and University Hospital Carl Gustav Carus, UCC Section Medical Systems Biology, TU Dresden, Dresden, Germany*
OLGA ALEXANDRA SIDOROVA • *Medical Faculty and University Hospital Carl Gustav Carus, UCC Section Medical Systems Biology, TU Dresden, Dresden, Germany*
GABRIELLE M. SIEGERS • *Department of Oncology, University of Alberta, Edmonton, AB, Canada*
MUKHAYYO SULTONOVA • *Department of Biology, University of Prince Edward Island, Charlottetown, PE, Canada*
DURAN SÜRÜN • *Medical Faculty and University Hospital Carl Gustav Carus, UCC Section Medical Systems Biology, TU Dresden, Dresden, Germany*
MAHDIS SADAT TABATABAEI • *Department of Chemistry, University of Prince Edward Island, Charlottetown, PE, Canada*
CATHERINE A. TAYLOR • *Atlantic Cancer Research Institute, Hôtel-Dieu Pavilion, Moncton, NB, Canada*
SUZANNE M. VERCAUTEREN • *BC Children's Hospital Research Institute, Vancouver, BC, Canada; Division of Hematopathology, Department of Pathology and Laboratory Medicine, BC Children's Hospital, Vancouver, BC, Canada*
ALICIA M. VILORIA-PETIT • *Department of Biomedical Sciences, Ontario Veterinary College, University of Guelph, Guelph, ON, Canada*
MARTIN WABITSCH • *Department of Paediatric Endocrinology and Diabetes, ULM University Hospital, Ulm, Germany*
TIMOTHY J. WALKER • *Division of Cancer Biology and Genetics, Cancer Research Institute, Queen's University, Kingston, ON, Canada*
LOUISA L. WIEDE • *Immunology and Infectious Diseases Group, Division of BioMedical Sciences, Memorial University of Newfoundland and Labrador, St. John's, NL, Canada*
XI XUE • *Department of Ocean Sciences, Memorial University of Newfoundland, St. John's, NL, Canada*

Chapter 1

Overview of the Maintenance of Authentic Cancer Cell Lines

Sherri L. Christian

Abstract

The proper maintenance of cancer cell lines is critical to maintain the integrity of any experiment. Proper maintenance includes handling of cell lines and maintaining records on usage and confirmation of identity. Cross-contamination of cancer cell lines has been well documented by the ATCC and must be avoided at all costs. In addition, contamination by *Mycoplasma* sp. is a well-known issue, which, if left unchecked, can affect all data collected as this contamination can affect cell line metabolism and has uncharacterized effects. Considerations and specific recommendations for the proper maintenance of cancer cell lines are presented here.

Key words Cancer cell lines, Cross-contamination, Cell line identification, Mycoplasm

1 Introduction

Studies of cancer rely heavily on cancer cell lines as these can be propagated indefinitely and can be genetically manipulated relatively easily. In addition, using the same cell line allows different studies to be directly compared. However, each type or sub-type of cancer can be different in a myriad of ways, and cell lines do not fully recapitulate the intra- or inter-tumor heterogeneity seen in patient tumors. Nevertheless, the use of cancer cell lines to unravel basic mechanisms of oncogenesis is fundamental to advancing our understanding of cancer. Thus, it is critically important that cell lines are maintained in such a way that their integrity is uncompromised. This includes confirming cell line identity, preventing cross-contamination, and preventing and tracking mycoplasma contamination. Provided is a sample template for researchers to track the relevant information needed for proper cell line maintenance (*see* Fig. 1). The following is a brief discussion on the considerations for maintaining cancer cell lines. For specific protocols on culturing cancer cell lines, please *see* Chapters 2 and 3 of this book.

Sherri L. Christian (ed.), *Cancer Cell Biology: Methods and Protocols*,
Methods in Molecular Biology, vol. 2508, https://doi.org/10.1007/978-1-0716-2376-3_1,

Cell line name:___________________________

Source	**Company/lab:** **Catalogue number if commercial product:** **Date received**
Mycoplasma	**Status:** **Date tested:**
Cell identify confirmed	**Method:** **Date:**
Growth Characteristics	**Growth Rate:** **Appearance (attach photo if available):**
Growth Medium	
Freezedown conditions	**Media:** **Cell concentration:**
Notes	

Fig. 1 A sample template for tracking cell line information

2 Biosafety

All handling of cancer cell lines must be performed in a biosafety cabinet of level IIA or higher in order to protect both the user and the cells. It is critically important to know the biohazard level of the cell line in use in order to take the right level of precaution. Many cancer cell lines are known to be infected by human retroviruses, which can possibly infect the user if mishandled. Users should consult with the cell line repository (e.g., ATCC or DSMZ) or original source for the biohazard level of each cell line being used and handle the cancer cell line with the appropriate degree of care. Level 1 cell lines are very unlikely to cause harm to the user, and so, contaminated plastic consumables can be disposed of in the open air. In contrast, level 2 cell lines have the potential to cause harm to the user, and so, contaminated media or consumables must be decontaminated in the biosafety cabinet or kept in a closed container until decontamination is performed (i.e., by autoclave). All users must be trained in the proper operation of a biosafety cabinet by an experienced user prior to working with any cancer cell line.

3 Cell Line Authentication

It has well been documented that up to 30% of cell lines are contaminated with HeLa cells or other cell lines or are otherwise misidentified [1, 2]. This is a serious issue that calls into question any study that has not thoroughly authenticated the cell lines being used and is a likely source of irreproducibility of data. Moreover, many journals are now requiring cell line authentication within 5 years of the cell lines being used in order to prevent further corruption of the literature. This can be done by purchasing a cell line from a reputable repository, such as ATCC, which provides a certificate of authenticity, or by authenticating the cell line in question genetically and morphologically.

Genetic authentication is generally done via short tandem repeat (STR) profiling. This method analyzes polymorphic markers in the genome by PCR to generate a unique fingerprint. The fingerprint pattern is then compared to a database of known STR profiles. Many sequencing facilities, along with ATCC, offer STR profiling services, which can either require isolation of DNA by the source laboratory or a sample of the cells on filter paper. The specific STR profile and source material differ between service providers so the reader should consult their local provider directly for instruction.

Further authentication by analysis of morphology and growth curve should be performed regularly by the lab using the cancer cell line. Cancer cell lines have distinctive shapes at both high and low densities, and a change in morphology of the cells is often the first indication that cross-contamination has occurred. Similarly, each cancer cell line has a particular rate of growth, which, when combined with morphological analysis, can serve as a benchmark for a particular cell line during the culture period.

4 Avoiding Cross-Contamination

It is a common practice in many labs to culture multiple cancer cell lines at one time. This is an important practice by which labs take advantage of economies of scale. As well, culturing multiple cancer cell lines is a necessity when comparing different cell lines to one another or a genetically manipulated cell line to the parental cell line of origin. However, any time that two or more cell lines are in the biosafety cabinet at one time, there is a possibility for cross-contamination. This is because aerosols are easily generated during pipetting. These aerosols can persist in the air circulation for many minutes. Aerosols containing cancer cells can easily drop into an open culture dish, where those that are particularly robust with a short cell cycle can take over the culture given enough time.

The following strategies will significantly reduce the chances of cross-contamination:

(a) Aliquot culturing media in 50-mL aliquots in sterile tubes or bottles. Dedicate each one to an individual cell line.

(b) Have only one cancer cell line in the biosafety cabinet at one time.

(c) Even for biohazard level 1 cancer cell lines, keep all used plasticware in a covered container.

(d) Allow the biosafety cabinet to run empty for 5 min between cell lines to remove any aerosols that may have been generated.

(e) Make culturing media after running the biosafety cabinet empty for 10 min and/or treatment with UV light for 30 min and prior to manipulating any cells.

5 Genetic Drift

Genetic drift occurs when a cancer cell line is grown for too long in culture, allowing it to accumulate mutations that result in it being demonstrably different from the cell of origin (e.g., [3]). This can often be observed at the genetic or morphological level. However, as this occurs relatively slowly over time, it is often not apparent to the user who maintains the cells unless the cells are directly compared to the original cancer cell line. The solution to this issue is relatively straightforward: do not culture cancer cell lines for long periods of time.

Specifically, the following guidelines will help prevent genetic drift from creating new sub-lines of cancer cells.

(a) When a cancer cell line is first introduced to the lab, it should be immediately expanded and 10–15 vials of the lowest passage number possible stored in liquid nitrogen.
(b) Adherent cells should not be passaged more than 15 times and less if the cancer cell line demonstrates an unstable morphology, karyotype, or genotype.
(c) Suspension cells should not be continuously grown in culture more than 2–3 months and less if the cancer cell line demonstrates an unstable morphology, karyotype, or genotype.
(d) The lowest possible passage cell line should be used for experiments.
(e) Low passage stocks should be replenished from the lowest passage possible.

6 Contamination with Mycoplasmas

Mycoplasma are a species of bacteria that infect both murine and human cancer cell lines. They lack a cell wall and can pass through 0.45-μm filters due to their small size of 0.3–0.8 μm and flexibility [4]. *Mycoplasma* can grow both intra- and extracellularly and are resistant to most commonly used antibiotics. Furthermore, intracellular growth protects mycoplasmas from being targeted by mycoplasmacidal compounds. Although they are slow growing, *Mycoplasma*-contaminated cultures can reach 10^6–10^8 cells per ml [4]. While there are a relatively high number of different species, it has been reported that 90–95% of contaminations were due to one of the six species, namely, *M. orale*, *M. hyorhinis*, *M. arginini*, *M. fermentans*, *M. hominis*, or the related *Acholeplasma laidlawii* bacterium [4].

Contamination by mycoplasmas changes a myriad of characteristics of the infected cell line, from metabolism to gene expression [5–7]. It should go without saying that alterations to the fundamental characteristics of cancer cell line will impact all of the data acquired through its use. Therefore, vigilance against contamination is essential. However, as mycoplasmas are not visible under the light microscope and the effects of contamination can be relatively subtle, these contaminations can go undetected unless specific screening is performed.

Mycoplasma contamination is difficult to avoid entirely as humans are the natural host for three of the most common species, and others can be carried by a wide range of hosts, including bovine, swine, birds, and insects [4]. Aerosols generated during the manipulation of cancer cell lines and potential contamination from fetal bovine serum, commonly used for culture cells, are the likely source for most contaminations. Some labs promote culturing cells in the absence of common antibiotics, penicillin and streptomycin, as this will allow detection of mycoplasma contamination more readily. The more likely outcome, however, is that cultures are contaminated with fast-growing bacteria before they are contaminated by mycoplasmas, thereby keeping the culture free of mycoplasmas. In practice, culturing cancer cell lines in the absence of antibiotics tends not to be practical if cells are cultured for over 1 week.

Molecular detection, and elimination, if needed, of mycobacterium is relatively easy and inexpensive. However, regular and ongoing inspection of cultures is necessary to maintain cell lines mycoplasma free. The reader is directed to the in-depth report of *Mycoplasma* contamination by Drexler and Uphoff [4] for detailed recommendations on detection and elimination strategies.

(a) Cell lines should be tested for *Mycoplasma* contamination approximately 1 week after a new culture is started and before freezing down cells.

(b) Testing can be done using a number of methods. The most prominent methods are PCR-based analysis and enzymatic analysis. Both of these are easily and widely available commercially.

(c) If there are mycoplasma-negative aliquots of a cell line, then any mycoplasma-positive cultures should be discarded.

(d) It is possible to "cure" cultures using a choice of antimycotic, if there are no mycoplasma-free alternatives. Follow the manufacturer's instructions for concentration and duration of treatment.

Acknowledgments

SLC is a senior scientist with the Beatrice Hunter Cancer Research Institute. This work was funded by the Natural Sciences and Engineering Research Council of Canada and the Janeway Children's Hospital Foundation.

References

1. Capes-Davis A, Theodosopoulos G, Atkin I et al (2010) Check your cultures! A list of cross-contaminated or misidentified cell lines. Int J Cancer 127:1–8
2. Chatterjee R (2007) Cases of mistaken identity. Science 315:928–931
3. Esquenet M, Swinnen JV, Heyns W et al (1997) LNCaP prostatic adenocarcinoma cells derived from low and high passage numbers display divergent responses not only to androgens but also to retinoids. J Steroid Biochem Mol Biol 62: 391–399

4. Drexler HG, Uphoff CC (2002) Mycoplasma contamination of cell cultures: incidence, sources, effects, detection, elimination, prevention. Cytotechnology 39:75–90
5. Yu T, Wang Y, Zhang H et al (2016) Metabolomics reveals mycoplasma contamination interferes with the metabolism of PANC-1 cells. Anal Bioanal Chem 408:4267–4273
6. Feng N, Huang X, Jia Y (2019) Mycoplasma contamination affects cell characteristics and decreases the sensitivity of BV2 microglia to LPS stimulation. Cytotechnology 71:623–634
7. Kagemann G, Henrich B, Kuhn M et al (2005) Impact of mycoplasma hyorhinis infection on L-arginine metabolism: differential regulation of the human and murine iNOS gene. Biol Chem 386:1055–1063

Chapter 2

Culturing Suspension Cancer Cell Lines

Modeline N. Longjohn, Hong-Dien Phan, and Sherri L. Christian

Abstract

Suspension cell lines grow free-floating in the cell culture media without any attachment to the culture plate/vessel. Suspension cells typically mimic cells that exist in the circulation of multicellular animals such as mouse and humans. Generally, cell lines derived from the blood such as lymphocytes, megakaryocyte, and neutrophils grow in suspension. These cell lines can be used for experimental studies to understand the biology/biochemistry of cancer cells. In this chapter, procedures for working with suspension cell lines are provided, including protocols for thawing, culturing, and cryopreserving cancer cell lines. Importantly, this chapter demonstrates the best practices required to work with suspension cell lines, to minimize the risk of contaminations from adventitious microorganisms or from other cell lines.

Key words Cell culture, Cryopreservation, Culture media, Cell line, Cancer, Suspension cells

1 Introduction

Cell culture is an important laboratory technique used in cellular and molecular biology, providing the best platform for the growth of eukaryotic or prokaryotic cells in physiological conditions [1]. In general, suspension cells can be grown in culture in a straightforward way, with specific modifications, as required for the specific cell line type. Suspension cells in culture grow either as fluid-suspended single cells or as clusters (clump) of cells. Viable cells appear round and refractile, whereas dead cells appear smaller and darker [2]. Continued culturing of suspension cells depends on availability of space and nutrients. Hence, suspension cells need to be subcultured regularly, to provide a fresh source of nutrients. These nutrients are supplied by culturing cells in a balanced salt solution with a variable number of additions in the form of amino acids, vitamins, trace minerals, a pH indicator, serum, and if desired, antibiotics [3, 4]. Though antibiotic supplementation prevents contamination, they have been linked to changing gene expression and regulation; thus, their use is recommended with caution (in small amounts).

Sherri L. Christian (ed.), *Cancer Cell Biology: Methods and Protocols*,
Methods in Molecular Biology, vol. 2508, https://doi.org/10.1007/978-1-0716-2376-3_2,

Examples of media used for suspension cells include Roswell Park Memorial Institute 1640 (RPMI-1640), Minimal Essential Media (MEM), Dulbecco's Modified Eagle Medium (DMEM), and F12. These media are generally supplemented with additives include animal sera, (fetal bovine; FBS or fetal calf serum; FCS), pyruvate, glucose, and glutamate [5, 6]. The aim of the specific cell culture media and its supplementation is to mimic the in vivo environment, as much as possible, enough to enable the in vitro growth of these cancer cells.

To monitor the health of culturing cells, the media in the culture plate/vessel should be examined for microscopic evidence of microbial and fungal contaminations and changes in pH of media. A change in media color (determined by eye) from pink to yellow indicates a drop in pH (from neutral to acidic due to the release of metabolites by cells and/or bacterial contamination). Furthermore, a change in media consistency from clear to cloudy/turbid is usually indicative of too many cells (high to very high cell density) relative to available media. Cloudy/turbid media could also be due to microbial contamination. Another metric for monitoring health of cultures is cell morphology (shape and relative size) determined by regular examination under a light microscope (*see also* Chapter 1). The morphology of cancer cells should be regularly and carefully checked, as different cancer suspension cells typically have specific shape and consistency (*see* Fig. 1). Overall, microscopic evidence of cells of significantly different morphology from the typical culturing cells may be indicative of fungal or cell line cross contamination.

To ensure that healthy cells remain healthy and viable, cells need to be fed with fresh media. This can be done by simply changing/replacing media or by passaging/subculturing (diluting a small volume of cells in fresh media in a new culture plate/vessel), as required. The time between feeding and subculturing of cells varies between cell types and depends on cell density, growth/proliferation rate, and culture conditions. In this chapter, we present basic protocols and highlight when different requirements for a given cell line may be needed.

2 Materials

2.1 Thawing and Reviving Frozen Suspension Cancer Cell Lines

1. Cryovial containing frozen cells.
2. Cell culture media: 87.9% Roswell Park Memorial Institute (RPMI) 1640, 5–10% FBS (*see* **Note 1**), 1% penicillin/streptomycin (*see* **Note 2**), 1% 2-mercaptoethanol, 0.1% sodium pyruvate, pre-warmed in water bath set to 37 °C (*see* **Note 3**).
3. Disposable, sterile centrifuge or conical tubes.
4. 70% ethanol.

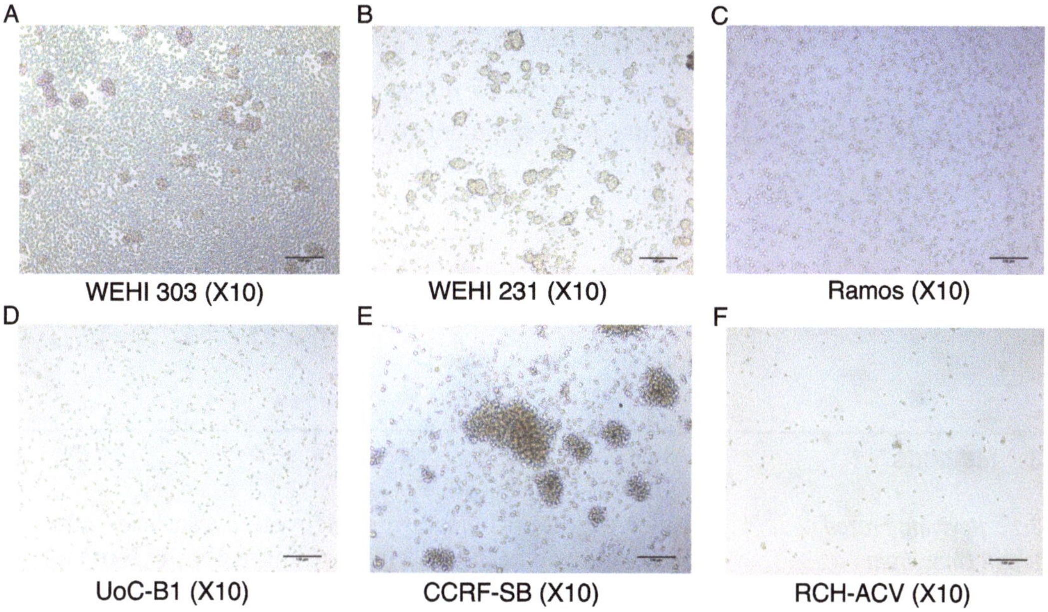

Fig. 1 Examples of different suspension cancer cell lines with different morphology after 24 h of culture. (**a**) WEHI 303 murine lymphoma cell line; (**b**) WEHI 231 murine lymphoma cell line; (**c**) Ramos, a human Burkitt's lymphoma cell line; (**d**) CCRF-SB; (**e**) UoC-B1; and (**f**) RCH-ACV, pediatric B cell acute lymphoblastic leukemia cell lines. X10 indicates objective used for magnification. Scale bar = 150 μm

5. Tissue-culture-treated dishes, plates, or flasks.
6. Pipette tips and pipettes.
7. Pipette aid.
8. Low-lint wipes.
9. 37 °C incubator with humidified atmosphere of 5% CO_2.
10. Biosafety cabinet, level II or greater.

2.2 Subculturing of Suspension Cell Lines

1. Culture vessels containing suspension cells.
2. Cell culture media, pre-warmed to 37 °C.
3. 37 °C incubator with humidified atmosphere of 5% CO_2.
4. Biosafety cabinet, level II or greater.
5. Reagents and equipment to determine cell viability and total cell counts (trypan blue and hemocytometer or cell counter).
6. Tissue culture vessels (round/flat bottom dishes, plates, or flasks).

2.3 Cryopreservation of Suspension Cell Lines

1. Culture vessels containing cultured cells in log phase of growth.
2. Cell culture media, pre-warmed to 37 °C.

3. Cryopreservation/freezing media: 5% dimethyl sulfoxide (DMSO) in 95% complete growth media (*see* **Note 4**).
4. Disposable, sterile 15-mL or 50-mL conical tubes.
5. Reagents and equipment to determine cell viability and total cell counts (e.g., hemocytometer, cell counter, and trypan blue).
6. Sterile cryogenic storage vials (i.e., cryovials).
7. Controlled rate freezing apparatus or isopropanol chamber (e.g., Nalgene's Mr. Frosty).
8. Liquid nitrogen storage Dewar.

3 Methods

3.1 Reviving Frozen B Cell Lines from Liquid Nitrogen Storage

1. Sterilize the hood by spraying the working surface with 70% ethanol and wiping down with low-lint wipes (*see* **Note 5**).
2. Dispense 20 mL of the appropriate pre-warmed media into a 100-mm culture plate using a 10-mL disposable pipette and pipetting aid.
3. Remove the cryovial/ampule from the liquid nitrogen storage. Spray the cryovial with 70% ethanol and wipe with low-lint wipes, and carefully loosen and unseal the cryovial to allow any liquid nitrogen to escape (*see* **Note 6**).
4. Reseal cryovial and thaw frozen cells rapidly (<1 min) in a 37 °C water bath (*see* **Note** 7).
5. Spray the ampule with 70% ethanol and wipe it down before bringing it into the biosafety cabinet.
6. Transfer the entire volume of the thawed cell suspension from the ampule into the 100-mm culture plates. Add 1 mL of media from the 100-mm plate to the ampule and rinse back into the culture plate. Rinse at least twice, to ensure the recovery of as many remaining cells as possible (*see* **Notes 8** and **9**).
7. Examine the cells under a microscope to evaluate cell morphology. Place 100-mm culture plates in a 37 °C incubator, allowing the cells to recover from the freeze–thaw process for 16–24 h.
8. Transfer the suspension cells into a sterile 50-mL conical tube.
9. Centrifuge cells at 500 × *g* for 5 min to pellet cells.
10. Discard the supernatant (old media) by aspiration, resuspend the pellet in 10-mL fresh media, and transfer the cell suspension to a fresh 100-mm culture plate.

3.2 Subculturing of Suspension Cell Lines Grown in 100-mm Culture Plates

1. Sterilize the hood by wiping it down with 70% ethanol.
2. Wipe down the surface of materials (*see* **Note 5**) for culturing, and bring the sterile materials into the biosafety cabinet, including media bottles, pipettes, pipette aid, and culture plates.
3. Remove culture plate containing cells to be subcultured from the incubator and examine the morphology and confluence of the culture under a microscope. Look for any signs of micro-organismal contaminations (*see* **Note 10**) to ensure the absence of contaminants in the culture.
4. To dislodge the cells from the culture plate surface, hold the culture plate at a 45° angle. Using a 10-mL pipette tip attached to a pipette aid, collect 8 mL of the culture media and rinse the plate surface, flushing the surface toward the lower side of the plate. Repeat a few times, until the growing surface becomes clear (*see* **Note 9**).
5. Using a 10-mL pipette tip attached to a pipette aid, transfer cells in suspension into a 15-mL conical tube and pellet cells by centrifuging at $500 \times g$ for 5 min.
6. After centrifugation, discard supernatant by aspiration and resuspend pellet in at least 1-mL of cell culture media.
7. Ascertain cell viability and cell density using a hemocytometer or an automatic particle counter, in order to determine whether the cells are ready to be passaged, based on the characteristics of the culture including cell density and proliferation rate (*see* **Note 11**).
8. If using a hemocytometer, mix cell suspension thoroughly by pipetting up and down. Add 10 μL of cells to 10 μL of trypan blue and pipette 10 μL of mixture underneath a coverslip placed over the containing chamber of a hemocytometer. Count all cells that do not stain for trypan blue in quadrants (Q) 1–4 and determine cell density (*see* **Note 12** and Fig. 2).
9. Subculture cells if the original cell density is between 5×10^6 and 1×10^7 cells/mL. Dilute the cell suspension accordingly using a cell culture media to a final seeding concentration of at least 1×10^6 cells/mL (at least five-fold dilution).
10. For routine passaging and maintenance, add an appropriate volume of cell suspension to a pre-determined volume of medium in a new culture plate to achieve the desired seeding cell density (*see* **Note 13**).
11. For culturing cells in multiple plates, transfer all the cell suspension into a tube of the appropriate volume. Dilute the cell suspension to the desired cell density with an appropriate volume of the medium. Distribute equal volumes of the diluted

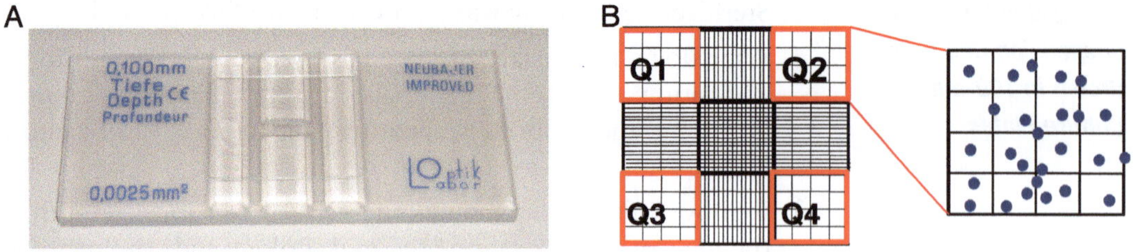

Fig. 2 Determining cell density manually, using a hemocytometer. (**a**) Neubauer improved hemocytometer with a coverslip placed over the top. Reproduced from Wikimedia commons, 2006, allowed under the Creative Commons CC0 copyright License. (**b**) Illustration of the laser etched lines on the surface of the counting area of the hemocytometer, with a schematic of cells in quadrant 2 (Q2)

cell suspension to the new plates to minimize the variation in cell density between plates.

12. Label the lid of the plate with the operator initials, date, split ratio/seeding cell density, cell line identifier, and passage number.
13. Place culture dish with cells into the CO_2 incubator to continue growing cells.

3.3 Cryopreservation of Suspension Cell Lines

1. Only preserve cultures containing healthy cells (based on trypan blue staining and inspection under a microscope), which are confirmed to be mycoplasma free (*see* Chapter 1).
2. Pre-label cryovials with the name of the cell line, freezing cell density, passage number, date, and researcher's initials.
3. Collect cells into a 15-mL or 50-mL conical tube as described in **step 4** in Subheading 3.2. Pellet cells by centrifugation at 500 × *g* for 5 min.
4. During centrifugation, make up working volume of cryopreservation/freezing media.
5. After centrifugation, discard supernatant and resuspend pellet in at least 1 mL of complete media. In a fresh Eppendorf tube, count cells as directed in **step 8** in Subheading 3.2 and determine volume of suspension that will result in the desired cell number (*see* **Note 12**).
6. Transfer desired volume to a new centrifugation tube and pellet cells by centrifugation (500 × *g* for 5 min).
7. Resuspend the cell pellet in desired volume (*see* **Note 14**) of cryopreservation/freezing media.
8. Carefully dispense 1 mL per vial of the cell suspension into the pre-labeled cryovials.
9. Transfer the cryovials into an appropriate freezing container and place in a −80 °C freezer overnight allowing the

temperature of the cryovials to drop slowly to the freezer temperature (*see* **Notes 15** and **16**).

10. Take out the frozen cryovials and rapidly move them to canes for storage in liquid N_2 container. Store frozen cryovials in the liquid or gas phase of liquid nitrogen as suggested by the cryovial manufacturer.

4 Notes

1. Sera for supplementation can be heat inactivated by incubating in a 56 °C water bath for 30 min. Heat inactivation inactivates complement, which can activate a cytolytic cascade and potentially kills or activates lymphocytes and macrophages [7]. B lymphocytes are especially vulnerable as they express or secrete antibodies that can activate the complement cascade.
2. Though antibiotic supplementation prevents contamination, they have been linked to changing gene expression and regulation; thus, their use is recommended with caution.
3. Temperature of water baths is monitored using either an attached thermometer or an internal control system.
4. Designate an aliquot of DMSO for cell culture and open only in a biosafety cabinet of level II or higher.
5. The surface of pipette tip containers, media bottles, and pipette aids should be sterilized by spraying with 70% ethanol and wiping down with low-lint wipes.
6. If any liquid nitrogen is escaping, you will see fog/vapor/fumes and hear a slight hissing sound.
7. Cells in the cryovials are frozen, and the thawing process (by gentle swishing motions of the ampule in a pre-warmed water bath) should take the cells from a solid to semi-liquid state (a small solid still visible).
8. Alternatively, thawed cells can be added to 10 mL of complete media and then pelleted by centrifuging at 500 × *g* for 5 min, followed by resuspending pellet in complete media and adding to culture dish.
9. Avoid creating bubbles as interactions of cells with bubbles can kill the cells and distort microscopic view of cells.
10. Signs of contamination include changes in media color (from pink to yellow), significant change in cell morphology, the presence of multiple cell morphologies in the same culture, and change in media consistency (clear to cloudy/turbid).

11. Counting cells to determine cell density is important for making decisions during cell culturing. Cells at medium or high density (most areas on the plate are covered with cells) have nearly or completely used up the nutrient supply in the media. Hence, the cells need to be subcultured for instance in a 1:1 ratio = twofold dilution (1-part cells, i.e., 5-mL cells from 100-mm culture plate into a fresh plate +5-mL fresh pre-warmed complete growth media). Cell doubling time is how long it takes for the cells to divide and double to fill the surface of the new culture plate/vessel. Cell doubling time can be estimated by counting the number of days between subculturing and comparing to cell number at each time point. Cell doubling time varies between cell lines and together with cell density per time determines the frequency of media changes and subculturing. For instance, Burkitt's lymphoma cell line Ramos has a double time of 2–3 days and should be subcultured or media changed every 2–3 days.
12. Trypan blue stains dead cells blue. Calculate cell count as follows: *Average trypan blue negative cells per quadrant* × *dilution factor* × 10^4 = *amount of cells/mL.*

$$\text{Desired volume} = \frac{\text{Density desired}}{\text{Density of cells determined}} \times \text{total resuspension volume}$$

13. Each cell culture vessel/plate has maximum volume of media that should be used. For instance, the recommended volume capacity for a 100-mm dish is 11–16.5 mL. If the capacity is exceeded, it must be for a specific reason and for a limited time only.
14. For freezing down cells, it is ideal to freeze at least 2×10^6 cells/mL, but up to 5×10^6 cells/mL is possible.
15. Cells should be transferred to −80 °C as fast as possible as exposure to DMSO at room temperature for prolonged periods will reduce cell viability.
16. Isopropanol surrounding cells placed into specially designed freezing containers will cool cryovials at a constant rate optimal for cell preservation (approximately −1 °C/min). Alternatively, styrofoam racks from a 15-mL conical tube packaging can be placed below and on top of tubes to enclose the tube fully to control the temperature drop.

Acknowledgments

SLC is a senior scientist with the Beatrice Hunter Cancer Research Institute. This work was funded by the Natural Sciences and

Engineering Research Council of Canada and the Janeway Children's Hospital Foundation. MNL was supported by a Memorial University School of Graduate studies fellowship and a Cancer Research Training program award from the Beatrice Hunter Cancer Research Institute with funds provided by GIVETOLIVE Cancer Studentship. We thank Dr. Sukhinder Cheema, Peter Isesele, and Abeer A. Awad for the help with the cell micrographs and Willow R. B. Squires for helpful comments on the text.

References

1. Segeritz C-P, Vallier L (2017) Cell culture: growing cells as model systems in vitro. In: Jalali M, Saldanha FYL, Jalali M (eds) Basic science methods for clinical researchers. Elsevier, pp 151–172
2. Freshney RI (2016) Quantification. In: Capes-Davis A, Gregory C, Przyborski S (eds) Culture of animal cells: a manual of basic technique and specialized applications. Wiley, Hoboken, pp 483–512
3. Ryu AH, Eckalbar WL, Kreimer A et al (2017) Use antibiotics in cell culture with caution: genome-wide identification of antibiotic-induced changes in gene expression and regulation. Sci Rep 7:1–9
4. Verhoeckx K, Cotter P, Kleiveland C et al (2015) General introduction to cells, cell lines and cell culture. In: Verhoeckx K, Cotter P, López-Expósito I et al (eds) The impact of food bioactives on health: in vitro and ex vivo models. Springer, Cham, pp 83–93
5. Ackermann T, Tardito S (2019) Cell culture medium formulation and its implications in cancer metabolism. Trends Cancer 5:329–332
6. Muraglia A, Nguyen VT, Nardini M et al (2017) Culture medium supplements derived from human platelet and plasma: cell commitment and proliferation support. Front Bioeng Biotechnol 5:1–15
7. Triglia RP, Linscott WD (1980) Titers of nine complement components, conglutinin and C3b-inactivator in adult and fetal bovine sera. Mol Immunol 17:741–748

Chapter 3

Culture of Adherent Cancer Cell Lines

Vanessa Noel and Mark D. Berry

Abstract

Adherent cell lines grow attached to the surface of a cell culture vessel. Due to the adherent nature of the cells, enzymes, such as trypsin, are required to lift the cells from the cell culture vessel for harvesting or subculturing. Many cancer cell lines are adherent, rendering adherent cell culture a critical experimental method in the fields of cell biology, biochemistry, and cancer research. In this chapter, we outline the protocols for culturing and maintaining adherent cells. We detail the procedures for preparing cell culture medium, thawing and reviving frozen adherent cells, subculturing adherent cells, freezing cells, and counting cells. Most notably, we outline the best techniques and practices for optimal growth of healthy adherent cells while diminishing the risk of contamination.

Key words Cell culture, Adherent cell, Cell line, Culture medium, Cancer

1 Introduction

Cell culture is a laboratory technique in which cells are removed from an animal or plant and then grown and maintained under controlled conditions and is fundamental to research in the fields of biochemistry, biology, and medicine [1].

The term "adherent cells" refers to cells that must attach to a surface to grow and survive. Due to the adherence of the cells, the confluency can be easily visualized and checked under a microscope. Confluency refers to the percentage of the surface area of the cell culture vessel that is covered by the adherent cells. This allows for easy identification of when the cells require passaging, the process by which cells are divided into more culture vessels than they originate from, hence decreasing the confluency/cell density. A limitation of adherent cell culture, however, is that growth is limited to the surface area of the cell culture vessel.

Adherent cells must be regularly examined under a microscope to ensure cells are healthy. If growth medium containing a pH indicator (typically phenol red) is used, then changes in the color of the medium provide a ready visual indication of a change in pH,

Sherri L. Christian (ed.), *Cancer Cell Biology: Methods and Protocols*,
Methods in Molecular Biology, vol. 2508, https://doi.org/10.1007/978-1-0716-2376-3_3,

which may indicate a number of factors including bacterial contamination, over-confluence of the cells, and "spent" medium (see below). An increase in the opacity (or cloudiness) of the medium is usually indicative of a microbial contamination.

To maintain healthy cell growth, the adherent cells need to be regularly fed with fresh medium to prevent depletion of trace nutrients and buildup of metabolic by-products. These by-products are often (although not exclusively) acidic in nature and so are accompanied by a change in the color of medium that contains a pH indicator, a simple indicator of "spent" medium. Replacement of medium is readily accomplished with adherent cells by simply aspirating the spent medium (most often through a vacuum aspiration line) and replacing it with fresh medium via a disposable pipette.

Cell passaging involves transferring a portion of healthy cells to a new cell culture vessel with fresh growth medium to promote continuous cell growth. Adherent cells need to be regularly passaged to prevent overcrowding, a situation that promotes altered growth and metabolic characteristics of the cells and introduces a readily avoidable source of variance in experimental parameters. In extreme situations, overcrowding can lead to induction of cell death and also contribute to phenotypic drift of cell lines. In contrast, maintaining cells at too low of a density can also lead to alterations in growth and metabolic characteristics and cell death. Most commercial sources of cells provide recommendations on optimal cell densities, and these should be closely adhered to. In general, we find that maintenance of cells at a confluency of 25–90% is appropriate for adherent cancer cell lines. The time between the medium changes and passaging of the cells will vary between different cell lines and should be experimentally determined, although most commercial suppliers provide information on expected cell doubling time and medium replacement recommendations.

2 Materials

2.1 Preparing Cell Culture Medium from Powder

1. 2-L beaker.
2. Magnetic stir plate and stir bar.
3. Deionized water.
4. Cell line recommended powdered medium (*see* **Note 1**).
5. Sodium bicarbonate.
6. pH meter.
7. 1 M HCl and 1 M NaOH to adjust pH.
8. Sterile 500-mL media bottles.
9. Sterile filtration apparatus.

10. Penicillin/streptomycin (5000 units/mL).
11. Fetal bovine serum (FBS).
12. Other medium adjuncts as required (*see* **Note 2**).
13. Biosafety cabinet of level II or greater.

2.2 Thawing and Reviving Frozen Adherent Cancer Cell Lines

1. Cryovial containing the frozen cancer cells.
2. Complete cell culture growth medium (cell line dependent) pre-warmed in a water bath set to 37 °C.
3. 70% ethanol for disinfecting.
4. Low-lint wipes.
5. Pipettes and pipette tips.
6. Pipette aid.
7. Disposable, sterile centrifuge tubes.
8. Centrifuge.
9. Aspirator.
10. Sterile tissue culture flasks, dishes, or plates.
11. 37 °C incubator with humidified atmosphere and regulatable medical-grade CO_2 supply (generally 5% in the incubator) (*see* **Note 3**).
12. Biosafety cabinet of level II or greater.

2.3 Passaging Adherent Cancer Cell Lines

1. Vessel containing the adherent cells (85–90% confluent).
2. Complete growth medium: (cell line dependent) pre-warmed in 37 °C water bath (*see* **Note 4**).
3. Aspirator.
4. Pipette aid.
5. Pipettes and pipette tips.
6. 0.25% trypsin-EDTA (pre-warmed in 37 °C water bath).
7. 37 °C incubator with humidified atmosphere of 5% CO_2.
8. Disposable, sterile centrifuge tubes.
9. Centrifuge.
10. Sterile tissue culture flasks, dishes, or plates.
11. Inverted microscope.
12. Biosafety cabinet of level II or greater.

2.4 Counting Adherent Cancer Cells

1. Materials required for passaging cells (*see* Subheading 2.2).
2. Trypan blue.
3. Hemocytometer.
4. Handheld cell counter.
5. Inverted microscope.

2.5 Freezing Adherent Cancer Cell Lines

1. Vessel containing adherent cancer cells.
2. Cell culture growth medium (pre-warmed in 37 °C water bath).
3. Disposable, sterile centrifuge tubes.
4. Freezing medium: 90–95% complete growth medium supplemented with 5–10% dimethyl sulfoxide (*see* **Note 5**).
5. Materials required to determine cell count (*see* Subheading 2.4).
6. Sterile cryovials.
7. Freezing apparatus or isopropanol chamber (e.g., Mr. Frosty).
8. Liquid nitrogen storage device (normally a liquid nitrogen Dewar).

3 Methods

3.1 Preparation of 1 L of Powdered Cell Culture Medium

1. Measure 900 mL of deionized water and add ~800 mL to a 2-L beaker. Place the beaker on a magnetic stir plate and add a stir bar.
2. Carefully open the medium packet by cutting along the top of the packet between the notches on the packet. Do *not* discard the cutoff top of the packet.
3. Slowly add the contents of the package to the 2-L beaker while the water is being stirred. It is critical that this be done slowly and care taken not to aerosolize the contents of the package.
4. Rinse the medium package at least twice using ~30 mL of the unused deionized water each time. Carefully squeeze the package together while the water is still inside to remove residual powder from all inner surfaces, and gently run a spatula down the seams of the packet to ensure all residual powder is captured. Pour the resultant solution into the 2-L beaker.
5. With gloved hands, carefully separate the panels of the cutoff medium packet top, and over the top of the 2-L beaker, carefully pour some of the residual distilled water between the panels to capture any residual medium powder directly into the beaker.
6. If not already a component of the powdered medium, add 2.0 g of analytical grade sodium bicarbonate to the stirring solution (*see* **Note 6**).
7. Using a calibrated pH meter, measure the pH of the medium and adjust to pH 7.6 by adding drops of 1.0 M HCl or 1.0 M NaOH as required (*see* **Note 7**).

8. Bring the medium solution up to 0.89 L by adding deionized water (*see* **Note 8**).
9. Transfer the medium to a biological safety cabinet (BSC) and divide it into two 445-mL aliquots.
10. Using a bottle-top 0.2-μm filter, filter sterilize each 445-mL aliquot directly into individual, sterile, 500-mL media bottles.
11. Add 50 mL (assuming 10% FBS is required) of FBS to each bottle (*see* **Note 9**).
12. Add 5 mL (assuming 1% penicillin/streptomycin is required) of penicillin/streptomycin (5000 Units/mL) to each bottle (*see* **Note 10**).
13. Prepared medium can then be stored in a refrigerator at 2–8 °C for 2–3 weeks.

3.2 Thawing and Reviving Frozen Adherent Cancer Cell Lines from Liquid Nitrogen Storage

1. Warm cell culture medium in 37 °C water bath (*see* **Note 11**).
2. Sterilize BSC by spraying the surface with a 70% ethanol solution and wiping down with low-lint wipes (*see* **Note 12**).
3. Obtain the cryovial containing the desired adherent cancer cells from liquid nitrogen storage and place the vial inside the BSC. Sterilize the outside of the cryovial by spraying with 70% ethanol and wiping with a low-lint wipe. Briefly (<5 s) loosen the cap on the cryovial to allow any liquid nitrogen to escape.
4. Reseal the cryovial and allow cells to thaw. This can be done by gently swirling the sealed cryovial in a 37 °C water bath until only a small amount of sample remains frozen in the vial (<1 min).
5. In the BSC, add 5 mL of pre-warmed medium to a 10-mL centrifuge tube. Transfer the contents of the cryovial into the same centrifuge tube.
6. Centrifuge at 400 × *g* for 5 min.
7. While the cells are centrifuging, add fresh growth medium to the final culturing vessel. For a 100-mm diameter cell plate, we recommend 9-mL growth medium; for a T25 culture flask, 5 mL; and for a T75 culture flask, 20 mL.
8. After the centrifugation, gently discard the supernatant using an aspirator without disturbing the cell pellet.
9. Resuspend the cell pellet in 1 mL of fresh growth medium. Mix the solution by pipetting up and down several times (trituration).
10. Transfer the resuspended cell solution to the medium-containing culture vessel, and gently swirl the container to ensure uniform distribution of cells. Label the outside of the container with the name of the cell line, passage number, researcher's initials, and the date (*see* **Note 13**).

11. Examine the cells under a microscope to check the uniformity of cell distribution.
12. Place the culture vessel in a 37 °C incubator with humidified atmosphere of 5% CO_2.
13. Cells should then be checked under the microscope daily for health (morphology), confluency, and signs of microorganism contamination (*see* **Note 14**).

3.3 Passaging of Adherent Cancer Cell Lines

1. Place growth medium and 0.25% trypsin-EDTA solution in 37 °C water bath.
2. Sterilize the BSC by spraying with 70% ethanol and wiping down with a low-lint wipe.
3. Wipe down external surfaces of all materials needed for culturing, including pipette boxes, media bottles, pipette aid, and cell culture plates/flasks, using 70% ethanol.
4. Remove the cell plate/flask from the incubator and examine the morphology and confluency of the cells under a microscope. If the cells are 85–90% confluent, they are ready to be passaged.
5. In the BSC, hold the plate at a 30–45° angle and aspirate the medium from the cells carefully so as to not disrupt the adherent cells.
6. For a T75 flask or 100-mm diameter dish, add 2 mL of 0.25% trypsin-EDTA solution and gently rock the culture container to ensure complete coverage of the growth surface of the plate. For a T25 flask, add 0.5 mL. Place the plate/flask back in the 37 °C incubator for 4 min (*see* **Note 15**).
7. Check under a microscope to ensure cells have detached from the plate (*see* **Note 16**). Greater than 90% of cells should be detached before proceeding to the next step. If necessary, return cells to incubator for additional 2 min.
8. Add 5 mL of growth medium to the plate to terminate the trypsin reaction. Then, holding the plate at a 30–45° angle, pipette up and down several times to rinse down the surface of the plate and collect all detached cells.
9. Transfer the cell solution to a 10-mL centrifuge tube.
10. Centrifuge at 400 × *g* for 5 min.
11. Discard the supernatant by gently aspirating taking care to not disturb the cell pellet.
12. Resuspend the cell pellet in 1 mL of fresh growth medium by trituration.
13. Add the desired volume of fresh medium to a fresh cell culture plate(s)/flask(s) (*see* Subheading 3.2 for recommended volumes). Add the desired aliquot of the 1-mL cell suspension

(e.g., 200 μL for a 1:5 split). Label the lid of the plate with the name of the cell line, passage number, researcher's initials, split ratio, and date (*see* **Note 17**).

14. Gently swirl the plate to evenly distribute the cells and examine the cells under a microscope.
15. Place the plate in a 37 °C incubator with humidified atmosphere of 5% CO_2.

3.4 Counting Adherent Cancer Cells

1. Follows **steps 1–12** of Subheading 3.3 (or **steps 1–9** of Subheading 3.2) to obtain a resuspended cell solution.
2. Obtain a hemocytometer and clean the surface of the hemocytometer and the coverslip with 70% ethanol and then wipe down with a low-lint wipe.
3. In a clean Eppendorf tube, add 10 μL of resuspended cell solution and 10–90 μL of trypan blue. Pipette up and down to mix (*see* **Note 18**).
4. Place the coverslip over the chamber of the hemocytometer and add 10 μL of the cell–trypan blue mixture in the chamber underneath the coverslip.
5. Under a microscope, count all cells that are not stained for trypan blue in quadrants 1–4 (*see* Fig. 1), and determine the cell density using the equation below (*see* **Note 19**):

$$\text{Cell number}\left(\text{amount of}\frac{\text{cells}}{\text{mL}}\right) = \left(\frac{A+B+C+D}{4}\right) \times \text{trypan blue dilution factor} \times 10,000$$

6. Dilute an appropriate aliquot of the original (no trypan blue added) cell suspension with growth medium to give the required cell density.
7. Clean the coverslip and hemocytometer using 70% ethanol and low-lint wipes immediately after use.

3.5 Freezing of Adherent Cancer Cell Lines

1. Obtain a cryovial and label with the name of the cell line, passage number, concentration of the cells, date, and researcher's initials.
2. Collect cells in a centrifuge tube as described in Subheading 3.3, **steps 1–10**.
3. As the centrifuge is running, make up the freezing medium (90–95% complete growth medium and 5–10% DMSO dependent on cell line).
4. After centrifugation, discard the supernatant and resuspend the cell pellet in 1 mL of freezing medium (*see* **Note 20**).
5. Transfer the 1-mL aliquot of resuspended cell solution into the labelled cryovial.

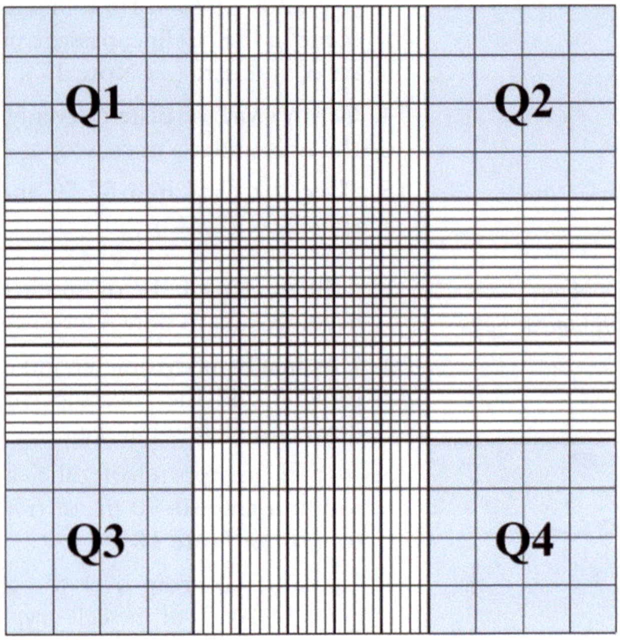

Fig. 1 Hemocytometer grid used for counting cells. The cells in Q1–Q4 that are not stained with trypan blue are counted manually using a handheld cell counter. The cell density can then be calculated as described in **step 5** of Subheading 3.4

6. Place the cryovial into a freezing apparatus and place it in a −80 °C freezer overnight (*see* **Note 21**).
7. The following day, quickly transfer the cryovial to the liquid nitrogen storage (*see* **Note 22**).

4 Notes

1. Cell culture medium is available in a wide variety of styles, from pre-made medium to powdered medium, medium plus/minus various nutrients and other growth-promoting additives, and medium plus/minus a pH-sensitive dye. Pre-made liquid medium is easier and requires less handling to use but is considerably more expensive (approximately $50 per liter) than powdered medium (approximately $6 per liter). We routinely use powdered medium containing phenol red pH indicator because of the increased cost-efficiency. Some examples of powdered media that are commonly used for adherent cancer cell lines include Roswell Park Memorial Institute (RPMI) 1640, Dulbecco's Modified Eagle Medium (DMEM), and DMEM-F12.
2. Depending on the medium and cell line being used, other additives/adjuncts may be required (e.g., INS-1 insulinoma

cells are grown in RPMI 1640 medium that requires further supplementation with sodium pyruvate and HEPES; other cell lines are grown in DMEM that requires additional glucose supplementation). See the manufacturer's guide for exact requirements.

3. The function of the CO_2 atmosphere is to provide appropriate buffering capacity following dissolution in the aqueous growth medium via the H_2CO_3–HCO_3^- –CO_2 buffer system. A vast majority of cells are grown in a 5% CO_2 atmosphere. At higher CO_2 concentrations the pH of the growth medium will decrease, as such, cell lines that prefer more acidic environments may require an increased CO_2 level for a given growth medium preparation. For a more complete discussion of CO_2 concentration and buffering in cell culture, see https://www.phe-culturecollections.org.uk/news/ecacc-news/co2-concentration-and-ph-control-in-the-cell-culture-laboratory.aspx (accessed May 25, 2021).
4. Complete growth medium is cell line dependent and is defined as cell culture medium supplemented with all required additives and adjuncts, for example, 89% RPMI 1640, 10% FBS, and 1% penicillin/streptomycin. See cell line supplier recommendations for the specific medium and additives required.
5. Typically, freezing medium is composed of 90–95% complete growth medium and 5–10% DMSO; however, this is subject to change depending on the cell line being used. The cell line supplier recommendation should be checked prior to preparing the freezing medium for exact requirements as some cell lines may require additional supplements or a modified growth medium base. For example, HL-60 cells undergo differentiation in the presence of DMSO, and so an alternate freezing medium is required for such cell lines.
6. The buffering capacity of almost all commercial media is based on the CO_2–HCO_3^- buffer system, and it is critical that the manufacturer's instructions are followed with respect to ensuring appropriate bicarbonate levels are present.
7. When preparing medium, adjust the pH to approximately 0.2 units above the required final pH as the sterilization process, addition of adjuncts, and use at 37 °C and 5% CO_2 will alter the pH. Setting the pH to 7.6 will result in a medium pH of 7.4 following sterilization and sterile adjunct additions, at 37 °C and 5% CO_2.
8. The medium volume may have to be adjusted depending on the type of medium being used and what adjuncts need to be added after sterile filtration. For example, if no additives are required, then the medium should be brought up to a final volume of 1 L.

9. FBS routinely comes in 500-mL bottles. We sterile aliquot this into sterile 50-mL Falcon tubes immediately on receipt (in a BSC) and store frozen at −20 °C.
10. Similar to FBS, we sterile aliquot penicillin/streptomycin into 10-mL Falcon tubes immediately on arrival and store at −20 °C.
11. Some researchers prefer to warm the medium in the cell culture incubator, with a "cracked" (loosened) cap, to ensure that appropriate buffering is in place by warming the medium in a 5% CO_2 environment. We have also successfully used such a technique without any contamination of cultures or medium occurring. We recommend, however, that this only be done where a single research group is using the incubator. In shared (or very large research group) facilities, where maintenance of common techniques is difficult to ensure, we recommend pre-warming solutions in fully sealed containers in a water bath.
12. With all procedures occurring in a BSC, care should be taken to avoid working over the top of open containers. Containers that have caps should always have the cap replaced immediately, even if just loosely, to ensure the contents are not open to the environment. The external surfaces of all sealed containers should be sprayed with 70% ethanol (plant spray bottles from the local hardware store are excellent for this purpose) immediately on placement in the BSC. Open containers (e.g., beakers) should have the external surfaces wiped with a lint-free wipe soaked in 70% ethanol.
13. Except in situations where cells have an extremely short culture life, cells should be allowed to grow to >80% confluence and passaged at least once before conducting an experiment. If a specific seeding density is required during routine maintenance, then incorporate the appropriate steps from Subheading 3.4 to obtain a cell count from the 1-mL cell suspension.
14. To check the confluency of adherent cells, estimate the percentage of the surface area of the cell culture vessel that is covered by the adherent cells. The morphology of a cell refers to the external characteristics, such as the shape, size, and structure of a cell. The presence of multiple cell morphologies in the same culture is indicative of contamination.
15. Some cell lines may require a longer or a shorter incubation period. It is also important to ensure that the trypsin reaction is not prolonged as prolonged trypsin exposure will initiate cell death.
16. To promote cell lifting, lightly tap the plate against the heel of the hand to dislodge the cells. This should only be done when using a flask (T25 or T75) as the cell culture vessel.

17. The split ratio will depend on when the cells are needed for an experiment. As a general rule of thumb, split cells 1:2 for an experiment in approximately 1–2 days, split cells 1:5 for an experiment in 2–4 days, and split cells 1:10 for an experiment in 4–6 days. This will vary depending on the cell line being used as some cells grow faster than others.
18. The exact dilution factor with trypan blue will depend on the density of the cells and the size of the container from which they were obtained. In general, cell counts of 20–100 per hemocytometer quadrant should be aimed for.
19. The "10,000" term in the equation is the dilution factor for the volume of one quadrant compared to the total volume added and is provided by the manufacturer.
20. When freezing down cells, aim to freeze at a concentration between 2×10^6 cells/mL and 5×10^6 cells/mL.
21. There are a number of commercially available freezing apparatuses (e.g., Mr. Frosty), most of which are based on an isopropanol bath to give a slow freeze and minimize the risk of cellular ice crystal development. These apparatuses allow for the temperature of the cryovial to slowly drop to the temperature of the freezer at a rate of approximately −1 °C/min.
22. Cells should always be stored in the vapor phase of liquid nitrogen, and appropriate PPE (face shield, thermally insulated gloves) always used.

Reference

1. Baust JM, Buehring GC, Campbell L et al (2017) Best practices in cell culture: an overview. In vitro cellular & developmental biology. Animal 53(8):669–672. https://doi.org/10.1007/s11626-017-0177-7

Chapter 4

Breast Cancer Xenograft Murine Models

Wasundara Fernando, Krysta M. Coyle, and Paola Marcato

Abstract

Mice are used as model organisms to understand the pathological basis of a variety of human diseases, including breast cancer. Both immunocompetent and immunocompromised mouse models are used depending on the scope of the study. Immunocompetent models allow the study of the impact of the immune system in murine models of mammary cancer, while immunodeficient mice serve as ideal host organisms to understand the behavior of human breast cancers within a biological system. Xenografting of human breast cancer cells into immunocompromised mouse models continues to be the most used fundamental animal model in preclinical breast cancer research. These in vivo models allow critical understanding of tumor biology and assessment of novel treatments, a necessary prelude to testing new drugs in the clinic. In this chapter, we provide detailed methodology for the use of non-obese diabetic (NOD) severe combined immunodeficient (SCID) mice in several breast cancer xenografting procedures, including established cell lines and patient-derived xenografts (PDXs).

Key words Immunodeficient, Xenografting, PDX, Mouse models, Tumors

1 Introduction

The advancement of knowledge in the processes of cancer development and progression and the discovery of novel and improved treatments for cancer relies on having established and well-characterized murine tumor models. Demonstrated efficacy in mouse tumor studies is often a requirement for a novel drug or treatment strategy to advance to the clinical testing; hence, murine tumor models are an essential facet of cancer research.

Preclinical breast cancer research studies utilize both immunocompetent and immunocompromised mouse models. There are a wide variety of immunocompetent models for breast cancer, which encompass two broad approaches: spontaneous syngeneic models and genetically engineered transgenic models. Although we include some description of syngeneic models of breast cancer, this article focuses predominantly on methods and protocols for human breast tumor xenografting in immunocompromised mice. These

Sherri L. Christian (ed.), *Cancer Cell Biology: Methods and Protocols*,
Methods in Molecular Biology, vol. 2508, https://doi.org/10.1007/978-1-0716-2376-3_4,

immunocompromised mouse models are essential for working with human tumors in an in vivo setting, since they will not reject the human tumor cell lines or patient-derived tissues. The use of human cell line–derived xenografts in immunodeficient animal models is a common, relatively inexpensive technique. Additional advantages include that the xenografts often develop in a short time and tumors grow relatively synchronously.

Immunocompromised models include the athymic nude mouse or variants of the non-obese diabetic (NOD) severe combined immunodeficient (SCID) mouse. Since most strains of nude mice are "leaky" and have a few T cells [1], more immunodeficient models, such as the NOD-SCID mouse, are more commonly used in human tumor xenografting experiments. CB17-SCID mice are unable to complete V(D)J recombination due to a defect in protein kinase, DNA-activated, catalytic polypeptide (PRKDC) [2]. As a result, they lack mature T and B cells and are similarly unable to reject allografts or xenografts. CB17-SCID mice were crossed with NOD mice [3], which, as a result of inbreeding, have defects in complement activity, T and B lymphocytes, and dendritic cells. The NOD models also display reduced activation of NK cells and defects in macrophage maturation as compared to the SCID mouse [2, 4]. The resultant NOD-SCID mouse thus supports engraftment of numerous cell types. The NOD-SCID mouse has been crossed with an IL-2 receptor (IL2R) γ-chain deficient mouse resulting in the NOD-SCID IL2Rg−/− (colloquially referred to as NOG or NSG) mouse with even more severe impairment of immune function [5]. These highly immunodeficient models are particularly useful for establishing patient-derived xenografts (PDXs) [6]. Other models of immunodeficiency include recombination activating gene (RAG)1- and RAG2-knockout mice; these may be better models for the study of radiation as they have intact DNA repair mechanisms [7].

1.1 Grafting of Breast Cancer Cell Lines

Following in vitro culture of human cell lines and assay, the xenografting of cell lines is a common technique to study breast tumor growth dynamics that more closely mimics the human scenario and allows for the interactions of tumor cells with mammary stroma. There are several factors to consider when choosing a cell line for xenografting. Not all breast cancer cell lines are tumorigenic, even within the immunocompromised NOD-SCID model [8]. Additionally, estrogen-sensitive cell lines require the addition of estrogen in order to engraft [9]. Orthotopic implantation in the mammary fat pad is ideal for breast cancer cell lines, and this can increase the take rate over subcutaneous implantation [10]. Metastasis of breast cancer cell lines is rare and occurs most often to the lung [11]; in contrast, the most common metastatic site in human breast cancer is the bone [12]. The importance of the immune system in promoting the metastatic cascade is a likely explanation for the rarity of metastases in immunocompromised hosts [13]. More complex experiments have resulted in a series of MDA-MB-231 clones, which metastasize to different tissues of the mouse [14]; additionally, SUM1315MO2 will preferentially metastasize to human bone grafts over mouse bone [15].

Syngeneic mouse models of breast cancer are typically derived from spontaneous tumors, which arise in murine mammary tissue, and their use combines in vitro and in vivo methods. The most well-studied is the 4T1 cell line model, which, along with several sister lines, was derived from a spontaneous tumor in a BALB/cfC3H mouse. The 4T1 variant is a thioguanine-resistant cell line, which spontaneously metastasizes in vivo to the lung and liver. The sister lines fail at various points in the in vivo metastatic cascade; thus, the model has been utilized as a model for metastasis and the epithelial-to-mesenchymal transition [16]. While syngeneic models are useful to understand essential elements of tumor biology, there are only a few cell lines available [17]. This makes it difficult to fully represent the wide molecular heterogeneity found in human breast cancer. Additionally, syngeneic cell lines are typically expanded in vitro prior to implantation; thus, a large number of cells implanted in mice do not represent the early stages of breast cancer progression and rarely model intratumoral heterogeneity.

1.2 Xenografting Patient-Derived Tumors

An alternative to cell line xenografts is PDXs. PDXs can be established from implantation of cells or pieces of tissue from human tumors [18]. Successive passaging of breast cancer PDXs in mice results in gradual replacement of the human mammary stroma with mouse mammary stroma [19]. Not all tumor types support engraftment as PDXs. Much like triple-negative breast cancer (TNBC) is overrepresented in available cell lines, triple-negative patient tumors have a higher take rate than hormone-sensitive tumors [20, 21]. Additionally, basal-like TNBCs engraft more readily than claudin-low xenografts, making it challenging to fully represent the heterogeneity of human breast cancer with xenografts alone [6]. Similar to cell line xenografts, PDXs rarely metastasize [22]; however, PDXs present several advantages over cell line xenografts. PDXs display more fidelity with human tumors in terms of intratumoral heterogeneity and tumor architecture and are passaged less often than in vitro cell culture. This allows the initial characteristics of the clinical sample to be preserved more reliably.

Many human cell lines and PDX models exist, which can recapitulate much of the intertumoral heterogeneity in breast cancer. However, these models are not entirely realistic, and there are several pitfalls associated with their exclusive use. First, there is a species mismatch between the stromal cells of the mammary epithelium and the xenografted cancer cells, which may limit cross-talk between the tumor and the supportive stroma. This can be addressed by clearing of the mammary fat pad and humanization with immortalized human fibroblasts [23], but this is most typical in studies of normal mammary development and may inhibit the growth of cancerous xenografts [6]. Second, research is unable to investigate the interactions between the human tumor cells and host immune cells. The lack of a chronic inflammatory milieu means that studies of these tumor models are not complete mimics of human cancer. Third, the implantation of large numbers of cells does not imitate the multi-step progression of spontaneous human cancer. There are several approaches in immunocompetent models, which can address some of these limitations. Regardless, as discussed above, there are many advantages to using human xenograft models. The successful implantation of human cell lines and PDX models provides valuable information for understanding the factors involved in tumor establishment, cancer progression, and drug responses. Herein, we describe these methods.

The growth kinetics of tumors in mice depends on a variety of factors such as cell number, mouse strain, and the type of cells [11, 24, 25]. The methods we describe will reliably produce tumors in NOD-SCID mice; however, changing the mouse strain will likely require a change in the number of cells implanted. It is also important to note that each mouse model of breast cancer is aimed at providing a specific background for the experiment; hence, the most appropriate model needs to be selected, with consideration of multiple factors [26]. The route of administration of cells into the mice also plays a role in the successful grafting of the cells [24, 26, 27]. This methods chapter provides fundamental guidelines for xenografting and expanding PDXs and cell line–derived cells in mice. However, these protocols can be modified to fit the specific needs of different experiments. Undoubtedly, mouse models are a vital component of breast cancer research progressing toward the clinical testing phase. These methods need to be well controlled and optimized for the best results.

2 Materials

1. 1-mL insulin syringes.
2. 15-mL and 50-mL conical tubes.
3. 225 units/mL collagenase type 3 in sterile Hanks' balanced salt solution (HBSS) (*see* **Note 1**).
4. 4-0 Dexon II.
5. 6-cm and 10-cm sterile cell culture dishes.
6. 70-μm cell strainers.
7. Breast cancer cell lines (*see* Table 1).
8. Complete cell growth medium compatible with breast cancer cell lines (*see* Table 1).

Table 1
Cell line in vitro culture specifications and optimum number for injections

Name of the cell line	Molecular subtype	Receptor expression ER	PR	HER2	In vitro culture specifications	Optimum number of cells for injection per mouse	References
BT20	Basal	–	–	–	RPMI-1640 with 10% FBS	5×10^6	[28]
Du4475	Basal/ immunomodulatory	–	–	–	RPMI-1640 with 10% FBS	5×10^6	[29]
HCC1187	Basal/ immunomodulatory	–	–	–	RPMI-1640 with 10% FBS	5×10^6	[28]
HCC1806	Basal	–	–	–	RPMI-1640 with 10% FBS	1×10^3	[29]
HCC1937	Basal	–	–	–	RPMI-1640 with 10% FBS	5×10^6	[28]
HCC38	Mixed claudin low/basal	–	–	–	RPMI-1640 with 10% FBS	5×10^6	[30]
HCC70	Basal	–	–	–	RPMI-1640 with 10% FBS	5×10^6	[28]
MDA-MB-231	Claudin low	–	–	–	DMEM with 10% FBS	2×10^6	[28]
MDA-MB-436	Claudin low	–	–	–	DMEM with 10% FBS	5×10^6	[28]
MDA-MB-453	Luminal	–	–	–	DMEM with 10% FBS	5×10^6	[29]
MDA-MB-468	Basal	–	–	–	DMEM with 10% FBS	2×10^6	[28]
SUM149	Basal	–	–	–	F12 with 5% FBS, 10 mM HEPES, 1 μg/mL hydrocortisone, 5 μg/mL insulin	2×10^6	[28]
SUM159	Claudin low	–	–	–	F12 with 5% FBS, 10 mM HEPES, 1 μg/mL hydrocortisone, 5 μg/mL insulin	2×10^6	[28]
4 T1	N/A (murine origin)	–	–	N/A	DMEM with 10% FBS	1×10^4	[31]

Anti-Anti antibiotic-antimycotic, *DMEM* Dulbecco's Modified Eagle Medium, *ER* estrogen receptor, *FBS* fetal bovine serum, *HEPES* 4-(2-hydroxyethyl)-1-piperazineethanesulfonic acid, *HER2* Human epidermal growth factor-2, *N/A* not applicable, *PR* progesterone receptor, *RPMI* Roswell Park Memorial Institute medium

9. Supplemented Dulbecco's Modified Eagle Medium (DMEM): DMEM, 10% fetal bovine serum (FBS).
10. Disinfected bead sterilizer.
11. Disinfected cart.
12. Disinfected circulating heating pad.
13. Disinfected hibitane, alcohol, and iodine bottles.
14. Disinfected portable anesthetic machine with medium-sized nose cone.
15. Fetal bovine serum (FBS).
16. Hair clippers.
17. HEPES: 4-(2-hydroxyethyl)-1-piperazineethanesulfonic acid.
18. High concentration (HC) Matrigel (*see* **Notes 2** and **3**).
19. Human breast cancer tissue or PDX.
20. Hydrocortisone.
21. Insulin.
22. Meloxicam.
23. Non-obese diabetic severe combined immunodeficient (NOD-SCID) female mice.
24. Peroxigard.
25. Red blood cell lysis buffer: 157-mM ammonium chloride (NH_4Cl) and 20-mM potassium bicarbonate ($KHCO_3$) (*see* **Note 4**).
26. Scalpel blades.
27. Slow-release estrogen pellets.
28. Sterile cold phosphate-buffered saline (PBS).
29. Sterile gloves, masks, and sterile gowns.
30. Sterile surgical pack (scissors, tweezers, needles holder, wound clip dispenser loaded with wound clips, cotton swabs, and gauze surgical gowns for mice).
31. Trypsin.
32. Two disinfected round weights to hold anesthetic tube in place.

3 Methods

3.1 Personnel Preparation for Procedures

All the procedures that involve immunodeficient mice take place in a biosafety cabinet using aseptic techniques. The personnel who perform the techniques should put on a surgical cap, a mask, and a sterile gown. Hands should be sterile gloved.

3.2 Protocol 1: Passage of Patient-Derived Xenografts as Tumors Pieces and Freezing

Once the PDX is established in a mouse [6], it can be passaged several times in mice:

1. Cut the PDX into ~3 × 3 × 3 mm^3 chunks and store in supplemented DMEM and 5% dimethyl sulfoxide (DMSO) at each passaging.
2. Thaw a frozen vial from liquid nitrogen for transplantation using a 37 °C water bath.
3. Spray the tube down with ethanol and place in cell culture hood.
4. Remove existing freezing media and add 1-mL supplemented DMEM to the tube.
5. Mix several times and then dump media and tissue into a 6-cm cell culture dish, and move the PDX tumor piece repeatedly to new dishes containing fully supplemented DMEM media. This will gradually remove the DMSO from the PDX tumor piece.
6. After the final wash, transfer the PDX piece into a fresh 6-cm cell culture dish containing DMEM, and place the dish on ice if transplanting into a new mouse (passaging).

3.3 Protocol 2: Xenograft of Patient-Derived Xenografts into Mice

Steps 1–10 except **step 5** are performed by the assistant. Two people (surgeon and assistant) can effectively perform this procedure. This procedure should be performed in the biosafety cabinet (*see* Fig. 1):

1. Turn the surgery light on.
2. Disinfect the working surface of the biosafety cabinet with Peroxigard 1:16 and place the heating pad on it.

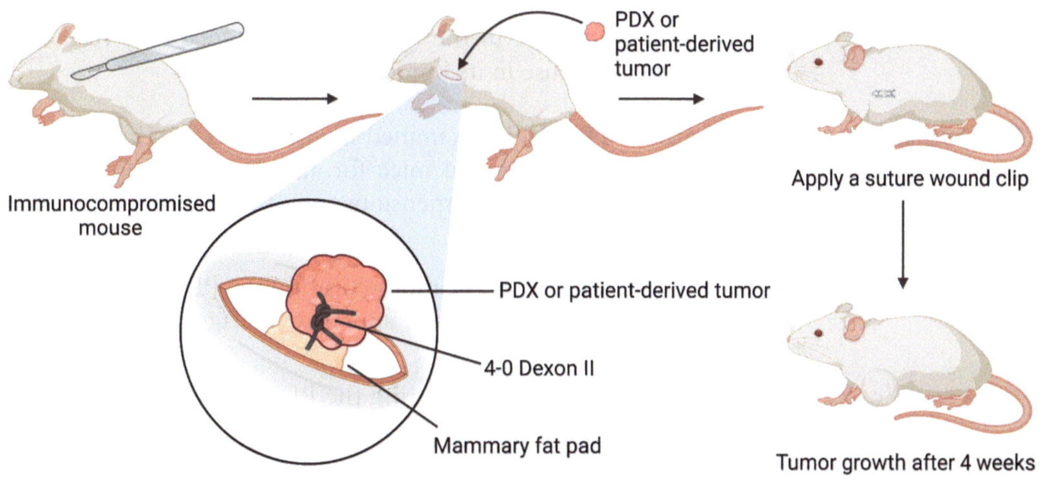

Fig. 1 Implanting patient-derived xenograft (PDX) in non-obese diabetic severe combined immunodeficient (NOD-SCID) female mice

3. Pour Peroxigard into a small bucket to facilitate disinfection of gloved hands.
4. Open the sterile surgical pack inside the biosafety cabinet.
5. Place a sterile towel.
6. Sterile recovery cages must be available.
7. Dip gloved hands in Peroxigard before handling the mouse.
8. Anesthetize the mouse.
9. Prepare the surgical site by clipping the hair on the lateral shoulder area followed by scrubbing the skin with hibitane, alcohol, and iodine.
10. Place the mouse on the sterile sheet/towel. Use a round weight on the anesthetic tube so that it is held in position.
11. Make a 0.5-cm incision on the lateral shoulder area.
12. Bluntly dissect the skin from the underlying tissue.
13. The mammary fat pad appears lighter in color compared to the surrounding tissue and is often adhered to the skin.
14. The assistant should hand the breast cancer tissue to the surgeon.
15. Suture the small piece of tissue to the mammary fat pad using 4-0 Dexon II.
16. Place the slow-release estrogen pellet in the subcutaneous pocket that has already been created (this step is not necessary when passaging PDXs).
17. Apply a suture wound clip (if using a wound clip, will need to remove clip in 7–10 days).
18. Inject meloxicam (5 mg/kg body weight) subcutaneously to provide analgesia.
19. Place the mouse in the recovery cage. Once fully recovered, place the mouse in its home cage.

3.4 Protocol 3: Generation of Single-Cell Suspensions from Patient-Derived Xenografts

Cells from PDX can be examined in vitro or can be injected back into immunocompromised mice for additional experiments. The generation of single-cell suspensions from PDX is done as explained below (*see* Fig. 2):

1. Before euthanizing the mouse bearing the PDX, make sure the incubator is set at 37 °C, and thaw collagenase tubes in the water bath at 37 °C.
2. Euthanize the mouse bearing the PDX.
3. Harvest the tumor using aseptic techniques (in the biosafety cabinet).

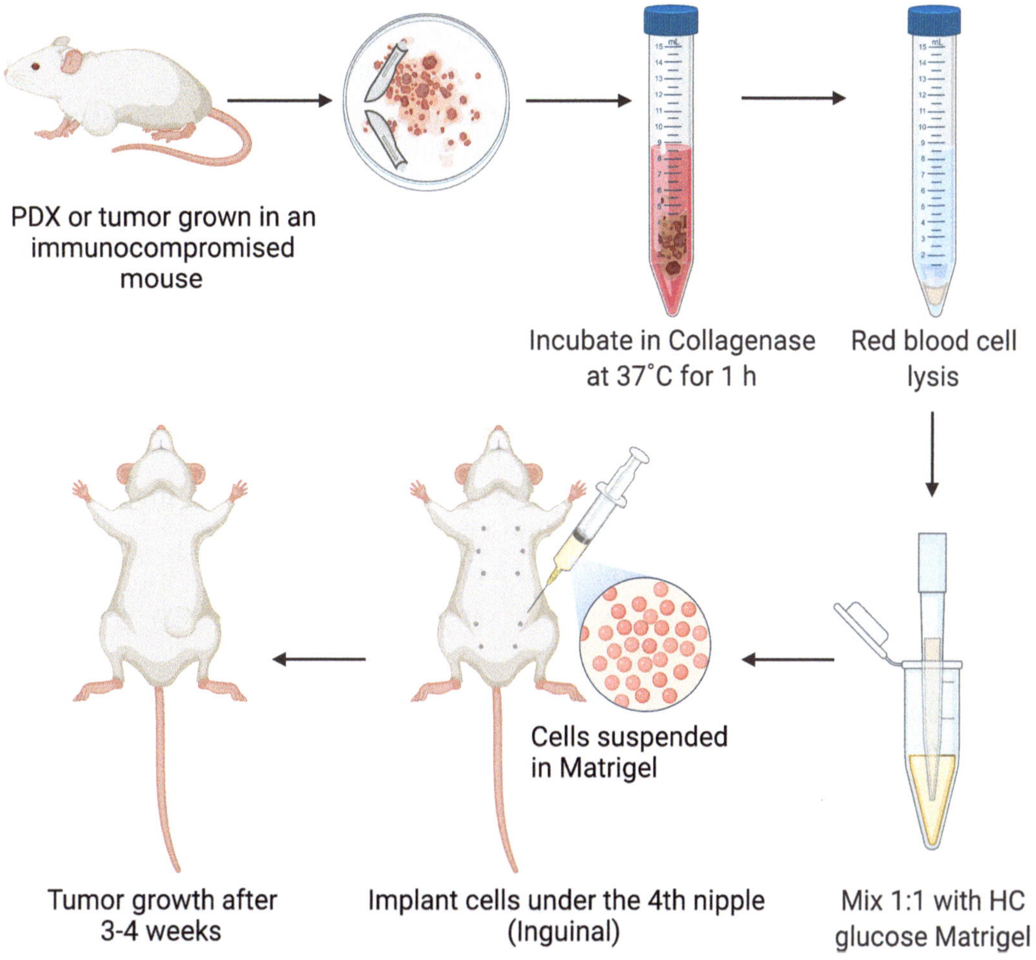

Fig. 2 Xenografting patient-derived xenograft (PDX)-derived cells in NOD-SCID female mice

4. Chop the tumor in a 6-cm dish with two blades, using a dicing technique. Continue to chop until there are no chunks larger than 1 mm^3. However, most of the tumor should be minced smaller than 1 mm^3 (*see* **Note 5**).
5. Using one of the blades, scoop up the diced tumor and put in a 15-mL conical tube containing 10-mL collagenase type 3. Close tightly and shake very well, making sure the tumor bits are well resuspended, and place on tube rotator in a 37 °C incubator for 1 h. Efficient digestion of the tumor tissue requires the constant mechanical motion of the tube rotator.
6. Check on the tube a couple of times, sometimes chunks collect at the bottom, and give it a good shake and put back on end-over-end.

7. Take the tube out, and strain the digested materials in a 6-cm dish using a 70-μm cell strainer. Collect the strained supernatant, and pellet the cells at 500g for 5 min.
8. Remove the supernatant very carefully. The cellular pellet is very soft at this stage, and it is easy to lose the pellet by suction; therefore, leave a 100 μL or so supernatant on top of the pellet.
9. Resuspend the pellet with 5 mL of 1× red blood cell lysis buffer, and incubate for 5 min at room temperature.
10. Spin at 500g for 5 min at 4 °C.
11. Aspirate the supernatant. Pellet is no longer soft now, so most of the supernatant should be removed. Check pellet to make sure it is no longer red.
12. Resuspend pellet with a 10-mL cold PBS.
13. Strain into a 6-cm dish over a 70-μm cell strainer. Collect strained cells into a new 15-mL conical tube.
14. Spin at 500g for 5 min at 4 °C.
15. The pelleted cells can be used for in vitro assays or to implant into new mice.

3.5 Protocol 4: Xenografting of Cells Derived from Cell Lines and the Single-Cellular Suspensions Prepared from PDXs

This protocol details the xenografting procedure for cells derived from cell lines and the single-cellular suspensions prepared from PDXs (*see* Fig. 3 and **Note 6**):

1. Culture breast cancer cells in 150-mm dishes.
2. Rinse the cell monolayers with sterile cold PBS and add 2-mL trypsin. Incubate the cells in a 37 °C incubator for 2–3 min.
3. Add 6-mL complete media (neutralize trypsin) and combine cells from 5 dishes in a 50-mL tube; spin down the cells at 500 × *g* for 5 min at 4 °C.
4. Combine cells of all the dishes in one 50-mL tubes and wash the cells in a 10-mL cold sterile PBS. Resuspend the resulting pellet in cold sterile PBS.
5. Count cells by trypan blue exclusion method. Calculate the required cell density to achieve the desired number of cells per mouse (*see* Table 1). Double the required cell density and resuspend the cell pellet in cold sterile PBS to achieve this density (*see* **Note 7**).
6. Withdraw 500 μL of the cell suspension using a sterile cold pipette tip, and carefully mix the cells with 500 μL of high concentration Matrigel without generating air bubbles while mixing.

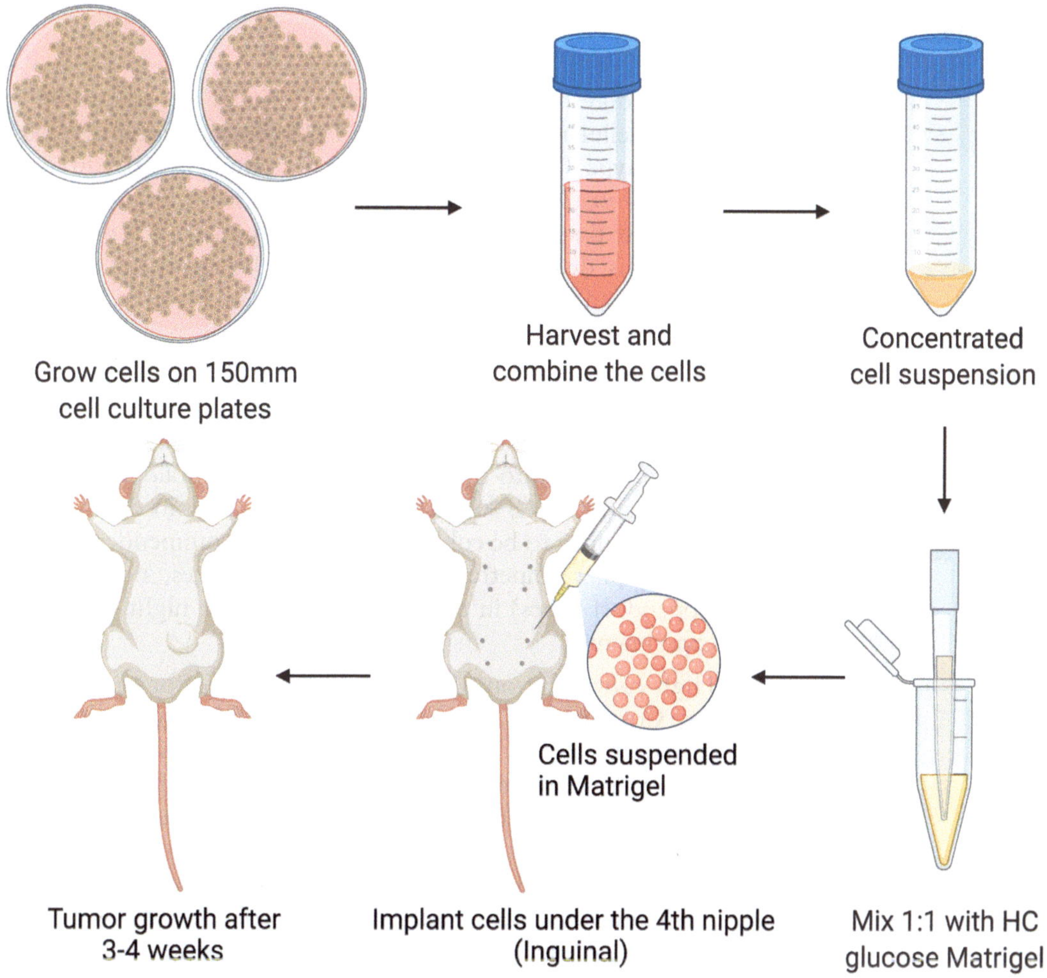

Fig. 3 Xenografting cell line–derived cells in non-obese diabetic severe combined immunodeficient (NOC-SCID) female mice

7. Place five mice in the induction chamber and sedate the mice with isoflurane (0.8 g/L O_2 & isoflurane at 3).
8. Closely observe the mice, and when they are under anesthesia, take the mice out of the induction chamber, inject the cell and high concentration Matrigel mixture, and place carefully under the skin into the fifth mammary fat pad very slowly (inguinal). No need to shave the fur.
9. Perform the ear punch for identification purposes and record the body weights.
10. Transfer the mouse into the recovery cage and monitor for recovery. Once fully recovered, place the mouse in its home cage.

4 Notes

1. Tissue digestion is a crucial step during the preparation of single-cellular suspensions from PDX passaged in mice. Collagenase type 3 (Bioshop Inc. Reference number COL006) from *Clostridium histolyticum* is commonly used for tissue digestion of breast tissues [32]. The lyophilized powder of collagenase type 3 is dissolved in sterile Hanks' balanced salt solution (HBSS), and preferably 10-mL aliquots are made to be stored at −20 °C.
2. Matrigel remains as a viscous liquid at 2–8 °C. Therefore, it is important to strictly maintain the temperature between 2 °C and 8 °C while working with Matrigel. It forms a solid gel plug after injecting into the mice, and it remains solid at temperatures below 2 °C. Any lab supplies that come in contact with Matrigel need to be cold; therefore, it is recommended to keep the supplies (pipette tips, stripettes, syringes, needles, and Eppendorf tubes) in the −20 °C freezer the night before the experiment.
3. High concentration Matrigel (recommended for tumor studies, Corning Inc. Reference number 354262) needs to be stored at −20 °C (preferably in 500 μL aliquots). Two days before the experiments, thaw a 10-mL bottle of high concentration Matrigel at 4 °C. The bottle can be placed in ice and thawed overnight in a 4 °C cold room. The day before the experiments, high concentration Matrigel aliquots need to be made under aseptic conditions, under cold conditions.
4. Red blood cell lysis is important when preparing single-cellular suspensions from PDXs and cell line–derived tumors. The pH of the red blood cell lysis buffer should be maintained at 7.4. The recipe for 250 mL of the buffer is given below:
   (a) 2.1 g NH_4Cl
   (b) 0.5 g $KHCO_3$
   (c) Adjust pH to 7.4
5. Do not over mince the PDX, it may negatively affect the viability of the PDX cells and reduce overall yield.
6. The day before the experiments that involve high concentration Matrigel, put PBS at 4 °C and the syringes and 1000-, 100-, and 10-μL pipette tips into the −20 °C freezer. It is important to also routinely check the cells for mycoplasma contamination [33]. At least 4–5 h before the planned cell implantation, the high concentration Matrigel aliquots need to be slowly thawed at 4 °C. The cells need to be maintained at 4 °C during the processing steps, prior to injection into mice.

7. After combining the cells from several culture plates for injections, the cell pellet is significantly large. Therefore, take the volume of the cell pellet into account when resuspending the cell pellet in PBS.

Acknowledgments

PM is a senior scientist of the Beatrice Hunter Cancer Research Institute (BHCRI). WF is funded by a project grant to PM from the Canadian Institutes of Health Research (CIHR, PJT 162313). KMC is funded by a postdoctoral fellowship from the CIHR. Figures were created with BioRender.com.

References

1. Manjili MH (2011) Revisiting cancer immunoediting by understanding cancer immune complexity. J Pathol 224:5–9
2. Bosma GC, Custer RP, Bosma MJ (1983) A severe combined immunodeficiency mutation in the mouse. Nature 301:527–530
3. Shultz LD, Schweitzer PA, Christianson SW et al (1995) Multiple defects in innate and adaptive immunologic function in NOD/LtSz-scid mice. J Immunol 154:180–191
4. Prochazka M, Gaskins HR, Shultz LD et al (1992) The nonobese diabetic scid mouse: model for spontaneous thymomagenesis associated with immunodeficiency. Proc Natl Acad Sci U S A 89:3290–3294
5. Ito M, Hiramatsu H, Kobayashi K et al (2002) NOD/SCID/γcnull mouse: an excellent recipient mouse model for engraftment of human cells. Blood 100:3175–3182
6. Zhang X, Claerhout S, Prat A et al (2013) A renewable tissue resource of phenotypically stable, biologically and ethnically diverse, patient-derived human breast cancer xenograft models. Cancer Res 73:4885–4897
7. Shultz LD, Ishikawa F, Greiner DL (2007) Humanized mice in translational biomedical research. Nat Rev Immunol 7:118–130
8. Holliday DL, Speirs V (2011) Choosing the right cell line for breast cancer research. Breast Cancer Res 13(4):215
9. Clarke R (1996) Human breast cancer cell line xenografts as models of breast cancer – the immunobiologies of recipient mice and the characteristics of several tumorigenic cell lines. Breast Cancer Res Treat 39:69–86
10. Price JE, Polyzos A, Dan Zhang R et al (1990) Tumorigenicity and metastasis of human breast carcinoma cell lines in nude mice. Cancer Res 50(3):717–721
11. Price JE (1996) Metastasis from human breast cancer cell lines. Breast Cancer Res Treat 39:93–102
12. Manders K, van de Poll-Franse LV, Creemers GJ et al (2006) Clinical management of women with metastatic breast cancer: a descriptive study according to age group. BMC Cancer 6:179
13. Kitamura T, Qian BZ, Pollard JW (2015) Immune cell promotion of metastasis. Nat Rev Immunol 15:73–86
14. Kang Y, Siegel PM, Shu W et al (2003) A multigenic program mediating breast cancer metastasis to bone. Cancer Cell 3:537–549
15. Kuperwasser C, Dessain S, Bierbaum BE et al (2005) A mouse model of human breast cancer metastasis to human bone. Cancer Res 65:6130–6138
16. Yang J, Mani SA, Donaher JL et al (2004) Twist, a master regulator of morphogenesis, plays an essential role in tumor metastasis. Cell 117:927–939
17. Galli S, Colombo L, Vanzuli S et al (2000) Characterization of a fibroblastoid mammary carcinoma cell line (LM2) originated from a mouse adenocarcinoma. Int J Oncol 17:1259–1265
18. Tentler JJ, Tan AC, Weekes CD et al (2012) Patient-derived tumour xenografts as models for oncology drug development. Nat Rev Clin Oncol 9:338–350

19. Cassidy JW, Caldas C, Bruna A (2015) Maintaining tumor heterogeneity in patient-derived tumor xenografts. Cancer Res 75:2963–2968
20. Bruna A, Rueda OM, Greenwood W et al (2016) A biobank of breast cancer explants with preserved intra-tumor heterogeneity to screen anticancer compounds. Cell 167:260–274, e22
21. Whittle JR, Lewis MT, Lindeman GJ et al (2015) Patient-derived xenograft models of breast cancer and their predictive power. Breast Cancer Res 17:1–13
22. Paez-Ribes M, Man S, Xu P et al (2016) Development of patient derived xenograft models of overt spontaneous breast cancer metastasis: a cautionary note. PLoS One 11(6):e0158034
23. Brill B, Boecher N, Groner B et al (2008) A sparing procedure to clear the mouse mammary fat pad of epithelial components for transplantation analysis. Lab Anim 42:104–110
24. Kim IS, Baek SH (2010) Mouse models for breast cancer metastasis. Biochem Biophys Res Commun 394:443–447
25. Fantozzi A, Christofori G (2006) Mouse models of breast cancer metastasis. Breast Cancer Res 8:1–11
26. Borowsky AD (2011) Choosing a mouse model: experimental biology in context-the utility and limitations of mouse models of breast cancer. Cold Spring Harb Perspect Biol 3:1–16
27. Gomez-Cuadrado L, Tracey N, Ma R et al (2017) Mouse models of metastasis: progress and prospects. Dis Model Mech 10:1061–1074
28. Prat A, Parker JS, Karginova O et al (2010) Phenotypic and molecular characterization of the claudin-low intrinsic subtype of breast cancer. Breast Cancer Res 12:1–18
29. Riaz M, van Jaarsveld MTM, Hollestelle A et al (2013) MiRNA expression profiling of 51 human breast cancer cell lines reveals subtype and driver mutation-specific miRNAs. Breast Cancer Res 15:R33
30. Prat A, Karginova O, Parker JS et al (2013) Characterization of cell lines derived from breast cancers and normal mammary tissues for the study of the intrinsic molecular subtypes. Breast Cancer Res Treat 142:237–255
31. Fernando W, Coyle K, Marcato P et al (2019) Phloridzin docosahexaenoate, a novel fatty acid ester of a plant polyphenol, inhibits mammary carcinoma cell metastasis. Cancer Lett 465:68–81
32. Shi L, Ermis R, Garcia A et al (2010) Degradation of human collagen isoforms by Clostridium collagenase and the effects of degradation products on cell migration. Int Wound J 7:87–95
33. Nikfarjam L, Farzaneh P (2012) Prevention and detection of mycoplasma contamination in cell culture. Cell J 13:203–212

Chapter 5

Processing and Cryopreservation of Blood, Cancer Tissues, and Cancer Cells for Viable Biobanking

Jennifer A. Chan and Suzanne M. Vercauteren

Abstract

Biorepositories of fresh frozen and formalin-fixed paraffin-embedded tissues have been foundational to many molecular cancer research studies. Collections of these materials, however, do not enable the establishment of short-term cultures, cell lines, or patient-derived xenograft models for functional studies. Also, intact dissociated cells that are required for some single-cell analyses cannot be obtained from these material types. Adding viable tumor banking to the repertoire of routine cancer biobanking would increase the value of samples collected. This chapter outlines procedures for processing and storing blood and tissue specimens viably in order to expand the future utility of the samples collected. We provide practical tips that can be used by banks and other researchers seeking to incorporate the cryopreservation of viable materials as part of their overall biobanking strategies.

Key words Biobanking, Cryopreservation, Cryoprotection, Peripheral blood mononuclear cells, PBMC, Cancer

1 Introduction

High-quality biospecimens are critical for enabling cancer research—whether toward finding new biomarkers, identifying new targets for therapy, or defining the genomic, transcriptomic, and proteomic landscapes of cancers. To this end, practice of tumor banking (the systematic process of collecting, preserving, storing, and supplying tumor samples and associated data) has become an essential part of the cancer research endeavor. A challenge for banks, however, is that the future use of the samples is often not known at the time of collection and processing. One goal in banking, therefore, is to consider all possible future uses and preserve the samples in a state or states that can be used for as many downstream applications as possible.

Traditionally, cancer-related biobanking has used a variety of preservation methods to meet the needs of histology and the needs

Sherri L. Christian (ed.), *Cancer Cell Biology: Methods and Protocols*,
Methods in Molecular Biology, vol. 2508, https://doi.org/10.1007/978-1-0716-2376-3_5,

of a range of molecular analyses that can be performed on derivative DNA, RNA, or protein extracted from the samples. This includes the processing and storage of tumor samples as formalin-fixed paraffin-embedded tissue, flash frozen tissue, and frozen whole blood or blood fractions—when possible from longitudinal collections that represent not only the primary site or diagnostic time point but also those of later relapses. The collection of additional sample types such as urine and cerebrospinal fluid is additionally important for research toward the discovery of new biomarkers and liquid biopsies.

Recently, technical advancements in translational cancer research and the need for more patient-derived models have driven the desire for the collection and distribution of viable samples to be added to the repertoire of the modern biobank. Viably banked samples are valuable for enabling patient-derived model generation (including generation of cell lines, organoid cultures, and patient-derived xenografts), drug screening, and pre-clinical testing [1–4]. Viable samples also allow for later separation of cells into different compartments such as infiltrating immune cells and neoplastic cells and for special preparation of cells for some single-cell molecular analyses that require intact whole cells.

Although these types of studies could be supported by the immediate release and use of the freshly collected material by dedicated investigator teams, in many cases, the costs and workflow implications are prohibitive for systematic implementation of up-front xenografting or culturing for most samples collected. A practical alternative to immediate use is to collect and store samples cryoprotected as either small tissue fragments or dissociated cells. The stored cryoprotected frozen samples may be thawed and re-aminated later—thus allowing for the more costly and labor-intensive work of culturing, xenografting, or preparation for single-cell analyses to be performed only on the samples that are desired and with a timeframe that can be tailored to the scheduling/staffing of the investigator.

The most common methods of cryopreservation employ a variety of components such as cell culture media, serum or albumin, and dimethyl sulfoxide (DMSO) as the main cryoprotective agent (typically with a final concentration of ~10%)—in a solution that is freshly prepared on the day of use. DMSO-containing media have an extensive history of use as a cryoprotectant for viability/cell culture and have also been shown to work well with modern single-cell molecular analysis techniques [5–9]. In addition to DMSO-based methods, other non-DMSO cryoprotectants and methods such as vitrification have also been used to preserve viability [10]. Here, we describe methods that use DMSO as the cryoprotectant.

The following protocols support live biobanking of blood and a wide range of tissues/cells. The optimal cryoprotective media for live biobanking of cells/tissues, however, can vary by cell/tissue type; so users may wish to further optimize parameters with respect to enzymes, duration of incubations, and cryoprotectant for different types of sample. We focus on technical procedures for processing and storing two main material types: liquid tumor samples (e.g., from peripheral blood draws and bone marrow aspirates) and solid tumor samples (e.g., from tissue biopsies, resections, and rapid autopsies).

1.1 Peripheral Blood and Bone Marrow Aspirates

The gold standard technique for the isolation and storage of human peripheral blood mononuclear cells (PBMCs) is density centrifugation with Ficoll-Paque followed by freezing in a cryopreservation medium containing DMSO at temperatures of −135° or below (liquid nitrogen tank). Non-neoplastic blood samples and most hematological malignancies can be frozen using this method as the malignant cells separate with the mononuclear cells. However, a few malignancies consisting of malignant neutrophils, eosinophils, and basophils as the dominant population cannot be cryopreserved using this method.

1.2 Solid Tumor Tissue

Tumor tissue may be cryopreserved as either tissue fragments (which can later be dissociated to single cells post-thaw, if desired) or dissociated cells (where dissociation is performed prior to freezing). Although cryopreservation of solid tumors may be more effective as dissociated cells rather than tissue fragments, possibly owing to greater contact/diffusion of the cryoprotectant into dissociated cells, both methods enable the isolation of viable individual cells, allowing for subsequent xenografting, culturing of tumor cells/organoids, or separation/culturing of cellular compartments such as neoplastic cells and tumor-infiltrating lymphocytes. The choice to cryopreserve tissue fragments or dissociated cells or both may consider staffing, number of samples coming to the bank, time constraints, available equipment, cost, and other factors (*see* **Note 1**).

2 Materials

2.1 General Materials and Equipment

1. 2.0-mL cryovials.
2. Pipettors.
3. Aerosol-resistant pipet tips.
4. Controlled rate freezing container (e.g., Cool Cell LX or Nalgene Mr. Frosty; *see* **Note 2**).
5. −80° freezer.
6. Liquid nitrogen vapor-phase freezer.

2.2 Additional Materials and Equipment for Bone Marrow Aspirate or Peripheral Blood

1. Tabletop centrifuge with swinging rotor and tube adapters.
2. 1–10-mL pipets.
3. Heat-inactivated fetal bovine serum (FBS), prepared by heating FBS to 56 °C for 30 min (for preparation of wash buffer and freeze media).
4. Wash buffer: calcium- and magnesium-free PBS (137-mM NaCl, 2.7-mM KCl, 8-mM Na_2HPO_4, 2-mM KH_2PO; pH 7.4) + 2% heat-inactivated FBS.
5. Density gradient medium: Ficoll-Paque Plus (GE Health Care) or Lymphoprep (Stemcell Technologies).
6. Freeze media for PBMCs/marrow: 90% heat-inactivated FBS + 10% DMSO (*see* **Notes 3** and **4**).
7. 15-mL and 50-mL conical tubes.
8. Hemocytometer or automated cell counter.
9. Trypan blue.

2.3 Additional Materials and Equipment for Tissue

1. Scalpel blades or fine spring scissors.
2. Forceps.
3. Sterile dish (petri dish or tissue culture dish) for dissecting.
4. Freeze media for tissue/cells: DMEM + 12.5% BSA + 10% DMSO; freshly prepared (*see* **Notes 3** and **4**).

2.4 Additional Materials and Equipment for Dissociating Tissue to Individual Cells

1. Human tumor dissociation kit (Miltenyi Biotec) (*see* **Note 6**).
2. GentleMACS Octo Dissociator with heaters (Miltenyi Biotec) (*see* **Note 7**).
3. DMEM.
4. GentleMACS C-tubes (Miltenyi Biotec, cat# 130-093-237).
5. 70-μm cell strainers.
6. Red blood cell (RBC) lysis buffer: 15-mM NH_4Cl, 1-mM $KHCO_3$, 10-mM EDTA (*see* **Note 5**).
7. 50-mL conical tube.
8. Hemocytometer or automated cell counter.
9. Trypan blue.

3 Methods

Obtain all required ethics approvals and consents prior to working with these protocols. All patient specimens are considered potentially infectious and must be handled in an appropriate, approved laboratory setting and in a biosafety cabinet. Personal protective equipment (gloves, lab coat, and goggles) should be worn. For samples known to be infectious, additional safety measures may be required; consult your local safety office for guidance. Any blood or tissue remaining after completion of processing should be disposed of in accordance with biosafety guidelines.

Fresh samples should be processed as soon as possible for maximal viability. Tissue samples should be kept moist or in media and chilled at 4 °C or on wet ice during holding and transport to the lab (*see* **Note 8**). Carry out all procedures at room temperature unless otherwise noted. Maintain sterile technique throughout the protocols.

3.1 Peripheral Blood Processing for Plasma and Viable PBMCS

The protocol here includes both plasma and PBMC processing from the same blood sample (*see* **Note 9**). Fresh, anticoagulated, non-stabilized whole blood (e.g., sodium citrate, acid citrate dextrose, EDTA, heparin tubes) is the acceptable inputs for viable PBMCs (*see* **Note 10**). Blood samples should be processed as soon as possible upon receipt and within 4 h of collection (*see* **Note 11**).

1. Fractionate the whole blood by centrifuging at 1500 × *g* for 10 min at room temperature. This will separate the plasma from other blood components (*see* Fig. 1). The upper layer is the plasma layer and should be clear to pale yellow in color. The second layer is a narrow grayish white interface band representing the "buffy coat" or leukocyte fraction. The third or bottom layer is dark red and consists of the erythrocytes or red blood cells. If the plasma fraction is cloudy/turbid, repeat the spin at 1500 × *g* for 10 min; otherwise proceed to the next step.
2. Using a pipet, carefully collect the plasma layer down to 1–2 mm from the buffy coat layer. If the plasma will be banked for general use, aliquot the plasma into 0.5-mL or 1.0-mL aliquots in cryovials, and then place the cryovials in −80 °C freezer for storage.

   If the plasma is intended for use in circulating tumor DNA (ctDNA) studies, temporarily place the collected plasma in a 15-mL conical tube on wet ice while proceeding with the rest of the steps in the PBMC preparation; the plasma will be returned to later in **step 13** for additional centrifugation prior to freezing.
3. Once the plasma layer has been removed, dilute the remaining blood layers in wash buffer such that the final volume is 2× the original blood specimen volume (including the plasma layer). Use 15- or 50-mL falcon tubes depending on the starting blood specimen volume.
4. In a separate tube, pour density gradient media (the same volume as undiluted blood), and then carefully layer the diluted blood on top of the cushion of density gradient media (*see* **Note 12**).

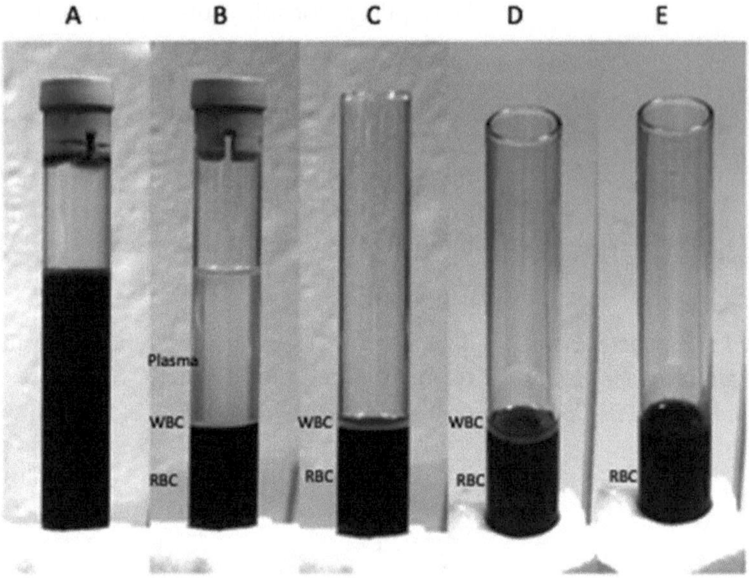

Fig. 1 Appearance of blood samples during recovery of white blood cells. (**a**) Whole blood. (**b**) Post-centrifugation white blood cells (WBC), red blood cells (RBCs), and plasma. (**c**) Post plasma removal. (**d**) Top view of buffy coat WBC. (**e**) Top view of sample after WBC removal

5. Centrifuge at 400 × *g* in a swing bucket without brake for 30–40 min at 22 °C. After centrifugation, the mononuclear cells are in the interface between the density gradient media (bottom layer) and the buffer (upper layer).
6. Collect the mononuclear cells using a disposable transfer pipette and transfer them into a new 15-mL conical tube. Add wash buffer to the PBMC tube to bring the volume up to 15 mL, and then gently invert the tube 5 times to resuspend.
7. Centrifuge at 500 × *g* for 5 min at 4 °C, and then carefully remove the supernatant with a pipet.
8. Add a new wash buffer depending on the size of the cell pellet, mix gently to resuspend, and count using trypan blue with an automated cell counter or a manual method of cell counting.
9. In order to deplete platelets, add fresh wash buffer and resuspend the cells again by gently inverting the tube 5 times. Centrifuge at 100 × *g* for 10 min at 4 °C, and then carefully remove with a pipet and discard the supernatant that contains the platelets. The mononuclear cells will remain as a pellet at the bottom of the tube.
10. Resuspend the pellet in freeze media to a concentration between 5×10^6 per mL to 20×10^6 cells per mL (*see* **Note 13**).

11. Mix and quickly transfer cell suspension into labeled cryovials.
12. Place the vials in a controlled rate freezing container, and place container in −80 °C freezer. After overnight storage at −80 °C, move the vials to vapor phase of liquid nitrogen for long-term storage (*see* **Notes 14** and **15**).
13. If the purpose of collecting plasma is for ctDNA analysis, additional clarifications of the plasma may be performed: Centrifuge the plasma for 10 min at 1600 × *g* at 4 °C. Carefully collect the plasma supernatant with a pipet, taking care not to disturb any pellet, and transfer to a new 15-mL tube. Centrifuge the plasma for another 10 min at 1600 × *g* at 4 °C, and then collect the plasma supernatant with a pipet. Aliquot the plasma into 0.5-mL or 1.0-mL aliquots in cryovials. Place the cryovials in −80 °C freezer for storage.

3.2 Bone Marrow Processing

1. Measure the specimen volume. Dilute the bone marrow to four times the volume in wash buffer. Use either 15- or 50-mL Falcon tubes depending on the sample size. Gently mix 3–4 times.
2. In a separate tube, pour density gradient media (amount equal to half the volume of the diluted bone marrow specimen) and carefully layer the diluted specimen on the cushion of density gradient media.
3. Centrifuge at 400 × *g* in a swing bucket without brake for 30–40 min at room temperature (22 °C).
4. After centrifugation, the mononuclear cells are in the layer between the density gradient media (bottom layer) and the buffer (upper layer). With a transfer pipette, collect the mononuclear cells and transfer them into a new conical tube containing wash buffer. Wash the cells by gently mixing and centrifuging again at 500 × *g* for 5 min at 4 °C.
5. Carefully remove the supernatant, add a new wash buffer depending on the size of the cell pellet, and mix gently to resuspend (a small pellet requires 1 mL of wash buffer, a large pellet 5–10 mL). Count the cells using trypan blue with an automated cell counter or use a manual method of cell counting.
6. In order to deplete platelets, resuspend the cells in a wash buffer and centrifuge at 100 × *g* for 10 min at 4 °C, and then carefully remove and discard the supernatant that contains the platelets. Mononuclear cells will remain as a pellet at the bottom of the tube.
7. Resuspend the cells (pellet) in freeze media to have 1×10^7 cells per vial (*see* **Note 13**).

8. Mix and quickly transfer cell suspension into cryovials. Place the vials in controlled rate freezing container, and then place container in −80 °C freezer. For long-term storage (>1 month), move the vials to vapor phase of liquid nitrogen storage (*see* **Notes 14** and **15**).

3.3 Tissue Processing for Frozen Viable Tissue Fragments

1. Prepare freeze media: If using DMEM, 12.5% BSA, and 10% DMSO as the freezing media, prepare this fresh. If using ready-to-use freeze media such as CryoStor CS10, take it out of the fridge and keep it on ice.
2. Transfer tissue to a petri dish and keep bathed in chilled HBSS, DMEM, or other sterile buffered media). Use forceps and scalpel or fine spring scissors to dissect and remove frankly necrotic or cauterized areas.
3. Mince the tissue using scalpels or scissors into fine fragments 1 × 1 × 1 mm or smaller.
4. Aliquot fragments into cryovials (*see* **Note 16**).
5. Pipet 1-mL freeze media per vial (*see* **Notes 3** and **4**). Gently mix.
6. Pre-freeze: incubate the tissue in freeze media for 5 min on ice, flicking tubes occasionally to mix.
7. Place the vials in a controlled rate freezing container and place the container in −80 °C freezer.
8. Transfer cells from freezing container to vapor phase in LN2 freezer (*see* **Notes 14** and **15**).
9. To thaw and use cryopreserved tissue, thaw rapidly at 37 °C for 1–2 min, rinse immediately in DMSO-free media, and use per downstream application (*see* **Note 17**).

3.4 Tissue Processing for Dissociated Cells

1. Place the tissue in a sterile petri dish with 4.7-mL media (*see* **Note 18**).
2. Hold the tissue with forceps and mince with scalpel or fine spring scissors until pieces are as small as possible (less than 1-mm fragments).
3. Using a pipet, transfer the media and minced tumor to C-tube. Add 200-ul enzyme H, 100-ul enzyme R, and 25-ul enzyme A into the tube, and then securely close the lid of the C-tube (*see* **Note 6**).
4. Turn on GentleMACS Octo Dissociator with heaters. Place C-tube upside down on GentleMACS Octo Dissociator and perform automated dissociated on the GentleMACS instrument per manufacturer's protocol using the "37C_h_TDK_1" program. The automatic dissociation will be completed in approximately 1 h (*see* **Note 7**).

5. When dissociation program is complete, remove C-tube and turn off the machine.
6. Open C-tube, add additional 10-mL DMEM into the tube to dilute the sample.
7. Transfer sample (15 mL) to a 50-mL tube to a 70-uM cell strainer to collect single-cell suspension. Let the sample pass through by gravity. Rinse the strainer with another 10-mL DMEM.
8. Centrifuge at 300 × *g* for 5 min at 4 °C.
9. Carefully remove the supernatant from the pellet.
10. Resuspend sample in a 5-mL cold RBC lysis buffer and incubate for 2–3 min on ice with occasional flicking or tapping to gently mix. Add 10-mL cold DMEM to quench.
11. Centrifuge at 300 × *g* for 5 min at 4 °C, and then remove the supernatant (*see* **Note 19**).
12. Resuspend the pellet in 10-mL DMEM.
13. Use 10 μl of the suspension to count cells with trypan blue and a hemocytometer or an automated cell counter.
14. Centrifuge at 300 × *g* for 5 min at 4 °C, discard the supernatant, and then resuspend the pellet in appropriate amount of cryopreservation media to result in a concentration of 1×10^7 cells per mL.
15. Quickly transfer cell suspension into labeled cryovials. Place the vials in controlled rate freezing container, and then place freezing container in −80 °C freezer. After overnight storage at −80 °C, move the vials to vapor phase of liquid nitrogen for long-term storage (*see* **Notes 14** and **15**).

4 Notes

1. Cryopreservation as tissue fragments is less technically demanding and faster (20–30 min for typical processing time) and requires fewer reagents up front than preparing and freezing dissociated cells (approximately 2 h). Cryopreservation as tissue fragments is the easier method to implement in the setting of pathology departments. Freezing as tissue fragments is also useful in a typical wet lab setting in situations where personnel may be limited or where cases arrive late in the day or after hours.
2. The ideal cooling rate of −1 °C per minute can be achieved using electronic programmable controlled rate freezing units. Such units, however, can be costly and difficult for workflow in settings where samples arrive frequently and at unpredictable

times. Freezing units such as Mr. Frosty (which requires filling with isopropanol) or Corning CoolCell units, which do not need alcohol, are practical and cost effective. With the latter two methods, cryovials containing the materials to be frozen are placed inside the containers, and the containers are then placed into a −80 °C freezer or dry ice locker for several hours prior to transfer to liquid nitrogen. Some freeze media such as CELLBANKER/STEM-CELLBANKER may be effective for cryopreservation even without controlled rate freezing; this could be a consideration for cryoprotectant selection in settings where controlled rate freezing is not available.

3. The requirement to make up fresh freeze media immediately before use may be challenging for workflow and may be a source of variability, depending on the volume of samples and the availability of resources and staffing. We and others have found that commercially available ready-to-use cryopreservation media such as CryoStor CS10 is a useful alternative to freshly prepared media [4, 11]. This is particularly convenient in facilities such as pathology departments, where cryovials pre-aliquoted with the ready-to-use media can be kept refrigerated and close at hand, obviating the need for special preparation at the time of banking. Ready-to-use media include the CryoStor series of cryopreservation media products (BioLife Solutions) and CELLBANKER/STEM-CELLBANKER solutions (AMSBIO), both of which can be used to store cells or tissues. Recently, several DMSO-free ready-to-use commercial cryopreservation media have also been developed such as CELLBANKER DMSO-free (AMSBIO), STEM-CELLBANKER DMSO-free (AMSBIO), and CryoSOfree DMSO-free (Sigma-Aldrich) that use alternative cryoprotectants in their formulations.

4. Although serum-containing freeze media are effective and the mainstay of cryopreservation of PBMC and bone marrow preparations, some may prefer to use a serum-free defined freezing medium. The consideration of whether to use serum in the freeze media or not depends on factors such as cell type and may also take into account considerations such as international shipping, where animal-derived substances such as serum may complicate import/export procedures, depending on the country. Serum also has the potential for variable quality with different sources and preparations. Another consideration is the effect of serum on the cells. For brain and other solid tumors, we prefer to use serum-free defined freeze media to avoid exposing the viable cells to serum, as serum is a strong inducer of differentiation and can affect the phenotype of the cancer stem-like cells.

5. A 10× stock of RBC lysis solution can be prepared easily and inexpensively in the lab as follows: Dissolve 90-g NH4Cl (0.155 M), 10-g KHCO3 (0.01 M), and 370-mg EDTA (0.1 mM) in one liter of ddH2O. Adjust pH to 7.3–7.4. Filter sterilize through a 0.22 micron filter. The 10× RBC lysis solution can be stored at 4 °C for 6 months. Dilute 1:10 in ddH2O before using as 1× working solution. Refrigerate the working solution until use. Many commercially available products also exist for the osmotic lysis of RBCs; these are commonly based on similar ammonium chloride solutions and come as either ready-to-use 1× solution or a 10× stock that needs to be diluted with water before use.
6. For dissociation of tissues to single cells, the optimal dissociation media or enzymes will vary and may be optimized by the protocol user. The final selection of conditions such as dissociation enzymes and cryoprotectant will depend on factors that can include cell type, the need for GMP or animal-free formulations in some instances, and practical considerations such as available equipment and staffing. We use Miltenyi Gentle-MACS enzyme kits and Miltenyi octo dissociator with heaters due to its reproducibility, ease of use, and decreased hand-on time for lab personnel. We have found that the Human Tumor Dissociation Kit (Miltenyi #130-095-929) and the Brain Tissue Dissociation Kit (P) (Miltenyi #130-095-942) cover most of the solid tumor sample types that we collect in our bank.
7. If using the Miltenyi system, follow the appropriate program for the specific kit per manufacturer's protocol (e.g., program 37C_h_TDK_1 for human tumor dissociation kit and 37C_BTDK_1 for brain tumor dissociation kit). Dissociation may also be performed with the same enzyme kits, but without the automated dissociator instrument by substituting alternating cycles of incubation and manual pipet trituration (for example, 37 °C incubation × 15 min followed by pipetting up and down 20 times, followed by another cycle of incubation and trituration before passing the material through a 70-μm strainer). A useful reference for tissue dissociation procedure for single-cell analyses without the use of an automated system is Ref. 6.
8. A variety of media can be used for temporary holding/transport of samples to the lab. Sterile PBS, HBSS, and RPMI are all commonly available and suitable for short-term transport of tissue samples. Hibernate A media may also be considered for brain cancer samples [8, 12] and reportedly improves viability. The requirement to prepare the media fresh, however, may make it less practical to use in some settings.

9. It is not necessary to separate and save plasma in order to obtain viable PBMCs from anticoagulated whole blood. We collect both plasma and PBMCs from our anticoagulated bloods, in keeping with the principle of maximizing future utility of all samples collected. If plasma is not desired, dilute the starting whole blood with an equal volume of PBS + 2% FBS, and then begin the peripheral blood processing protocol at **step 5**.
10. There are some considerations for selection of anticoagulant tube type. Acid citrate dextrose (ACD), EDTA, and heparinized blood tubes will all yield viable PBMC, but there are some differences in stability of PBMC phenotypes, with EDTA-anticoagulated blood deteriorating somewhat more quickly (after 24 h) and having greater changes in some phenotypic cell parameters compared to ACD or heparin anticoagulated blood [13–15]. On the other hand, heparin is a known PCR inhibitor, and ACD performance can be impacted by short draws. For studies examining cell-free DNA in plasma, EDTA has been suggested as an anticoagulant of choice for anticoagulated, non-stabilized blood over citrate or heparin [16, 17].
11. Delayed processing of blood affects not only viability of PBMCs but importantly also affects the suitability of plasma samples for molecular studies examining cell-free DNA [18, 19]. Always note collection and processing times in records. For samples that will have delayed processing, placing the tube at 4 °C can extend the time before lysis of cells and reduction of viability begins from 8 to 24 h. If processing will routinely be delayed beyond 6 h of collection, however, and if plasma is desired for cfDNA studies, the use of stabilized anticoagulated blood tubes (e.g., Streck BCT tubes or equivalent) for collection of plasma should be considered in addition to drawing the anticoagulated, non-stabilized blood for PBMC prep.
12. More recently, cell preparation tubes (CPTs) and SepMate tubes with Lymphoprep have been developed that make the gradient prep easier and more reproducible, especially when no trained personnel are available. Use of such specialized tubes results in comparable cell recovery, viability, PBMC population composition, and cell functionality, relative to the gold standard technique [20].
13. Cell concentration should be at least 5×10^6/mL and not more than 20×10^6 cells/mL in freezing media. Because cell counts can be variable, a practical approach to aliquoting is to determine the number of cells per vial based on the overall cell number: If the count is 20–30 million cells, freeze 5 million

cells per mL per vial. If the count is 40–100 million cells, freeze 8–10 million cells per mL per vial. If the count is >100 million, freeze 10–20 million cells per mL vial in freezing medium.

14. Once tissues or cells are frozen, it is critical to keep the frozen material from warming during transfers. Extract frozen vials from CoolCell or Mr. Frosty and place them on bed of dry ice to assist transfer to LN2 storage without thawing. Rapid transfer is important. Cryovial contents can rise from −80 to over −50 in less than 1 min when exposed to room temperature air and can negatively affect viability.
15. Temporary storage at −80 °C should be limited to short periods (a few days to weeks). To preserve viability of cells/ tissue and to preserve the integrity of biomolecules' cells/ tissues (regardless of whether flash frozen or cryoprotected frozen), long-term storage of viable samples should be performed in liquid nitrogen freezers, with storage temperatures at or below −135 °C. Although this can be achieved by storing vials in the liquid phase of LN2 (−196 °C), storage in the vapor phase above the liquid (temperatures typically between −140 °C and −180 °C) is preferred for safety and ease of access, as the temperatures are still low enough to stop cell activity, but the possibility of leaky vials exploding during removal is reduced.
16. Aliquoting of tissue fragments: We typically freeze 3–5 fragments, or about 2 mm^3 in aggregate of fine fragments, per vial so that each vial is sufficient for implantation into 3–5 mice when thawed. After collecting the fragments and placing into cryovials, if the dissection media are cloudy with cells or there are small fragments that are difficult to pick up, the residual dissection media with cells/fragments can be transferred into a 15-mL conical tube and centrifuged at 180 $g \times 5$ min at 4 °C. The fragments/pellet can be resuspended in freeze media and placed in another cryovial.
17. Thawed material may be rinsed by centrifuging, removing freeze media, and then adding fresh media. It is important to remove DMSO from the cells as soon as possible after thawing, as DMSO is toxic to cells when at 37 °C. Thawed tissue fragments may be subsequently dissociated to single cells using the same dissociation procedures described in Subheading 3.4.
18. The protocol for dissociated cells describes digestion in a 5-mL enzyme mix based on an input of 0.2- to 1.0-g tissue. If the tissue for dissociation is smaller than 0.2 g, use a 2.5-mL enzyme mix, and follow the rest of the protocol as written.
19. Inspect the pellet post-RBC lysis. If it is red, then RBC lysis (*see* **steps 10** and **11**) may be repeated before proceeding.

References

1. Abel L, Durmaz A, Hu R et al (2021) Impact of immediate cryopreservation on the establishment of patient derived xenografts from head and neck cancer patients. J Transl Med 19(1):180
2. Beato F, Reverón D, Dezsi KB et al (2021) Establishing a living biobank of patient-derived organoids of intraductal papillary mucinous neoplasms of the pancreas. Lab Investig 101(2):204–217
3. Tsai YH, Czerwinski M, Wu A et al (2018) A method for cryogenic preservation of human biopsy specimens and subsequent organoid culture. Cell Mol Gastroenterol Hepatol 6(2): 218–222
4. He A, Powell S, Kyle M et al (2020) Cryopreservation of viable human tissues: renewable resource for viable tissue, cell lines, and organoid development. Biopreserv Biobank 18(3): 222–227
5. Wohnhaas CT, Leparc GG, Fernandez-Albert F et al (2019) DMSO cryopreservation is the method of choice to preserve cells for droplet-based single-cell RNA sequencing. Sci Rep 9(1):10699
6. Slyper M, Porter CBM, Ashenberg O et al (2020) A single-cell and single-nucleus RNA--Seq toolbox for fresh and frozen human tumors. Nat Med 26(5):792–802
7. Jacob F, Ming GL, Song H (2020) Generation and biobanking of patient-derived glioblastoma organoids and their application in CAR T cell testing. Nat Protoc 15(12):4000–4033
8. Jacob F, Salinas RD, Zhang DY et al (2020) A patient-derived glioblastoma organoid model and biobank recapitulates inter- and intra-tumoral heterogeneity. Cell 180(1): 188–204
9. Leelatian N, Doxie DB, Greenplate AR et al (2017) Preparing viable single cells from human tissue and tumors for cytomic analysis. Curr Protoc Mol Biol 118:25C.1.1–25C.1.23
10. Bojic S, Murray A, Bentley BL et al (2021) Winter is coming: the future of cryopreservation. BMC Biol 19(1):56
11. Valyi-Nagy K, Betsou F, Susma A, Valyi-Nagy T (2021) Optimization of viable glioblastoma cryopreservation for establishment of primary tumor cell cultures. Biopreserv Biobank 19(1): 60–66
12. Lin GL, Monje M (2017) A protocol for rapid post-mortem cell culture of diffuse intrinsic pontine glioma (DIPG). J Vis Exp 121:55360
13. Mallone R, Mannering SI, Brooks-Worrell BM et al (2011) Isolation and preservation of peripheral blood mononuclear cells for analysis of islet antigen-reactive T cell responses: position statement of the T-Cell Workshop Committee of the Immunology of Diabetes Society. Clin Exp Immunol 163(1):33–49
14. Bull M, Lee D, Stucky J et al (2007) Defining blood processing parameters for optimal detection of cryopreserved antigen-specific responses for HIV vaccine trials. J Immunol Methods 322(1–2):57–69
15. McKenna KC, Beatty KM, Vicetti Miguel R, Bilonick RA (2009) Delayed processing of blood increases the frequency of activated CD11b+ CD15+ granulocytes which inhibit T cell function. J Immunol Methods 341(1–2):68–75
16. Lam NY, Rainer TH, Chiu RW, Lo YM (2004) EDTA is a better anticoagulant than heparin or citrate for delayed blood processing for plasma DNA analysis. Clin Chem 50(1):256–257
17. Barra GB, Santa Rita TH, de Almeida VJ et al (2015) EDTA-mediated inhibition of DNases protects circulating cell-free DNA from ex vivo degradation in blood samples. Clin Biochem 48(15):976–981
18. Parpart-Li S, Bartlett B, Popoli M et al (2017) The effect of preservative and temperature on the analysis of circulating tumor DNA. Clin Cancer Res 23(10):2471–2477
19. Risberg B, Tsui DWY, Biggs H et al (2018) Effects of collection and processing procedures on plasma circulating cell-free DNA from cancer patients. J Mol Diagn 20(6):883–892
20. Grievink HW, Luisman T, Kluft C et al (2016) Comparison of three isolation techniques for human peripheral blood mononuclear cells: cell recovery and viability, population composition, and cell functionality. Biopreserv Biobank 14(5):410–415

Chapter 6

Adipocyte–Breast Cancer Cell Co-Culture in Transwells

Soofiyah P. Ayaani, Martin Wabitsch, and Mark Christian

Abstract

Experiments with cell co-culture systems facilitate investigation of the effects of one cell population on another, when the cells are grown in close proximity. Here we describe co-culture of Simpson–Golabi–-Behmel syndrome (SGBS) adipocyte cells with the MCF-7 breast cancer cell line using the Corning® Transwell® 12-mm, 0.4-μm pore polyester membrane insert cell culture system. The SGBS adipocyte cell line, which was developed from cells taken from an infant with Simpson–Golabi–Behmel syndrome is comparable, both functionally and biochemically, to primary preadipocytes. The MCF-7 breast cancer cell line is an ER^+/PR^+ and $HER2^-$ line used very commonly in studies of breast malignancy. Consisting of insert supports with a permeable membrane 'floor,' which sit suspended in wells, the Corning® Transwell® co-culture system allows communication between physically separate cells cultured on the membrane and in the well beneath. This co-culture procedure described here can be applied to analyze the effects of cancer cells on the process of adipogenesis and the changes in cancer cells due to adipocyte-secreted factors.

Key words Simpson–Golabi–Behmel syndrome, SGBS, Adipocyte, MCF-7, Breast cancer, Cell co-culture, Corning® Transwell®

1 Introduction

Cell co-culture techniques allow the detailed study of interactions between different cell populations when these cells are grown in systems that allow them to exert influence on each other [1]. These techniques are essential for studying both natural and atypical interactions between cell populations and are also a means by which the culturing success for select cell populations can be improved [1]. Co-cultures may be simple and consist of two discrete cell populations or complex with multiple cell types and even whole tissue models. The technique can be used for analysis of proliferation and migration, as well as drug research [2], infection studies [3], and cancer experiments [4], to name a few. Examples of co-culture systems include culture dishes, hydrogels, Transwell®

Sherri L. Christian (ed.), *Cancer Cell Biology: Methods and Protocols*,
Methods in Molecular Biology, vol. 2508, https://doi.org/10.1007/978-1-0716-2376-3_6,

plates, 3D scaffolds, microfluidic platforms, and the xCELLigence® RTCA™ DP Cell Proliferation Assay, which allows for the real-time analysis of co-cultured cells [5].

Experiments may be performed with complete mixing of the different cell types or with populations remaining in physically distinct compartments. In the latter, a permeable membrane between the two allows movement of signaling molecules while preventing cell mixing. Various factors must be considered when co-culture experiments are being planned; for example, the (common) culture media to be used and the incubation environment should be appropriate for all cell types in the study, and the system used must allow the appropriate level of cell interaction. Membrane type must support growth of the cell type to be cultured; coating, for example, with gelatin, collagen, or fibronectin may be required. Furthermore, membrane pore size must also be carefully considered to prevent cell migration if this is not desired. If individual cell populations are to be examined after the co-culture ends, there should be no mixing of cells, or cell sorting should be undertaken to ensure cell contamination does not lead to inaccurate results.

The Corning® Transwell® permeable supports are inserts that have a membrane 'floor' and sit suspended in the wells of a plate. The membrane may be polycarbonate, polyester, or collagen-coated polytetrafluoroethylene and permits culture of both adherent and non-adherent cells [6]. When a separate population of cells is cultured in the well below and sufficient media are added to allow movement of signaling molecules between the two compartments, a co-culture system is created (*see* Fig. 1).

Here, the co-culture of SGBS adipocytes and MCF-7 breast cancer cells using the Corning® Transwell® 12-mm, 0.4-μm pore polyester membrane insert cell culture system will be described.

In 2001, Wabitsch and Brenner et al. [7] published their work describing the preadipocyte cell line they developed from the white subcutaneous adipose tissue of an infant with Simpson–Golabi–Behmel syndrome (SGBS). The cells are neither transformed nor immortalized, but can be proliferated to at least 50 generations [8]. The cell line is recognized as both functionally and

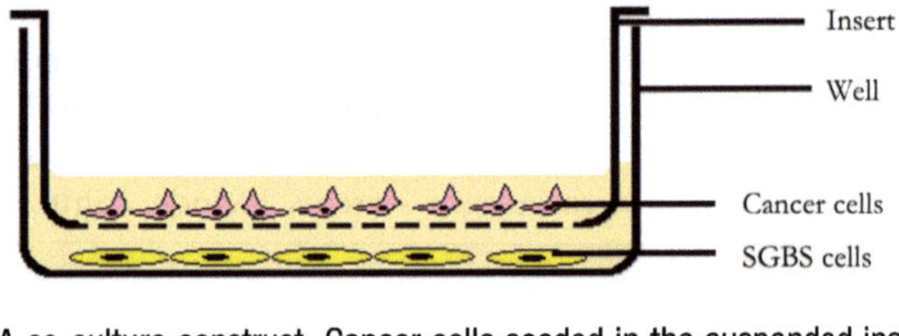

Fig. 1 A co-culture construct. Cancer cells seeded in the suspended insert and Simpson–Golabi–Behmel syndrome (SGBS) adipocytes seeded in the well below form a co-culture, with signaling molecules passing between the permeable membrane 'floor' of the insert

biochemically comparable to primary preadipocytes [9]. In contrast to pre-differentiated cells, which retain their proliferative properties, mature adipocytes do not multiply; they have lost their capacity for mitotic activity. For the purposes of laboratory work, therefore, preadipocytes are first expanded in number and then differentiated to obtain mature lipid-containing cells. They are adherent cells, cultured in DMEM/F12 media and 1% penicillin/streptomycin, with various further additives required dependent upon the stage of growth and differentiation.

The MCF-7 breast cancer cell line fits a *Luminal A* molecular sub-classification pattern, being ER^+/PR^+ and $HER2^-$. It was established in 1973, at the Michigan Cancer Foundation [10]. The epithelial-like adenocarcinoma cells were taken from a metastatic pleural effusion in a 69-year-old Caucasian female and immortalized. The cells are adherent and cultured in DMEM/F12 media with 10% fetal bovine serum (FBS), supplemented with 1% penicillin/streptomycin. They are the most commonly used breast cancer cell line in the world [10].

2 Materials

1. SGBS cells; frozen aliquots of 1×10^5 cells/mL.
2. MCF-7 cells; frozen aliquots of 1×10^5 cells/mL.
3. Trypsin–EDTA solution 1× (trypsin).
4. Phosphate-buffered saline (PBS).
5. Gelatin (from bovine skin) solution: 0.01% *w/v* gelatin, sterile water.
6. 4% paraformaldehyde (PFA).
7. 0F medium for SGBS cells: DMEM/F12 with 1% *v/v* penicillin/streptomycin, 3.3-mM pantothenate, and 1.7-mM biotin.
8. Quick differentiation (QD) medium for SGBS cells: 0F medium with 0.01-mg/mL apo-transferrin, 20-nM insulin, 100-nM cortisol, 0.2-nM triiodo-L-thyronine, 25-nM dexamethasone, 250-μM 3-isobutyl-1-methylxanthine, and 2-μM rosiglitazone (*see* Table 1).
9. 3-FC medium for SGBS cells: 0F medium with 0.01-mg/mL apo-transferrin, 20-nM insulin, 100-nM cortisol, and 0.2-nM triiodo-L-thyronine (*see* Table 1).
10. General medium for SGBS cells (GM_{SGBS}): 0F medium with 10% *v/v* fetal bovine serum.
11. General medium for MCF-7 cells ($GM_{MCF\text{-}7}$): DMEM/F12 with 10% *v/v* fetal bovine serum and 1% *v/v* penicillin/streptomycin.

Table 1
Media supplements for SGBS cell differentiation

Compound	QD	3-FC
Transferrin	0.01 mg/mL	0.01 mg/mL
Insulin	20 nM	20 nM
Cortisol	100 nM	100 nM
T_3	0.2 nM	–
Dexamethasone	25 nM	–
IBMX	250 μM	–
Rosiglitazone	2 μM	–

T_3 triiodo-L-thyronine, *IBMX* 3-isobutyl-1-methylxanthine

12. Oil Red O (ORO) stock solution: 500-mg powdered ORO in 200-mL 100% isopropanol.
13. T75 flask.
14. 6-well cell culture plate.
15. Corning® Transwell® 12-mm, 0.4-μm pore polyester membrane insert cell culture plate.

3 Methods

All steps are to be carried out at room temperature, unless otherwise specified, in a biosafety cabinet/laminar flow hood of the appropriate class. Media and PBS should be warmed to 37 °C.

3.1 Pre-differentiated SGBS Cell Retrieval

1. Partially thaw the frozen pre-differentiated SGBS cells in a water bath set to 37 °C by holding the vial in the water for approximately 30 s (*see* **Note 1**).
2. Add 1 mL of GM_{SGBS} to the vial to complete the thaw and then transfer the contents to a 50-mL centrifuge tube containing a further 20-mL GM_{SGBS}. Mix to ensure complete suspension (*see* **Note 2**).
3. Centrifuge the cells (200 *g*, 5 min), and then aspirate the supernatant to remove the diluted freezing medium. Resuspend the cell pellet in 10-mL fresh GM_{SGBS}.
4. Count the cells and then plate in a flask at a density of 2×10^5 cells/cm^2, and culture in an incubator (5% CO_2, 37 °C) to 80% confluence. Change medium twice per week (*see* **Note 3**).

3.2 Pre-differentiated SGBS Cell Subculture

1. When 80% confluence is achieved, aspirate medium from the flask and wash the cells with PBS warmed to 37 °C (*see* **Note 4**). Aspirate and discard the PBS.
2. Add 750-μL trypsin to a T75 flask (*see* **Note 5**), and incubate (5% CO_2, 37 °C, 5 min) to promote cell detachment.
3. Add 20-mL medium as a trypsin-neutralizer once the cells have lifted, and gently pipette the cell/trypsin/medium mix up and down to ensure cells are not clumped.
4. Centrifuge the cells (200 *g*, 10 min), and resuspend the pellets in fresh GM_{SGBS}. Divide the cells at a ratio of 1:3 to continue culturing.

3.3 SGBS Cell Differentiation

1. Using sterile forceps, transfer the inserts from a 6-well Corning® Transwell® plate to a standard 6-well plate and store at 4 °C, wrapped in plastic, until required (*see* Subheading 3.6).
2. Pipette 3-mL 0.01% gelatin solution into each well of the now-empty 6-well Corning® Transwell® plate and incubate at room temperature for 30 min. Aspirate and discard the gelatin solution and allow the plate to air dry in a biological safety cabinet (*see* **Note 6**).
3. Plate undifferentiated SGBS cells in the gelatin-coated plate (*see* **Note** 7), in GM_{SGBS}, at a density of 2×10^5 cells/well. Culture to 100% confluence and then leave for a further 48 h before commencing the differentiation process (*see* **Note 8**).
4. At 48 h, remove GM_{SGBS} and gently wash the cells twice with PBS. Once all the FBS-containing GM_{SGBS} and the PBS have been removed, add 3 mL of QD medium to each well (*see* **Note 9**). From the addition of QD (Day 0), cells require 14 days to differentiate (*see* Fig. 2) into fully mature adipocytes ready for experimentation (*see* **Note 10**).
5. On Day 4, switch the medium from QD to 3-FC (PBS washing not required), and change the 3-FC medium on Days 8 and 12 (*see* Fig. 2). The cells are ready for use from Day 14 (*see* Fig. 3; *see* Subheading 3.7 for Oil Red O Staining of Adipocytes Protocol).

3.4 MCF-7 Cell Retrieval

1. Partially thaw the frozen MCF-7 cells in a water bath set to 37 °C by holding the vial in the water for approximately 30 s (*see* **Note 1**).
2. Add 1 mL of $GM_{MCF\text{-}7}$ to the vial to complete the thaw and then transfer all to a 15-mL centrifuge tube containing a further 8-mL $GM_{MCF\text{-}7}$. Gently pipette up and down to ensure complete suspension.

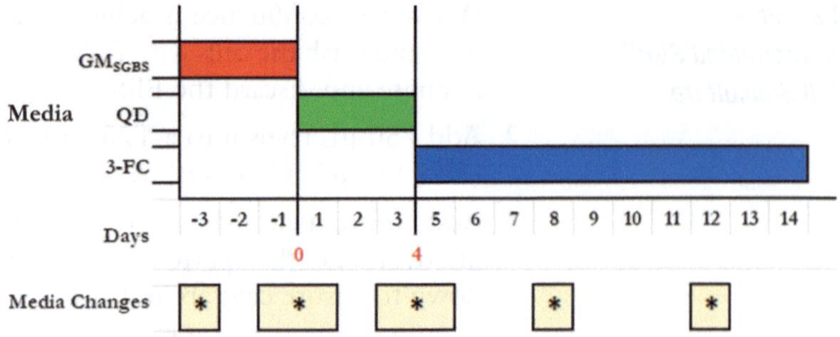

Fig. 2 Media used in the Simpson–Golabi–Behmel syndrome (SGBS) differentiation process. GM_{SGBS} is used during cell proliferation until Day 0 when differentiation is initiated by replacing GM_{SGBS} with Quick Differentiation (QD) media. On Day 4, 3-FC media replace QD. 3-FC is changed every 4 days, until Day 14 when cells are fully differentiated and ready for use

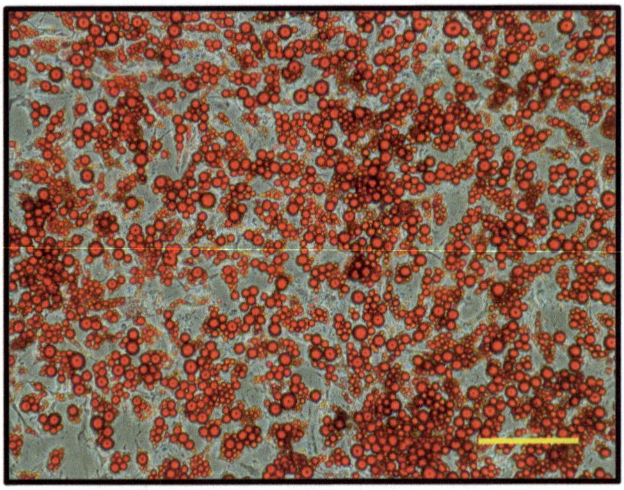

Scale bar: 0.1 mm

Fig. 3 Light microscopy image of Day 14 fully differentiated Simpson–Golabi–Behmel syndrome (SGBS) adipocytes stained with Oil Red O (ORO). Scale bar: 0.1 mm

3. Centrifuge the cells (250 *g*, 5 min), and then aspirate the supernatant to remove the diluted freezing medium. Resuspend the cell pellet in a 10-mL fresh GM_{MCF-7}.
4. Plate cells in a flask at a density of 2×10^5 cells/cm^2, and culture in an incubator (5% CO_2, 37 °C). Change medium the day after thawing and twice per week thereafter.
5. To maintain cells in the logarithmic growth phase, subculture when 80–85% confluence is achieved.

3.5 MCF-7 Cell Subculture

1. When 80–85% confluence is achieved, aspirate medium from the flask and wash the cells with PBS warmed to 37 °C. Aspirate and discard the PBS.
2. Add 1.5-mL trypsin to a T75 flask (*see* **Note 5**), and incubate (5% CO_2, 37 °C, 5 min) to promote cell detachment.
3. Add 7.5-mL $GM_{MCF\text{-}7}$ as a trypsin-neutralizer once the cells have lifted, and gently pipette the cell/trypsin/medium mix up and down to ensure cells are not clumped. Transfer to a 15-mL centrifuge tube.
4. Centrifuge the cells (200 *g*, 5 min), and resuspend the pellets in fresh $GM_{MCF\text{-}7}$. Divide the cells at a ratio of 1:4 to continue culturing.

3.6 Cell Co-Culture

1. Co-culture should be commenced once SGBS cells have been differentiated to maturity (*see* Subheading 3.3). On Day 14, replace the 3-FC medium with 2.5-mL/well GM_{SGBS} (*see* **Note 4**).
2. Incubate (5% CO_2, 37 °C, 1 h) six 24-mm Corning® Transwell® inserts (0.4-μm pore polyester membranes), placed in a standard 6-well plate, with 2-mL GM_{SGBS} per well as per manufacturer's instruction, to allow for a greater membrane permeability.
3. Post-incubation, aspirate the media, and seed cancer cells in the inserts (*see* Fig. 4), at a density of 4×10^4 cells/cm^2 (*see* **Note 11**).
4. Transfer these inserts to the base plate containing the differentiated SGBS cells, to commence co-culture (*see* Fig. 1). Continue the co-culture for as long as required. If media change is required, deconstruct the co-culture to prevent loss of the SGBS cells due to the shearing force of pipetting media.
5. After completion of the co-culture experiments, cells may be fixed and stained on, or removed using trypsin from, the inserts as required [11]. Trypsinization of cells from inserts is as per trypsinization from standard flasks and plates. The manufacturer suggests 600-μL trypsin for a single insert; however, a volume as small as 75-μL should be sufficient.

3.7 Oil Red O Staining of Adipocytes

Oil Red O (ORO) is a lysochrome that stains lipid material an orange-red color and may be used to stain and visualize lipid droplets in differentiated adipocytes (*see* Fig. 3).

1. The stock solution ORO is prepared by dissolving 500 mg of powdered ORO in 200-mL 100% isopropanol at 56 °C for 1 h. Protect the solution from exposure to light.
2. A working solution of ORO is prepared the day before the staining procedure is performed (*see* **Note 12**). Dilute the stock

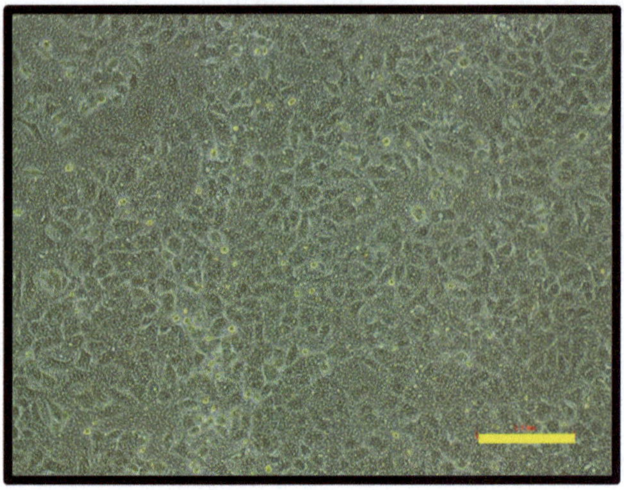

MCF-7 cells (pores just visible)

Scale bar: 0.1 mm

Fig. 4 Light microscopy image of MCF-7 breast cancer cells seeded in an insert (pores just visible). Scale bar: 0.1 mm

ORO solution at 3:2 stock:distilled water *v/v*, (e.g., 120-mL ORO stock + 80-mL water) and stir for 10 min to produce a working solution. Allow the solution to stand at room temperature overnight. Protect the solution from exposure to light.

3. When ready to stain, aspirate media from the SGBS adipocytes and wash the cells with warmed PBS.
4. Fix the adipocytes with 4% paraformaldehyde (PFA) for a minimum of 15 min.
5. Whist the cells are being fixed, pass the working ORO solution through a 0.22-μm filter to remove any precipitate.
6. After PFA aspiration and further PBS washing of adipocytes, add 500-μL of ORO working solution to each 24-mm well of the 6-well plate (adjust according to well size, as required, so that cells are completely covered), and incubate (room temperature, 60 min).
7. After incubation, remove the ORO solution and wash the cells first with 60% isopropanol (*see* **Note 13**) and then with PBS.
8. Capture images using an inverted cell culture microscope such as a Nikon Eclipse TS100F with DS-Fi2 digital color camera and DS-L3 camera control unit.

4 Notes

1. Be sure not to submerge the entire vial; as the water in the water bath is not sterile, submersion of the cap risks contamination of the cells. On lifting the vial out of the water bath, dry it off and then wipe with alcohol.
2. Ensure care is taken when pipetting and suspending to avoid excessive trauma to the cells.
3. SGBS cells are very sensitive to temperature changes; minimize the time they are out of the incubator.
4. It is very easy to shear cells off the surface when adding PBS or medium. Always gently pipette in, aiming for the wall of the flask or well.
5. Adjust trypsin volume appropriately to flask size, if different, such that all cells are just covered in a thin layer of trypsin.
6. Multiple plates may be gelatin-coated up to 4 weeks in advance if stored securely wrapped in plastic bags, at 4 °C. If the coating appears discolored or cracked, do not use the plates.
7. Although mature SGBS cells can be trypsinized and transferred between vessels with extreme care, it is easier and much less traumatic to the cells to differentiate them in the vessel of final experimentation. For this reason, it is easier to seed them in the well and differentiate, but mature cells can be seeded in inserts if needed. No specific change in protocol is needed, but extra care should be taken.
8. Packed SGBS cells differentiate better than cells at a lesser confluency. Further, a pause of 48 h allows the cells to move out of the proliferative phase of cell growth, making them more responsive to the chemical stimuli of differentiation.
9. FBS hinders cell differentiation. It is therefore *essential* that all traces are removed with careful but thorough washing with PBS. If cells are failing to differentiate, it may be that this washing is not thorough enough.
10. The presence of lipid droplets, usually seen from about Day 5–7 onwards, indicates successful commencement of the differentiation process. Formal assessment of lipid droplets can be undertaken with Oil Red O staining (*see* Subheading 3.7).
11. Seed the MCF-7 cells with the inserts in the regular 6-well plate; do not seed them placed in the Transwell® plate with the SGBS cells because the media movement will shear SGBS cells.
12. Achieving a precipitate-free stain can be challenging; the method described produces excellent staining. Preparing the ORO working solution the day before use and filtering on

the day of staining are essential steps. If a precipitate is seen to form prior to staining, filter the solution again. If precipitates form while staining, try shortening the stain time.

13. Keeping the washes with 60% isopropanol as short as possible prevents stripping the stain out of the cells. Incubation of stained cells with 100% isopropanol will remove the ORO from the cells which can be quantified using a spectrophotometer (absorbance 490–520 nm).

References

1. Goers L, Freemont P, Polizzi KM (2014) Co-culture systems and technologies: taking synthetic biology to the next level. J R Soc Interface 11(96):20140065
2. Wu MW, Huang SB, Lee GB (2010) Microfluidic cell culture systems for drug research. Lab Chip 10:939–956
3. Duell BL, Cripps AW, Schembri MA, Ulett GC (2011) Epithelial cell coculture models for studying infectious disease: benefits and limitations. Biomed Res Int 2011:852419. https://doi.org/10.1155/2011/852419
4. Cattaneo CM, Dijkstra KK, Franchi LF et al (2019) Tumor organoid–T–cell coculture systems. Nat Protoc 15:15–39
5. Dowling CM, Ors CH, Kiely PA (2014) Using real-time impedance-based assays to monitor the effects of fibroblast-derived media on the adhesion, proliferation, migration and invasion of colon cancer cells. Biosci Rep 34(4):e00126. https://doi.org/10.1042/BSR20140031
6. Corning. Transwell®, Snapwell®, Netwell, and Falcon® Permeable Supports. Accessed at https://www.corning.com/worldwide/en/products/life-sciences/products/permeable-supports/transwell-snapwell-netwell-falcon-permeable-supports.html. Accessed on 24 Apr 2021
7. Wabitsch M, Brenner RE, Melzner I et al (2001) Characterization of a human preadipocyte cell strain with high capacity for adipose differentiation. Int J Obes Relat Metab Disord 25(1):8–15
8. Fischer-Posovszky P, Newell FS, Wabitsch M et al (2008) Human SGBS cells – a unique tool for studies of human fat cell biology. Obes Facts 1:184–189
9. Rosenow A, Arrey TN, Bouwman FG (2010) Identification of novel human secreted proteins by using SGBS cells. J Proteome Res 9(10): 5389–5401
10. Holliday DL, Speirs V (2011) Choosing the right cell line for breast cancer research. Breast Cancer Res 13:215. https://doi.org/10.1186/bcr2889
11. Corning. Guidelines for Use. Transwell® Permeable Supports Including Snapwell™ and Netwell™ Inserts. Accessed at https://www.corning.com/worldwide/en/products/life-sciences/products/permeable-supports/transwell-guidelines.html. Accessed on 24 Apr 2021

Chapter 7

Three-Dimensional Co-Culture Method for Studying Interactions Between Adipocytes, Extracellular Matrix, and Cancer Cells

Emmanuel C. Asante, Nikitha K. Pallegar, Alicia M. Viloria-Petit, and Sherri L. Christian

Abstract

Three-dimensional (D) culture models are increasingly becoming the model of choice for studying different biological phenomena such as cell–cell interaction, drug resistance, and gene expression. These models include extracellular matrix (ECM) proteins that better model the in vivo conditions as it allows cells to have both cell–cell and cell–ECM contacts. In the context of the tumor microenvironment, there are additional types of cells present in addition to the ECM. Thus, an intermediate between 2D cell culture and in vivo mouse models can be desired to interrogate the interactions between multiple cell types under the influence of the ECM. Here we describe a 3D co-culture technique for studying breast cancer—adipocyte interactions. This technique could easily be modified to analyze interactions between other cancer cell types and different fibroblast-like cells.

Key words 3D culture, Breast cancer, Extracellular matrix, Adipocytes, Immunofluorescence

1 Introduction

Cell cultures have become an indispensable tool for studying different physiological and pathological processes in vitro. There are two kinds of cell culturing techniques: the conventional 2-dimensional (D) monolayer culture and 3D culture. In 2D culture, cells grow in monolayer on a flat rigid substrate usually made of plastic, while in 3D culture, cells grow in aggregates on a scaffold or matrix that mimics the in vivo extracellular matrix. [1]. Cells in 2D cell culture lack the 3D organization of cells between each other or with the extracellular matrix (ECM) as observed in organs and tissues in vivo [2]. In vitro 3D cell culture methods were thus developed to better mimic in vivo conditions and bridge the gap between in vitro and in vivo experiments [1, 3]. Unlike cells in 2D cultures, cells grown in 3D obtain a more physiological

Sherri L. Christian (ed.), *Cancer Cell Biology: Methods and Protocols*,
Methods in Molecular Biology, vol. 2508, https://doi.org/10.1007/978-1-0716-2376-3_7,

morphology, displaying aggregate structures with prevalent cell junctions. Moreover, cells in 3D obtain phenotypic heterogeneity with a varied cell proliferation rate, gene expression, and differentiation within one population [4]. Exposure to nutrients, growth factors, or drugs is also heterogeneous where cells on the outer side of a spheroid are more exposed compared to cells in the inner core, which is more similar to in vivo conditions. In addition, cells in 3D have greater viability and less susceptibility to external factors and show increased resistance to drug-induced stimuli [5, 6]. Due to its in vivo semblance, 3D culture models could potentially decrease the need for animal use in pre-clinical studies such as drug testing and pharmacological screening, gene and protein expression studies, and cancer research [7–10]. In cancer research, 3D models have been employed in studying tumor cell signaling pathways, tumor cell–ECM interaction, invasion, metastasis, and angiogenesis [1]. We have established a novel 3D co-culture system to co-culture adipocytes with triple-negative breast cancer cells in a manner that mimics an in vivo milieu as it includes a laminin-rich extracellular matrix [11]. Our model can be used to study adipocyte–breast cancer interaction or may be modified to study other cell types, including co-culture of other types of cancer with adipocytes or other fibroblast-like cells.

2 Materials

2.1 Adipogenesis

1. 3T3-L1 pre-adipocyte cell line.
2. DMEM/NCS: high-glucose Dulbecco's Modified Eagle Medium, 10% newborn calf serum (NCS), 1% sodium pyruvate, and 1% penicillin/streptomycin.
3. DMEM/FBS: high-glucose Dulbecco's Modified Eagle Medium, 10% fetal bovine serum (FBS), 1-mM sodium pyruvate, and 50-units/ml penicillin/streptomycin.
4. 0.5 M 3-isobutyl-1-methylxanthine (IBMX) stock.
5. 10-mM Dexamethasone (Dex) stock.
6. 10-mg/ml insulin stock.

2.2 Three-Dimensional Culture

1. MDA-MB-231 cell line.
2. Complete RPMI: RPMI 1640 medium with L-glutamine and 11-mM glucose, 10% FBS, and 1% penicillin/streptomycin.
3. Mammary epithelial basal media (co-culture media; PromoCell): Basal media plus 10-ng/ml epidermal growth factor, 5-μg/ml insulin, 0.5-μg/ml hydrocortisone, 0.4% bovine pituitary extract, and 2% Matrigel.
4. 0.05% Trypsin–EDTA.

5. Matrigel: a basement membrane matrix growth factor reduced, phenol red free, Cat# 356231, 9.1-mg/ml or similar, BD Bioscience (*see* **Note 1**).
6. 4-well chambered slides: e.g., Cat# 354104, Corning-Life Science, New York, NY, USA.

2.3 Fixation and Immunofluorescence (IF) Staining

1. 4% paraformaldehyde, pH 7.4 (*see* **Note 2**).
2. 1× Phosphate-buffered saline (PBS).
3. Permeabilization buffer: 0.5% Triton X-100 in 1× PBS (*see* **Note 2**).
4. PBS–glycine solution: 100-mM glycine in 1× PBS (*see* **Note 2**).
5. IF buffer: 0.1% BSA, 0.2% Triton X-100, 0.05% Tween 20, 1× PBS (*see* **Note 2**).
6. Blocking buffer: 10% donkey serum in IF buffer.
7. 5 mg/ml DAPI stock.
8. Primary and secondary antibodies, as per the experimental objectives.
9. Coverslip.
10. Anti-fade mounting media.
11. Silicone sealant (e.g., GE clear multi-purpose silicone sealant II).
12. Parafilm.
13. 20–36G needle.
14. 10-ml syringe.

3 Method

Culture media and cells must be prepared and handled using aseptic techniques. All procedures pertaining to cell culture should be carried out in a class II biosafety cabinet. All cell cultures are conducted in a humidified incubator at 37 C, in the presence of 5% CO_2 and 95% atmospheric air.

3.1 Adipogenic Differentiation of Pre-Adipocytes

1. Maintain 3T3-L1 cells in DMEM/NCS until they are approximately 60–70% confluent.
2. Seed 30,000 3T3-L1 cells in 1 ml of DMEM/NCS into a 4-well chambered slide and culture the cells until they are 100% confluent, approximately 48 h (*see* **Note 3**).
3. To initiate adipogenesis, add 1 ml of DMEM/FBS containing 0.5-mM IBMX and 1-μM Dex and incubate for 48 h at 37 °C with 5% CO_2.

4. After 48 h, replace the media with 1 ml of DMEM/FBS containing 10-mg/ml insulin and culture for additional 48 h at 37 °C with 5% CO_2.
5. On the sixth day, cells should be verified for the presence of lipid droplets in >60% of the cells visually under the microscope.

3.2 Three-Dimensional Co-Culture

1. After adipogenic differentiation, gently remove the media without disturbing the differentiated cells and then overlay 110 μl of Matrigel on the mature adipocytes (*see* **Note 4**).
2. Incubate the slide for 1 h at 37 °C to allow the Matrigel to solidify.
3. Meanwhile, trypsinize and count the MDA-MB-231 cells. Dilute the cells to 30,000 cells/ml in co-culture media containing 2% Matrigel. MDA-MB-231 cells are maintained in complete RPMI.
4. After 1 h, overlay the Matrigel with 450 μl of MDA-MB-231 cells containing 13,500 cells.
5. Maintain the co-culture for 5 days by changing media for every 2 days with co-culture media containing 2% Matrigel to allow colony formation.

3.3 Fixation

1. On day 5, remove culture media from the wells of the chambered slide and wash the co-cultured cells gently with 500 μl 1× PBS.
2. Fix cells with 800 μl of 4% paraformaldehyde for 20 min at room temperature (*see* **Note 5**).
3. Wash cells with 500 μl of 1× PBS for 15 min on an orbital shaker on low speed.
4. Repeat **step 3** two more times to remove any residual fixative. Proceed to Subheading 3.4 for staining or store at 4 °C until further use.
5. For storage, add 1 ml of 1× PBS to the wells, place the lid on chamber slide, seal it with parafilm, place the slide on the lid inside a 10-cm plastic culture dish surrounded with wet kimwipes, and seal the plate with parafilm. The cells can be stored for a week before staining.

3.4 Immunofluorescence Staining and Slide Mounting

1. Wash the cells with 1× PBS before adding permeabilization buffer, if continuing from Subheading 3.3, **step 5**; otherwise proceed directly to Subheading 3.4, **step 2**.
2. Permeabilize cells with 500 μl of permeabilization buffer for 10 min at 4 °C.
3. Wash cells with 500 μl of 1× PBS for 15 min on an orbital shaker on low speed.

4. Repeat **step 3** two more times to remove any residual permeabilization buffer.
5. Block with 500 μl of blocking buffer for 1 h at room temperature.
6. Dilute primary antibody of interest in blocking buffer, as per the manufacturer's instructions, and add 300 μl to each well.
7. Place the lid on chamber slide and seal it with parafilm, as directed in **step 5** in Subheading 3.3.
8. Incubate cells with the diluted antibody at 4 °C overnight.
9. Wash cells with 500 μl of IF buffer for 15 min on an orbital shaker at low speed.
10. Repeat **step 9** two more times to remove any residual antibody.
11. Dilute appropriate secondary antibody in blocking buffer, and incubate cells with 300 μl of the secondary antibody solution for 1 h at room temperature on an orbital shaker (*see* **Note 6**).
12. Repeat **steps 9** and **10** to remove any residual secondary antibody.
13. Counterstain cells with 300 μl of DAPI diluted 1:800 in 1× PBS at room temperature for 15 min on an orbital shaker at low speed.
14. Wash cells with 500 μl of 1× PBS for 5 min on a shaker at low speed.
15. Discard 1× PBS and separate the chamber from the slide according to the manufacturer's instructions (*see* Fig. 1).
16. Gently wipe the edges of the slide with delicate sorbent tissue without disturbing the Matrigel.
17. Gently place silicone sealant around the edges of the slide as shown in the figure below and wait for 3 min. Use a 20G needle and syringe to get even piping of silicone around the slide as shown in Fig. 2 (*see* **Note 7**).
18. Place 1–2 drops of anti-fade mounting solution on each section of the slide to cover the culture. Care must be taken to avoid bubbles as this might affect image acquisition. Bubbles can be removed by pipetting without disturbing the co-culture.
19. Allow the silicone to set for 2 min (*see* **Note 8**).
20. Place a coverslip over the posterior or anterior end of the slide at an angle of 25°. Gently lower the coverslip while using a forceps to apply gentle pressure at the back of the coverslip. Slide the forceps along as the coverslip is being lowered to cover the entire slide. This is to expel air and bubbles that may be created during the mounting procedure (*see* **Note 9**).
21. Place the slide in a dark chamber for at least 1 h to allow the silicone sealant to set (*see* **Note 10**).

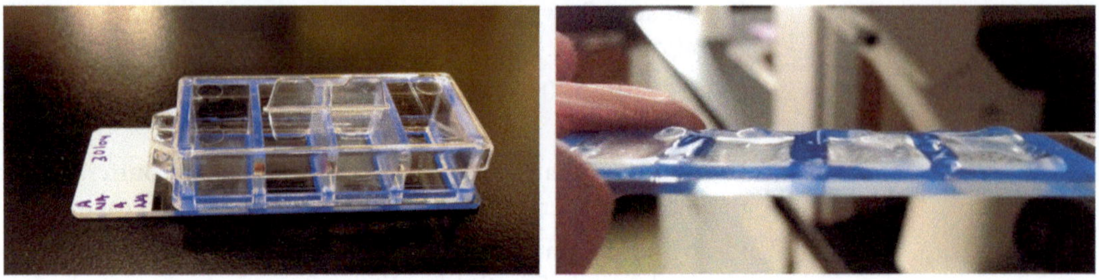

Fig. 1 Three-dimensional culture-containing chambered slide, (left) prior to and (right) after the detachment of the plastic chamber

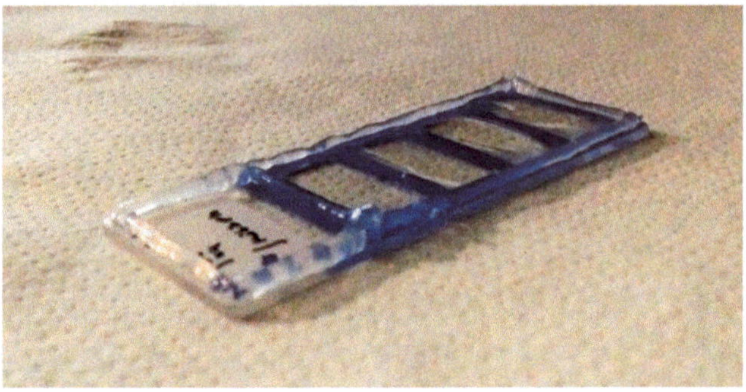

Fig. 2 Glass slide with silicone sealant placed around its edges following the detachment of the chamber

22. Let slides set overnight and at room temperature, and then image using a confocal microscope or store at 4 °C for up to 1 week. For example of images that can be obtained (*see* Fig. 3).

4 Notes

1. There can be lot-to-lot variation between batches of Matrigel. Try to obtain the same lot number between experiments or, if that is not possible, obtain a batch with a similar protein concentration.
2. Fixative and IF reagent can be stored at 4 °C up to 1 month. For optimum results, ensure that reagents do not exceed 1 month in storage.
3. Chambered slides of other dimensions can be substituted as long as volumes and cell numbers are adjusted according to the relative surface area of the chambers.
4. Matrigel solidifies at room temperature so it must always be kept on ice when in use. All the pipet tips and tubes used for Matrigel need to be pre-cooled before using. Aliquots of Matrigel can be

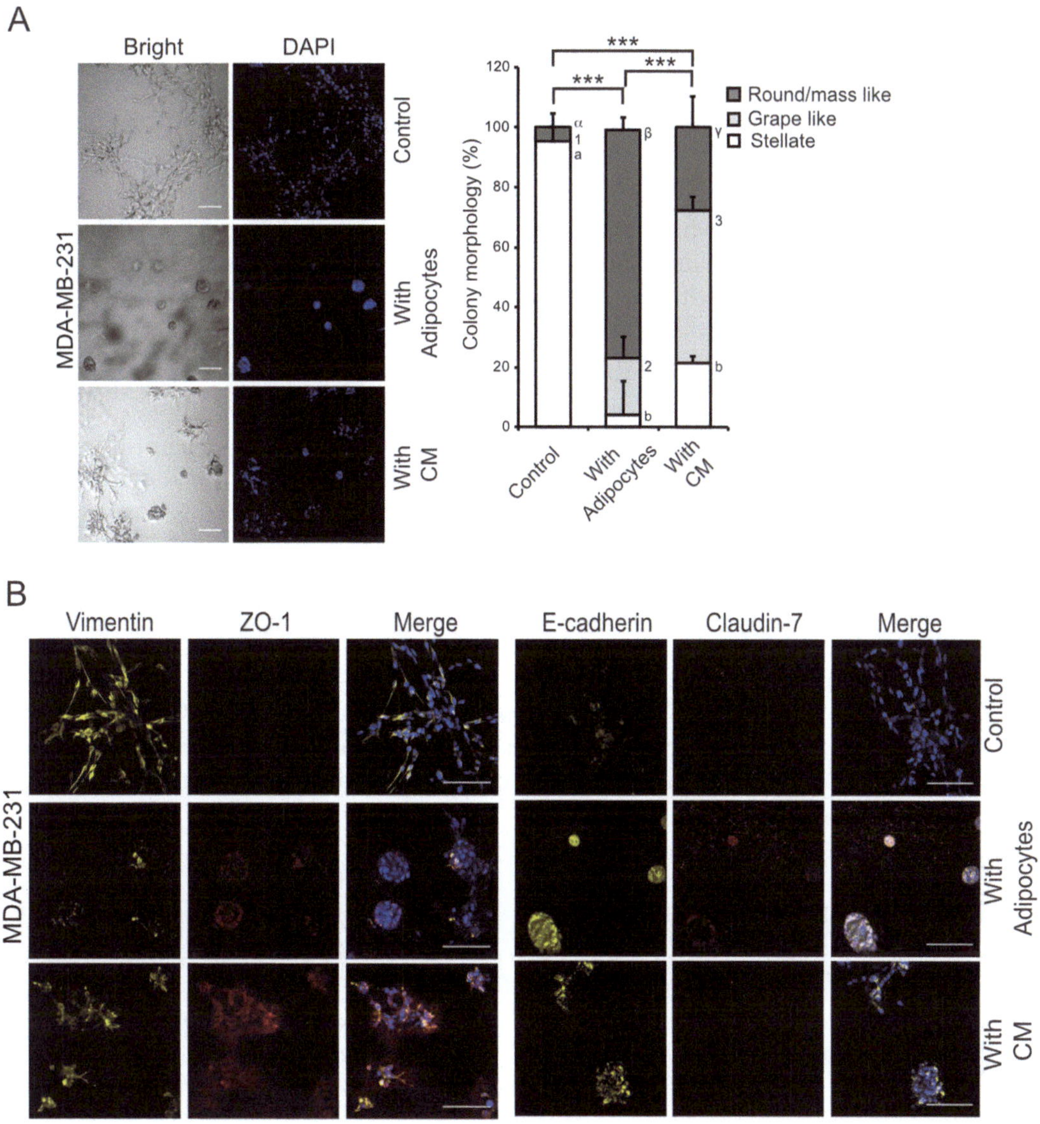

Fig. 3 (**a**) Representative images of MDA-MB-231 cells grown in the 3D co-culture system without adipocytes, with adipocytes, or with adipocyte-conditioned media (CM), $n = 3$. (**b**) Representative images of MDA-MB-231 cells cultured with or without adipocytes or CM and co-stained with anti-vimentin, anti-ZO-1, and DAPI (left panel) or with anti-E-cadherin, anti-claudin-7, and DAPI (right panel) (scale bar = 50 μm), $n = 3$. Reproduced from Pallegar (2019) with permission from Springer Nature

kept at −20 °C or 4 °C for long-term and short-term storage, respectively. Avoid multiple freeze–thaw cycles for best results.

5. Following fixation, cells can be stored at 4 °C for up to 1 week prior to immunofluorescence staining. When cells must be stored, add 1 ml of 1× PBS to each chamber and wrap the chambered slide with parafilm to prevent the cells from drying out.

6. Since most immunofluorescence secondary antibodies are conjugated with light-sensitive fluorophores, this and subsequent steps must be carried out in the dark.
7. Silicone piping may be placed surrounding the entire slide or just around the chamber area.
8. Do not exceed the time mentioned as the silicone will solidify, which prevents the coverslip from adhering properly to the slide. The silicone should be firm enough to hold the coverslip without compacting the Matrigel but soft enough so it doesn't resist the coverslip.
9. If air bubbles are trapped, re-apply the coverslip.
10. Clear nail polish can be applied around the edges of the slide after this step. This will seal the slide and prevent the anti-fade mounting solution from leaking out of the slide. The slide must be protected from light following this step point to prevent photobleaching.

Acknowledgments

SLC is a senior scientist with the Beatrice Hunter Cancer Research Institute. This work was funded by the Cancer Research Society. We thank Eman Elbakry for critical reading of the manuscript.

References

1. Ravi M, Paramesh V, Kaviya S, Anuradha E, Solomon FP (2015) 3D cell culture systems: advantages and applications. J Cell Physiol 230(1):16–26
2. Duval K, Grover H, Han L-H, Mou Y et al (2017) Modeling physiological events in 2D vs. 3D cell culture. Physiology 32(4): 266–277
3. Kenny PA, Lee GY, Myers CA, Neve RM et al (2007) The morphologies of breast cancer cell lines in three-dimensional assays correlate with their profiles of gene expression. Mol Oncol 1(1):84–96
4. Luca AC, Mersch S, Deenen R, Schmidt S et al (2013) Impact of the 3D microenvironment on phenotype, gene expression, and EGFR inhibition of colorectal cancer cell lines. PLoS One 8(3):e59689
5. Aljitawi OS, Li D, Xiao Y, Zhang D et al (2014) A novel three-dimensional stromal-based model for in vitro chemotherapy sensitivity testing of leukemia cells. Leuk Lymphoma 55(2):378–391
6. Fang Y, Eglen RM (2017) Three-dimensional cell cultures in drug discovery and development. SLAS Discov 22(5):456–472
7. Thoma CR, Zimmermann M, Agarkova I, Kelm JM, Krek W (2014) 3D cell culture systems modeling tumor growth determinants in cancer target discovery. Adv Drug Deliv Rev 69:29–41
8. Breslin S, O'Driscoll L (2013) Three-dimensional cell culture: the missing link in drug discovery. Drug Discov Today 18(5–6): 240–249
9. Birgersdotter A, Sandberg R, Ernberg I (2005) Gene expression perturbation in vitro—a growing case for three-dimensional (3D) culture systems. In: Seminars in cancer biology, vol 5. Elsevier, pp 405–412
10. Nath S, Devi GR (2016) Three-dimensional culture systems in cancer research: focus on

tumor spheroid model. Pharmacol Ther 163: 94–108

11. Pallegar NK, Garland CJ, Mahendralingam M, Viloria-Petit AM, Christian SL (2019) A novel 3-dimensional co-culture method reveals a partial mesenchymal to epithelial transition in breast cancer cells induced by adipocytes. J Mammary Gland Biol Neoplasia 24(1):85–97

Chapter 8

Modeling Collective Invasion and Single-Cell Mesenchymal Invasion in Three-Dimensional Matrigel–Collagen I Cultures

Erin M. Bell, Marcia L. Graves, Pamela M. Dean, Tate O. Goodman, and Calvin D. Roskelley

Abstract

Understanding the modes and mechanisms of tumor cell invasion is key to developing targeted therapies against metastatic disease. In vitro assays modeling tumor progression have primarily been optimized for studying classical single-cell migration through an epithelial–mesenchymal transition (EMT). Although experimental and clinical histopathological evidence has revealed that tumor invasion is plastic and that epithelial carcinomas can invade by a range of modes that vary from single, mesenchyme-like cells, all the way to cohesive, collective units, few in vitro assays have been designed to assess these modes specifically. Thus, we have developed a Matrigel–Collagen I overlay assay that is suitable for identifying and quantifying both collective and mesenchymal invasion. This three-dimensional (3D) culture assay utilizes the features of Matrigel and Collagen I to mimic the laminin-rich basement membrane and the stiff, fibrillar Collagen I tumor microenvironment allowing for spheroid invasion to be assessed at the interface between these two matrix components.

Key words Tumor invasion, 3D Culture, Matrigel, Collagen I, EMT, Collective invasion

1 Introduction

The classical mechanistic understanding of tumor cell invasion through an epithelial–mesenchymal transition (EMT) primarily came from experimental in vitro assays used in early developmental biology research, such as collagen gel culture, transwell migration, and two-dimensional (2D) culture wound healing scratch assays [1–4]. In the context of cancer biology, EMT describes the process whereby epithelial cells lose expression of key epithelial markers (namely, the adherens junction protein, E-cadherin) by an activation of mesenchymal transcriptional programs [5]. This loss of epithelial marker expression results in single cells breaking away from the primary tumor and migrating as single, mesenchymal cells. Yet, advances in in vivo lineage tracing, imaging technology,

Sherri L. Christian (ed.), *Cancer Cell Biology: Methods and Protocols*,
Methods in Molecular Biology, vol. 2508, https://doi.org/10.1007/978-1-0716-2376-3_8,

and three-dimensional (3D) culture have revealed that tumor cell invasion is plastic and can occur without an overt EMT, by multicellular modes termed collective invasion [6–10]. Collective invasion describes the process whereby epithelial tumor cells retain cell–cell adhesion (through maintenance of junctional E-cadherin) but exhibit dynamic, contractile interactions with the underlying extracellular matrix (ECM) and invade as a cohesive unit [8]. A recent consensus statement by The EMT International Association (TEMTIA) has adopted the term epithelial–mesenchymal plasticity (EMP) to describe conditions by which tumor cells exhibit hybrid epithelial–mesenchymal characteristics [9]. It is clear that tumor invasion is not a binary process and functional understanding of this phenotypic spectra is critical for the development of metastasis-targeted therapeutics.

The extracellular matrix (ECM), comprised of non-cellular proteoglycans and fibrous proteins (including laminin, fibronectin, and collagens), offers structural support for cells within tissues [11]. In addition to this scaffolding function, the ECM provides mechanical and biochemical cues that regulate tissue morphogenesis and gene expression as well as tissue-specific functions [11–13]. Hence, recapitulating the unique conditions of the ECM experimentally is necessary for modeling tumor cell invasion and further characterizing the mode and mechanisms driving such invasion.

Early three-dimensional culture assays utilized laminin-rich extract of Engelbreth–Holm–Swarm (EHS) mouse tumor tissue as a substrate for melanocytes—this extract was later commercialized as reconstituted basement membrane, Matrigel [14]. When cultured on or in Matrigel, cells form organized spheroid structures that can come to resemble the tissue of origin; for instance, endothelial cells form branching tube-like structures, and mammary epithelial cells, given appropriate growth factors, form duct-like structures capable of secreting milk [15, 12].

It is also important to consider the density and stiffness of the ECM. Stromal density has been shown to correlate with poor disease outcomes in breast cancer, for example [16]. Further, invasive carcinoma is often associated with an increase in tissue stiffness and fibrous Collagen I deposition [17–19]. Tumor cells interact with aligned Collagen I fibrils via surface receptors (i.e., integrins) promoting enhanced motility.

Combining the features of both Matrigel (laminin-rich basement membrane) and Collagen I (stiff, epithelial tumor microenvironment component) thus allows for modeling of an invasive tumor at the interface between the two ECM components (normal and tumor associated). Building upon a protocol from the lab of Andrew Ewald [20], we developed a Matrigel–Collagen I overlay assay, to assess modes of tumor invasion in 3D culture [21]. This assay is suitable for single-celled (mesenchymal-like) and multi-celled (epithelial collective) invasion—morphological differences indicative of different modes of invasion can be quantified and perturbed upon experimental alterations in protein expression.

2 Materials

2.1 Aliquoting and Storing Matrigel

1. Corning Matrigel Basement Membrane Matrix, 10 ml (Corning, catalogue #354234).
2. 1.5-ml microfuge tubes.
3. Class II biosafety cabinet (BSC) suitable for tissue culture.

2.2 Coating Coverslips with Matrigel

1. 18-mm diameter round No. 1 Micro Cover Glasses, 0.13–0.17-mm thickness.
2. Forceps, curved tapered ultrafine tip (Fisher, catalogue 12-000-131).
3. 70% Ethanol.
4. Standard tissue culture dish, 60 mm.
5. Polystyrene microplates, 12 wells.
6. Parafilm, double-sized roll.
7. Large orifice pipet tips, 1–200 μl (Fisher, catalogue 02707134 or similar).
8. Tissue culture incubator (humidified, 37 °C, 5% CO_2).

2.3 Plating Cells

1. Tumor cells of interest (i.e., MCF7).
2. Culture media: Dulbecco's Modified Eagle's Medium (DMEM)/F12, L-glutamine, 10% fetal bovine serum (FBS) for MCF7 cells (cell-type dependent).
3. Trypsin–EDTA 0.05%.
4. DMEM/nutrient mixture F12 Ham without $CaCl_2$ and phenol red, powder.
5. 15-ml Screw cap centrifuge tube.
6. Hemocytometer or automated cell counter.
7. Trypan blue.
8. PolyHEMA (Sigma, catalogue P3932).
9. 100-mm Tissue culture dish.
10. Dry oven or incubator at 37 °C.

2.4 Generating Fibrillar Collagen I Overlay

1. Cultrex Rat Tail Collagen I (Cultrex, catalogue 3440-100-01).
2. Dulbecco's Modified Eagle's Medium-High Glucose, powder.
3. Sodium bicarbonate.
4. Sodium hydroxide pellets.

5. Syringe and low-protein-binding syringe filter.
6. Distilled water, sterile.
7. Plastic pH strips, pH 5.5–8.0.
8. Invasion media: DMEM/F12, L-glutamine, 1% fetal bovine serum, 100 ng/ml EGF (epidermal growth factor) as a motility-inducing agent for MCF7 cells (cell-type dependent).

2.5 Imaging and Quantifying Spheroids

1. Inverted microscope with phase contrast and a humidified stage-top imaging chamber. The system we have consists of the following:
   (a) Inverted microscope (Leica DMI4000 B).
   (b) High-resolution digital monochrome camera (DFC345 FX).
   (c) 10×/0.25-na Ph1 objective lens (Leica Microsystems, catalogue 11506260).
   (d) Imaging software (AF6000 Leica Application Suite 3.1.0).
   (e) Humidified stage-top imaging chamber (Live Cell Instrument, catalogues CU-109, TC-L-Z013, TC-L-Z102, TC-L-Z101).
2. Chamlide 35-mm magnetic imaging chamber (Chamlide, CM-B18-1).
3. Image J or Fiji [22].
4. Adobe Illustrator or equivalent.

2.6 Permeabilizing and Staining Spheroids

1. 16% Paraformaldehyde (PFA).
2. OCT compound, if embedding gels.
3. Phosphate-buffered saline (PBS), 1×, without calcium chloride or magnesium chloride, pH 7.4: 137-mM NaCl, 2.7-mM KCl, 10-mM Na_2HPO_4, 1.8-mM KH_2PO_4.
4. Glycine solution: 100-mM glycine in 1× PBS.
5. Permeabilization solution: 0.3% Triton-X100 in 1× PBS (v/v).
6. Bovine serum albumin (BSA) solution: 1% bovine serum albumin (BSA) in 1× PBS (w/v).
7. Blocking solution: 10% normal serum from species in which secondary antibodies are raise, in bovine serum albumin solution (v/v).
8. Staining buffer: 1% BSA in 1× PBS (w/v).
9. Rhodamine phalloidin (Invitrogen, R415).
10. DAPI.
11. Pro Long Gold mounting media (Invitrogen, catalogue P36930).

12. Clear nail polish.
13. Parafilm.
14. Kimwipes.
15. Confocal microscope (i.e., Leica TCS SP5 II Basic VIS, Leica Microsystems).
16. 63× Oil immersion objective lens.

3 Methods

3.1 Aliquoting and Storing Matrigel

When working with Matrigel, it is essential that the Matrigel and all plastics that come in contact with it are kept on ice to avoid early, uneven polymerization of the extracellular matrix, which will result in considerable variation in spheroid formation and invasion across the culture:

1. Thaw the 10-ml stock bottle of Matrigel on ice at 4 °C overnight. Meanwhile, place sterile 1000-μl pipette tips and 1.5-ml microfuge tubes at −20 °C.
2. The following day, aliquot Matrigel on ice in a class II tissue culture biosafety cabinet (BSC) using pre-chilled pipette tips and microfuge tubes (*see* **Note 1**).
3. Store aliquots of Matrigel at −70 °C. Do not freeze-thaw Matrigel aliquots multiple times as this will affect its ability to polymerize efficiently.

3.2 Preparing Sterilized Coverslips

The procedure should be performed in a sterile environment in a class II BSC. Sterilized coverslips can be prepared in advance and parafilm-sealed multi-well tissue culture plates can be stored at −20 °C. Coverslips may also be sterilized the same day as coating, but ample time must be given to allow plates to cool on ice prior to applying Matrigel coating.

1. Place 18-mm coverslips in a 60-mm tissue culture dish containing 70% ethanol. Allow coverslips to soak for at least 15 min, ensuring that coverslips are not sticking together.
2. Using curved, ultrafine tip forceps, carefully remove one coverslip at a time from the ethanol wash. Lean the coverslip against the wall of a 12-well tissue culture plate (at roughly 45° angle). Leave the plate uncovered, allowing the ethanol to evaporate and the coverslip to fully dry, for at least 20 min; the plate lid can be placed on the edge of the dish, sterile side down, and angled down toward the working surface of the BSC (*see* Fig. 1a).
3. Once the coverslips are dry, lay them flat in the center of the well, taking care to ensure that they are not touching the edge

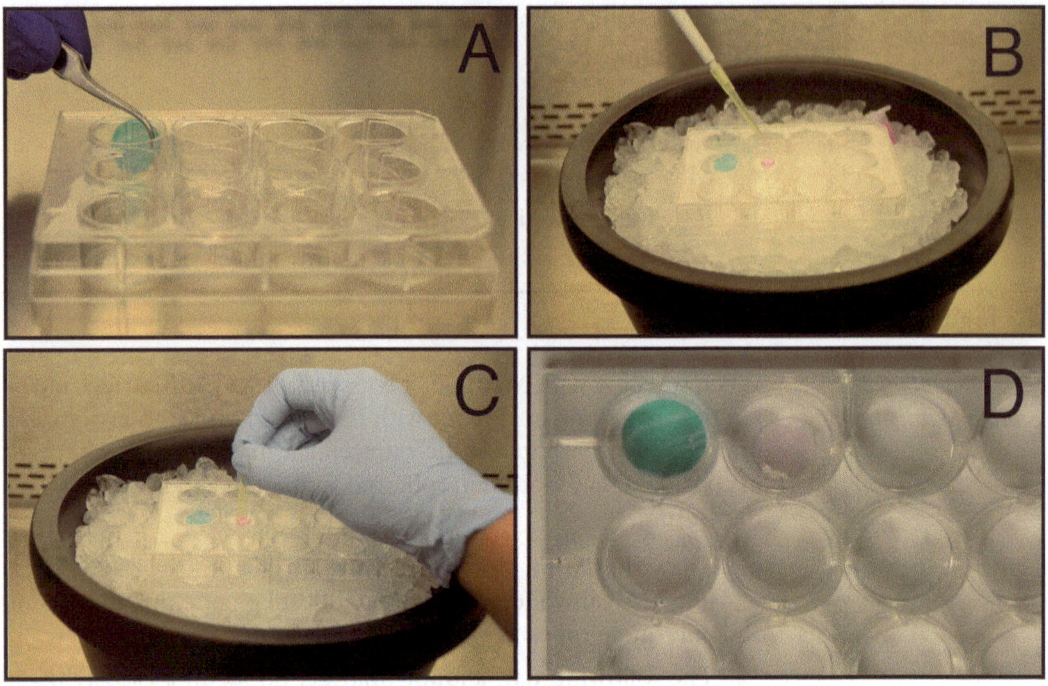

Fig. 1 Coating coverslips with Matrigel. When handling Matrigel ensure that the plate is kept on ice and all pipette tips are pre-chilled at −20 °C. (**a**) Coverslips (colored green for clarity) should be sterilized in 70% ethanol and allowed to dry by leaning the coverslip against the wall of a 12-well plate. (**b**) After ensuring that the coverslip is not touching the walls of the well, Matrigel is pipetted onto the center of the coverslip using a chilled large orifice pipette tip. (**c**) Using the wide bore, top end of a P-200 pipette tip, the Matrigel is spread around the coverslip in one circular motion. (**d**) Plated and spread Matrigel should not wick off the coverslip. The plate will now be placed at 37 °C to polymerize

of the well. Place the plate on ice for at least 1 h, or seal the plate with parafilm and store at −20 °C for use the following day (*see* **Note 2**).

3.3 Coating Sterilized Coverslips with Matrigel

Depending on downstream experimental endpoints and goals (i.e., immunofluorescence staining or growth versus invasion assays), this protocol can be slightly modified (*see* **Note 3**). All steps should be performed using sterile technique in a tissue culture BSC (*see* **Note 4**):

1. Calculate the amount of Matrigel required for the experiment—100-μl Matrigel or diluted Matrigel solution per 18-mm coverslip (approximately 40 μl/cm^2) (*see* **Note 3**).
2. Place the appropriate number of Matrigel aliquots on ice and thaw Matrigel at 4 °C overnight.
3. If diluting Matrigel 1:1 in regular maintenance culture media without serum (i.e., DMEM/F12 for MCF7 3D culture), aliquot the appropriate volume of cold culture media to a

pre-chilled microfuge tube, on ice. Carefully, avoiding the introduction of bubbles, pipette an equal volume of Matrigel into the media using a pre-chilled large orifice pipette tip. Mix gently by pipetting and keep on ice.

4. Ensure that the coverslips prepared previously are not touching the walls of the 12-well plate wells. While keeping the plate on ice and evenly levelled, use a large orifice P-200 pipette tip and gently pipette 100 μl of Matrigel solution (undiluted or diluted in base culture medium) onto the center on the coverslip. Quickly, using the top, wide bore end of a sterile P-200 pipette tip, spread the Matrigel over the coverslip in a single circular motion, ensuring that the Matrigel is not spread too close to the edge of the coverslip as it may wick to the side of the well and become unusable (*see* **Note 5**, Fig. 1b and c).
5. Once all coverslips have been coated, carefully move the plate to a 37 °C tissue culture incubator (humidified, 5% CO_2) and allow the Matrigel to polymerize for 30 min (undiluted Matrigel) or 1 h (Matrigel diluted up to 1:1 with culture media) (*see* Fig. 1d). Once the Matrigel is polymerized, it is important to rehydrate it with culture media. If polymerized Matrigel is left uncovered beyond 1 h, it may begin to dry out and crack. If cells are not ready to be plated after 1 h, add 1 ml of serum-free culture media into the well until the cells are ready to be plated.

3.4 Plating Single Cells on Top of Pre-polymerized Matrigel

Single cells can be plated on top of Matrigel as a means of assessing spheroid formation and invasion, particularly of the mesenchymal variety. Other matrix components (i.e., Collagen I) can be overlaid on top of the Matrigel to form a "sandwich assay" and allow for assessment of the effect of matrices on tumor spheroid invasion, particularly the collective variety (*see* Subheading 3.7) [21]. Plating cells on top of Matrigel, rather than within Matrigel as described in Subheading 3.5, has the advantage of allowing cells to be imaged in a single plane, which is beneficial for both live imaging and confocal microscopy of fixed, immunostained samples.

1. Rinse cells (up to 15 min for highly adhesive cell types) with calcium-free DMEM/F12. Trypsinize briefly with 0.05% Trypsin–EDTA and collect cells as usual. After centrifugation, resuspend cells in serum-free or low-serum media (i.e., 1% FBS, although this can be optimized for each cell type), and count live cells using a hemocytometer or automated cell counter (*see* **Note 6**).
2. Resuspend cells at a final concentration of 100,000 cells/ml. Cells should be thoroughly resuspended such that a single-cell suspension is maintained (*see* **Note 7**).
3. Gently pipette 1 ml of cell suspension on top of the Matrigel coating. It is important that this is done gently and dropwise

into the center of the coverslip to prevent damage to the Matrigel and allow for even distribution of the cells (*see* **Note 8**).

4. Incubate cells overnight in a 37 °C tissue culture incubator (humidified, 5% CO_2) to allow for spheroid formation. Depending on the cell type, more time may be required.

3.5 Plating Single Cells Within Diluted Matrigel

Plating cells within, as opposed to on top of, Matrigel allows for a more accurate assessment of cell proliferation within spheroids because it reduces the chance that an increase in spheroid size is due to aggregation of more than one cell, which can occur when cells are plated on top of Matrigel.

1. Trypsinize and collect cells as usual. Resuspend in cold low-serum media (2% FBS for a final diluted concentration of 1% FBS) (*see* **Note 6**).
2. Aliquot 50 μl of Matrigel into the appropriate number of pre-chilled microfuge tubes for the experiment (1 tube per coverslip).
3. Resuspend cells to 8000 cells/50 μl low-serum media, and keep diluted cells on ice (*see* **Note 7**).
4. Gently pipette 50 μl of cell suspension into the pre-aliquoted Matrigel and plate 80 μl of Matrigel cell suspension onto the center of a sterile coverslip, spreading it evenly with a pipette tip as described earlier. Incubate the coverslips in a 37 °C tissue culture incubator (humidified, 5% CO_2) for 1 h, and then carefully add 1-ml low-serum culture media to each well (*see* **Note 9**) and continue incubation for desired length of time.

3.6 Plating Pre-clustered Spheroids Within Diluted Matrigel

If cells do not readily form spheroids when plated on Matrigel, they can be pre-clustered by culturing on ultra-low attachment plates or tissue culture plates coated with polyhydroxyethylmethacrylate (PolyHEMA) [23].

1. Prepare a PolyHEMA stock solution in 95% ethanol (50 mg/ml). For optimal solubilization, add the desired volume of ethanol to a sterile conical centrifuge tube, and then add the PolyHEMA to the ethanol. This solution may need to be warmed in a 37 °C water bath overnight with gentle agitation to fully solubilize. PolyHEMA stock solutions can be stored at room temperature.
2. Prepare a PolyHEMA working solution in 95% ethanol (5–8 mg/ml). The working concentration may need to be optimized for the cell type.
3. Pipette 5 ml of PolyHEMA working solution into a 10-cm tissue culture dish. Scale the volume of PolyHEMA for the desired size of the tissue culture dish. For MCF7 cells, for example, proper cluster formation required PolyHEMA to be

coated at a minimum of 0.45 mg/cm^2. Place in a 37 °C dry oven for 2–3 days. Avoid moving the dishes until they have completely dried; it is essential to have a smooth coating. Coated dishes can be stored at 37 °C until needed.

4. Before plating cells on PolyHEMA-coated dishes, it is important to rehydrate the PolyHEMA coating. Rinse the coated dishes two times with 5-ml serum-free culture media (i.e., DMEM/F12). Replace the last wash with an additional 5 ml of serum-free media until you are ready to plate cells.
5. Trypsinize, collect, and resuspend cells in serum-free or low-serum culture media (may vary depending on the cell type). Plate up to 12 ml of desired cell concentration on PolyHEMA-coated 10-cm dishes (250,000 cells/ml for MCF7 cells; cell concentration will need to be optimized based on the desired spheroid size and cell type). Incubate the cells on PolyHEMA in a tissue culture incubator (37 °C, humidified, 5% CO_2) for 3–6 h or until the desired cell clusters have formed.
6. During the last hour of cell cluster formation, coat sterile, dry coverslips (*see* Subheading 3.2) with a 2.5% Matrigel solution. To do this, dilute Matrigel in ice cold, serum-free culture media to a final concentration of 2.5% v/v. Coat sterile, dry coverslips with 80 μl of the diluted Matrigel as described earlier, and incubate in the tissue culture incubator (37 °C, humidified, 5% CO_2) for 20 min.
7. Once cell clusters have formed, gently remove media and clustered cells from the dish and place in a 15-ml sterile conical tube on ice. Prepare a 5% Matrigel solution diluted in the same media formulation used to cluster cells, and gently mix an equal volume of the diluted Matrigel with the cell clusters (*see* **Note 9**).
8. Plate 1 ml of the 1:1 mixture of cell clusters and Matrigel solution onto coverslips that had been pre-coated with a thin layer of 2.5% Matrigel (as in **step 6** above).

3.7 Generating Fibrillar Collagen I Overlay on Matrigel (Sandwich Assay)

Dense, crosslinked, fibrillar Collagen I is a key component of the invasive tumor microenvironment [17–19]. Overlaying Collagen I on Matrigel cultures may thus more accurately model invasive conditions. Rat tail Collagen I is reconstituted in acetic acid solution as fibrils assemble at a neutral pH.

1. To preassemble Collagen I fibrils, first combine 75 μl of 10× DMEM (made from reconstituted powder and filter sterilized) and 25 μl of filter sterilized 1-M NaOH. Mix well by pipetting to ensure that the pH of the solution is consistent (pH 7); this will prevent localized precipitation of the Collagen I (*see* **Note 10**). Keep the tube on ice.

2. To the tube, add 650 μl of Collagen I (stock concentration ~5 mg/ml) and pipette to mix well, but avoid the introduction of bubbles. The solution should be reddish orange (neutral pH). The pH can be checked by spotting a small drop of the Collagen I solution onto pH paper. If necessary, adjust the pH by decreasing the amount of 1-M NaOH added and making up the volume with sterile ddH2O (*see* **Note 11**).
3. Let the Collagen I solution sit on ice for 2 h to allow for fibril formation.
4. Dilute the preassembled Collagen I 1:1 with 2× invasion media. This will yield a final Collagen I concentration of ~2.5 mg/ml, depending on the stock. Mix well by pipetting while avoiding the introduction of bubbles. Keep the diluted collagen solution on ice while preparing the coverslips.
5. Carefully remove media from spheroids on Matrigel-coated coverslips (*see* Subheading 3.4, **step 4**) using a P1000 pipette. Working quickly to avoid allowing the cells on Matrigel to dry out, add 100 μl of the Collagen I: media solution dropwise to the center of the coverslip. Up to 120 μl can be added, but limiting the volume added to just coat the coverslip prevents the Collagen I solution from wicking to the sides of the well. This wicking can result in the delicate Collagen I layer being pulled away from the Matrigel/cell layer.
6. Without adding media, carefully place the Collagen I–coated coverslips in the tissue culture incubator for 2 h (no more than 3 h) to allow for gelation of the Collagen I.
7. After 2 h, add invasion media (this requirement will vary depending on the cell type) to each well without disturbing the gel. Check the coverslips using a phase contrast microscope. If the pH was optimal, Collagen I fibrils should be visible as a meshwork of evenly dispersed-phase dark strands on the upper surface of the polymerized Matrigel (*see* Fig. 2, black arrowheads).
8. Depending on the cell line and experimental endpoint, spheroids can be maintained for a number of days in 3D culture (typically approximately 5 days) with regular media changes. As needed, every 2 days, gently aspirate off ~90% of the expired media using a P1000 pipette and replace with fresh culture media. Ensure that the pipette tip does not disrupt the Matrigel/Matrigel–Collagen I by pipetting into the side of the well.

3.8 Imaging Live Spheroids

Spheroid invasion can be monitored over time with the use of live imaging apparatus (including a humidified, CO_2 imaging chamber, stage warmer/incubator and motorized stage, preferably with Z-plane stabilization to maintain optimal focus on highly motile spheroids; *see* **Note 3**).

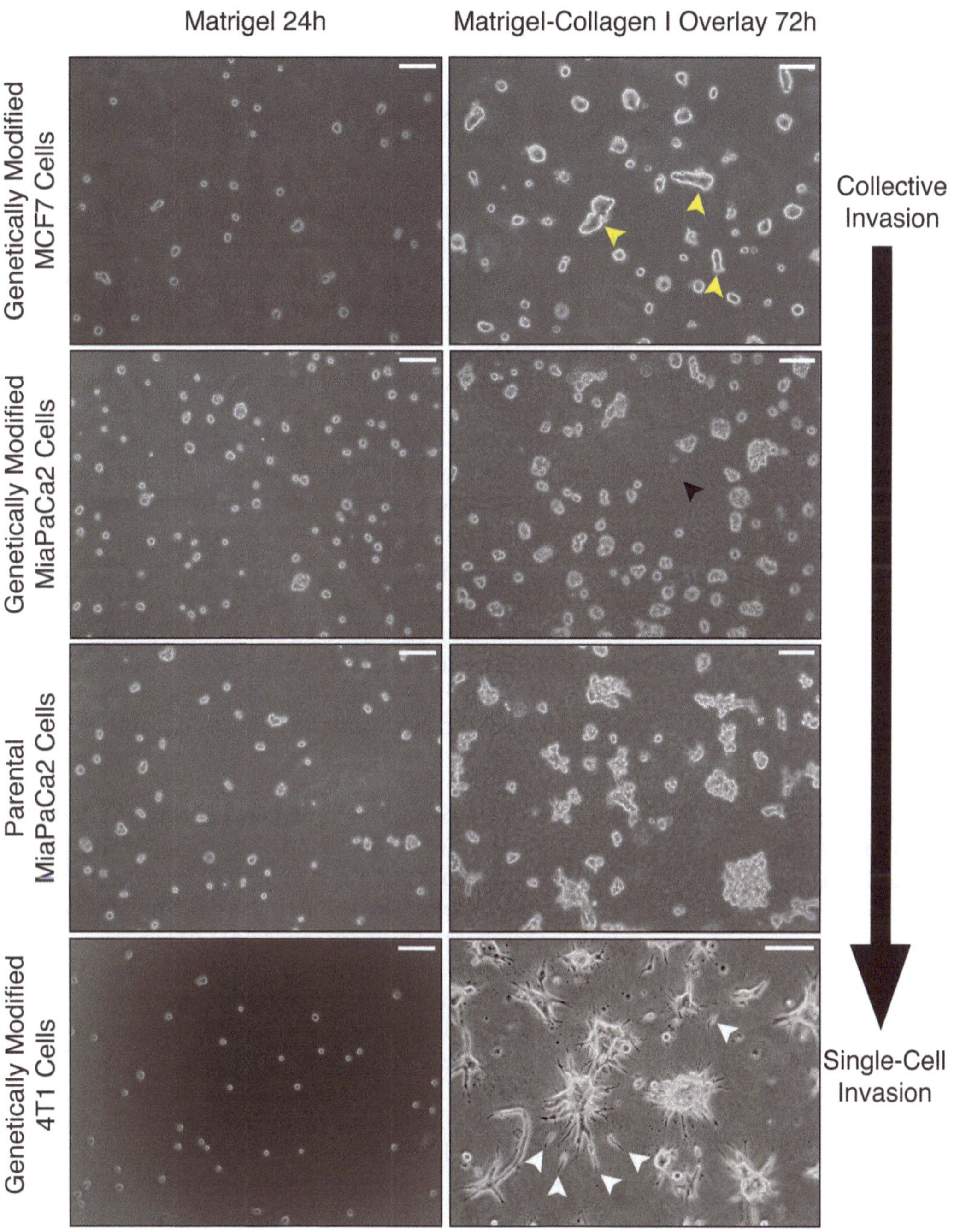

Fig. 2 Phase contrast images of live spheroids on Matrigel and in the Matrigel–Collagen I sandwich assay. Various cell types are shown after culture on top of Matrigel for 24 h. These spheroids were then overlaid with Collagen I and imaged 48 h later. A range of phenotypes are shown, from collective (yellow arrowhead) to quasi-mesenchymal to single-cell mesenchymal (white arrowhead) invading spheroids. Note the presence of phase dark Collagen I fibrils (black arrowhead). Scale bars represent 50 μm. Genetically modified 4T1 cell Matrigel–Collagen I overlay originally appeared in Fig. 7c of Graves et al. (2016).

1. To set up for live imaging, first prepare the imaging chamber so that the heat (37 °C), humidity (70%), and CO_2 (5%) levels equilibrate—this will take at least 1 h.
2. In the tissue culture BSC, disassemble and sterilize a Chamlide magnetic imaging chamber with 70% ethanol.
3. Using forceps, gently transfer a Matrigel/Collagen I coverslip from the 12-well plate to the imaging chamber bottom plate with cells/gel facing up.
4. Place the silicone O-ring in the chamber's main body and carefully lower it onto the bottom plate without allowing it to snap into place.
5. Fill the chamber with 900 μl of invasion media. Use a kimwipe to gently clean the bottom of the coverslip with 70% ethanol. Place the chamber on a dry kimwipe and check that no media are leaking and that a seal has formed. Place the transparent cover onto the chamber and transfer it to the microscope stage-top incubator, securing it in place.
6. Focus on a region in the center of the coverslip with multiple spheroids, and set the capture parameters to record one image every 5 min for 24 h (*see* **Note 12**). If using a motorized stage and a chamber slide stage adapter, set the positions (XYZ coordinates) as per the imaging software (*see* **Note 3** for multi-well chamber slides and the adjusted Matrigel/Collagen I volumes).

3.9 Quantifying Spheroid Invasion: Elongation Index in Collectively Invasive Spheroids

Spheroids can be imaged at given timepoints on any inverted phase contrast microscope with a 10× or 20× objective. Morphological differences between modes of invasion in this assay may be visually striking (i.e., cohesive elongated spheroids versus protrusive single cells from a loosely packed primary spheroid) (*see* Fig. 2). However, it may be beneficial to quantify the degree of invasion. To determine elongation index as a proxy for collective invasion and protrusion length as a proxy for mesenchymal-like invasion, multiple fields can be imaged at different timepoints (in this case, single spheroids will not be tracked over time, rather an average morphology will be determined; *see* Fig. 2). This can be done by simply imaging the coverslips in the 12-well plate.

To determine the elongation index of collectively invasive spheroids, measure the longest and shortest axis (length vs. width) and express this as a ratio of length/width [21] (*see* Fig. 3). This can be performed directly using the microscope software or by using Fiji [22].

1. Open a single TIFF image in Fiji.
2. Set the scale based on the objective lens used (10× is usually sufficient for quantification): (a) *Analyze* → *Set scale...*.

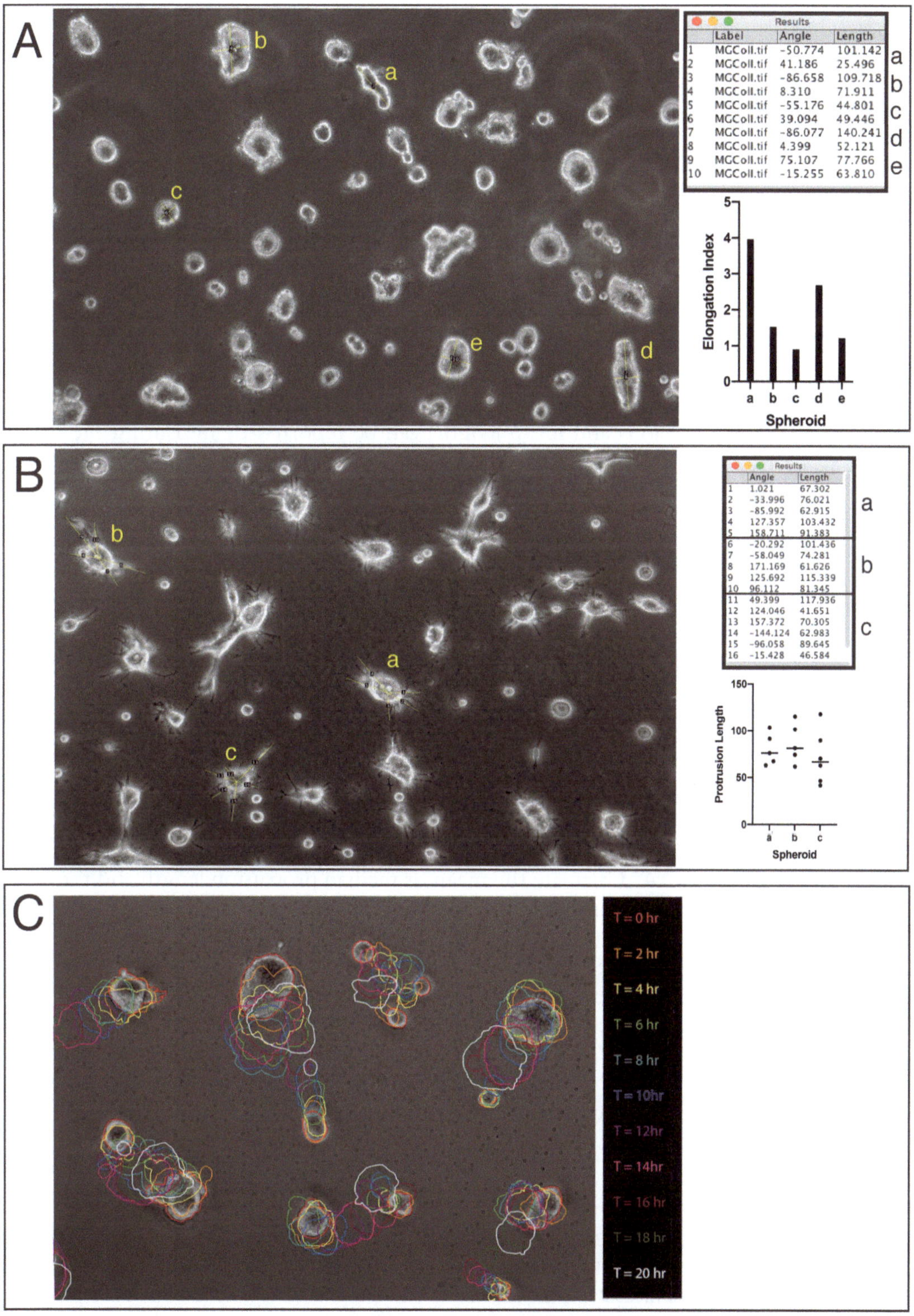

Fig. 3 Sample analyses for collective invasion (elongation index) and single-cell invasion (protrusion length) and visual representation of invasion over time. (**a**) The elongation index of collectively invasive MCF7 spheroids is measured using Fiji. Yellow lines on the phase contrast image show sample measurements. Elongation index = length/width of a spheroid. A number closer to 1 reflects low invasion (spheroid c),

(b) Enter the distance (pixels) and the known distance (μm). (c) Check the box for *Global* to apply this scale to subsequent images, and (d) select *OK*.

3. Select the *Straight Line* tool and draw a line along the longest axis of a spheroid: (a) *Analyze* → *Measure*. (b) Draw a line along the shortest axis of that spheroid. (c) *Analyze* → *Measure*, and (d) repeat with other spheroids in the field of view.
4. Repeat **steps 1–3** for the remaining images, then (a) select the Results table and save the .csv file, (b) *File* → *Save As*, and (c) calculate the elongation index (length/width) in Microsoft Excel or a statistical analysis software of choice.

3.10 Quantifying Spheroid Invasion: Protrusion Length in Mesenchymal-Type Invasive Spheroids

To quantify invasion in spheroids exhibiting sprouting, mesenchymal-type invasion, protrusion length (number of protrusions and mean length) can be reported by measuring the length from the centroid to each protrusion tip [24, 25] (*see* Fig. 3).

1. Repeat **steps 1** and **2** from Subheading 3.9.
2. Select the *Straight Line* tool and draw a line from the center of the spheroid to the first protrusion: (a) *Analyze* → *Measure*, (b) click the white selection point at the protrusion tip, and drag it to the tip of the next protrusion. This allows for measurement from the same center anchor point. (c) *Analyze* → *Measure*, and (d) repeat for all protrusions originating at that spheroid. Record the number of protrusions measured, and (e) repeat for all other spheroids, noting down the number of protrusions each time.
3. Select the Results table and save the .csv file by selecting *File* → *Save As*.
4. Calculate the mean or median protrusion length and the mean number of protrusions per condition.

3.11 Visual Representation of Spheroid Invasion over Time

If time-lapse images are recorded, it can be useful to generate a visual representation of spheroid invasion (shape and location change) over time. This can be done in Adobe Illustrator or a similar software, simply by drawing the outline of spheroids every × hours over the course of an experiment, depending on the length of the acquisition [26] (*see* Fig. 3).

Fig. 3 (continued) whereas a number greater than 1 reflects a higher invasion (spheroid a). (**b**) Protrusion length of mesenchymally invasive, genetically modified 4T1 cells is measured in Fiji. Yellow lines on the phase contrast image show sample measurements, originating from the center of the spheroid. Median protrusion length is reported for each spheroid. (**c**) Collectively invasive MCF7 spheroids were imaged over 20 h, and the spheroid perimeters were outlined every 2 h using Adobe Illustrator to show the invasive movement over time. Figure 3b, phase contrast image originally appeared in Movie S4 of Graves et al., 2016.

3.12 Permeabilizing and Staining Spheroids

To further characterize the modes of invasion or localization of relevant markers, it may be necessary to perform immunofluorescence staining (*see* **Note 13**). Using a P1000 pipette, aspirate off the media and replace with 1 ml of 4% PFA. Incubate at room temperature for 20 min. Fixation conditions will vary depending on the spheroid conditions as described in the steps in Subheadings 3.4, 3.5, 3.6, and 3.7 (*see* **Note 13**).

1. Using a P1000 pipette, aspirate off and safely dispose of the 4% PFA and replace with 1 ml of glycine buffer for 10 min.
2. Gently aspirate off the glycine buffer and replace with 1× PBS. Wash three times, 5 min each with 1× PBS.
3. Fill the wells of a new 12-well plate with 1 ml of permeabilization solution (*see* **Note 14**). Using forceps, carefully remove the coverslips from the original 12-well plate and place in wells of the new 12-well plate, cell-side up. The Collagen I layer will likely begin floating in this step; use forceps to gently peel off this layer and discard it. Incubate coverslips for 15 min at room temperature.
4. Gently aspirate off the permeabilization solution with a P1000 pipette and wash three times, 5 min each with 1× PBS at room temperature, without agitation.
5. Lay a few sheets of paper towel on a tray, label with coverslip conditions, and moisten with distilled water. Cut an appropriate-sized strip of parafilm and place the parafilm on the paper towel. Pipette 100-μl droplets of blocking solution onto the parafilm. Use normal serum derived from the species in which the desired secondary antibody was raised.
6. Using forceps, carefully remove the coverslips from the 12-well plate and place cell-side down onto the droplets of blocking solution. Cover with a plastic container to maintain humidity and incubate at room temperature for 30 min.
7. Meanwhile, prepare primary antibody dilutions in staining solution. Cut a new sheet of parafilm and pipette 60-μl droplets of antibody onto the parafilm.
8. After blocking, use forceps to transfer the coverslips to the primary antibody droplets, cell-side down. Cover the parafilm with a plastic container and wrap the tray with plastic wrap; incubate overnight at 4 °C.
9. The next day, fill the wells of a 12-well plate with 1 ml of 1× PBS. Using forceps, transfer coverslips to the wells, cell-side up, and wash three times, 5 min each with 1× PBS with gentle agitation on an orbital shaker, carefully aspirating with a P1000 pipette to avoid damaging the gel.

10. Meanwhile, prepare secondary antibody dilutions in staining solution. Cut a new sheet of parafilm and pipette 100-μl droplets of antibody onto the parafilm. Transfer coverslips to the secondary antibody droplets, cell-side down. Cover with a plastic container and incubate in the dark for 1 h at room temperature.
11. Fill the wells of a 12-well plate with 1 ml of 1× PBS. Using forceps, transfer coverslips to the wells, cell-side up, and wash one time for 5 min with 1× PBS in the dark, with gentle agitation. Replace the 1× PBS with 1:10000 DAPI (in 1× PBS) and incubate in the dark for 2 min.
12. Replace the DAPI with 1× PBS and wash three times, 5 min each, with gentle agitation.
13. Pipette 10–15 μl of Prolong Gold mounting media onto a clean microscope slide. Using forceps, remove a coverslip from the 12-well plate. Gently dab the side of the coverslip on a kimwipe to remove excess liquid. Place the coverslip cell-side down onto the mounting media. If the gel is too thick, push down gently with forceps to remove bubbles and allow the mounting media to cover the gel; alternatively add more mounting media (*see* **Note 15**).
14. Allow mounting media to harden for 2 h at room temperature in the dark, and then move to 4 °C overnight.
15. The next day, use clear nail polish to seal the edges of the coverslip. Store at 4 °C.
16. Image fixed and stained spheroids using a laser scanning confocal microscope equipped with a 63× oil objective, within 1 week (*see* Fig. 4).

4 Notes

1. Aliquots of 500 μl are ideal for setting up these experiments, but the volume can be adjusted depending on the scale of a typical experiment to prevent waste.
2. If Matrigel coating is not adhering well to glass coverslips, an acid wash can be performed prior to ethanol sterilization and coating. To do so, fill a beaker with 1-M HCl and add coverslips singly making sure that they are not stuck together. Heat the acid wash to 50–60 °C for 4–16 h with occasional agitation. Rinse thoroughly with three washes with fresh ddH_2O. Rinse in 95% ethanol and allow to dry before storing in a sealed container.

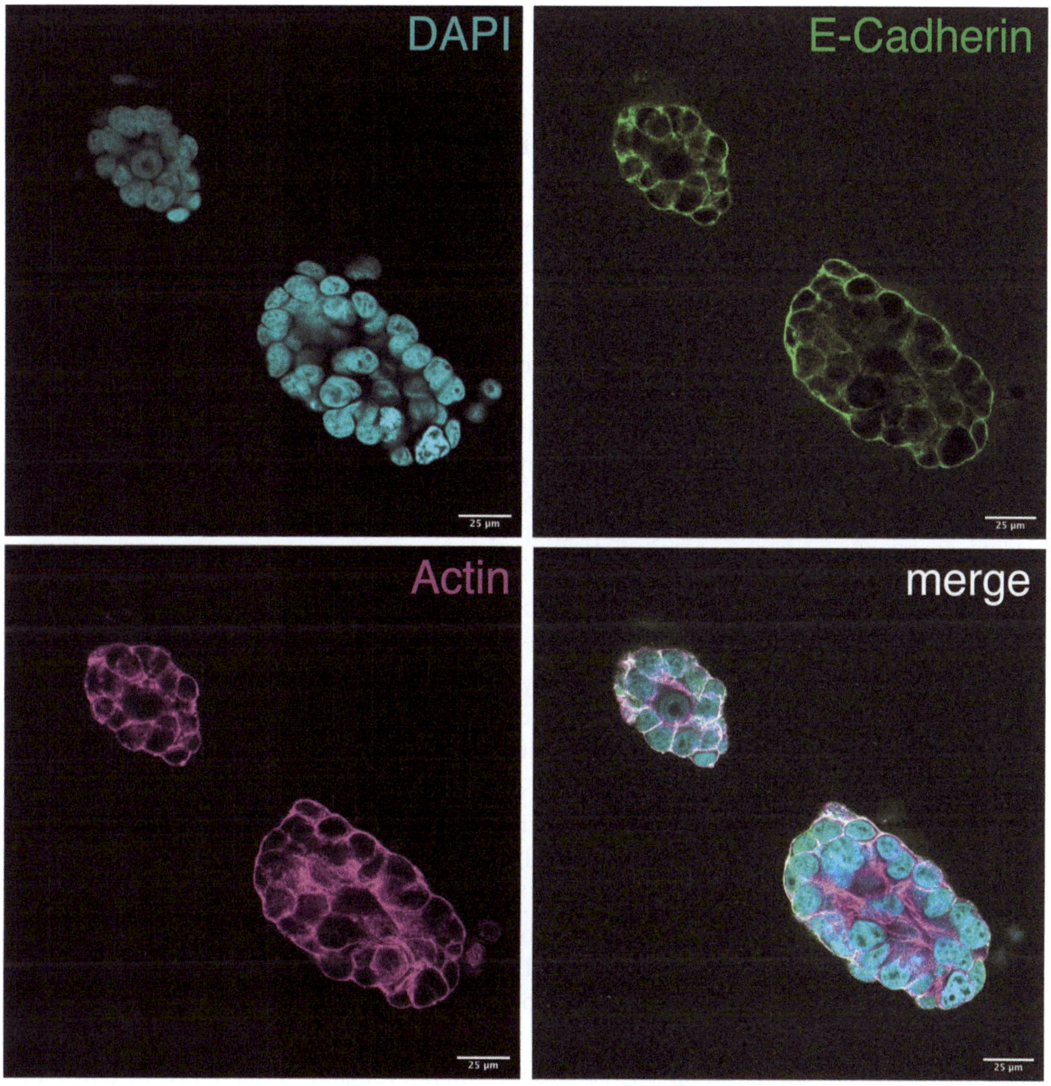

Fig. 4 Confocal imaging of 3D cultures. MCF7 spheroids cultured in a Matrigel–Collagen I overlay assay were immunostained for E-cadherin and actin and imaged using a Leica SP5 laser scanning confocal microscope equipped with a 63× oil immersion lens. Scale bars represent 25 μm

3. Undiluted Matrigel is required when doing immunofluorescence with 4% PFA fixation as diluted Matrigel will depolymerize upon fixation. However, diluted Matrigel can be functionally useful as it provides a less stiff substrate than undiluted Matrigel, which can facilitate the emergence of weakly invasive phenotypes. Glass coverslips generally work well for live imaging and immunocytochemistry (although some cell types, especially the more contractile mesenchymal types, will interact strongly with Collagen I and may be removed during staining). In this case, or if live imaging will

be performed using a microscope with a motorized stage, a multi-well chamber slide may be preferred. Matrigel and Collagen I volumes will need to be reduced accordingly (~40 μl/cm^2). In smaller format chamber slides, the small bore, tip end of the micropipette tip can be used to spread the Matrigel.

4. Matrigel must be carefully handled to avoid warming—pipette tips and tubes should be pre-chilled at −20 °C and work performed on ice.
5. Press down decisively with the wide bore, top end of the micropipette tip when spreading/dragging the Matrigel around the coverslip. A small volume of Matrigel will be removed by the pipette tip, but this is generally not a problem. Make sure that the coverslip does not lift up or that the plate is not tilted or bumped. If Matrigel spills off the coverslip and spreads onto the bottom of the well, it will tear during media changes and routine monitoring.
6. For this assay, it is best to test a few serum concentrations to determine the optimal growth conditions of each cell line used in 3D culture.
7. Cell number may need to be optimized for different cell types; low density is best when the object is to generate spheroids from single cells due to proliferation rather than due to aggregation as what occurs in a high-density setting.
8. In addition to damaging the Matrigel, harsh pipetting may result in the cells rolling off the Matrigel to the edges of the well.
9. Make sure that cells are suspended evenly in media (as single cells) before gently mixing with Matrigel. Vigorous mixing can cause bubbles to form, which disrupts the uniformity of the matrix and can be detrimental to later imaging.
10. Using DMEM with phenol red allows the pH to be constantly monitored. It is essential that the DMEM and NaOH are well mixed before Collagen I is added to facilitate pH uniformity during fibril formation.
11. The pH can differ between Collagen I lots and may fluctuate with older reagents. If the color of the final solution is more magenta (basic pH) than the reddish orange of normal culture media, fibrils will not form. In this case, new solutions should be prepared and the pH checked.
12. The Matrigel and Collagen I coating will be most consistent in the center of the well; hence, multiple fields should be captured from this region of the coverslip—the edge of the coverslip will likely have a thin, uneven coating so it is best to avoid imaging spheroids there.

13. Diluted Matrigel tends to depolymerize after fixation with 4% PFA. In the case that diluted Matrigel is required for the assay, methanol fixation or fixation with 2% PFA each for 20 min can prevent this; however, this fixation may not be suitable for every epitope of interest. Undiluted Matrigel (and Matrigel–Collagen I overlays) can be fixed in 4% PFA for 20–30 min without depolymerizing, and hence, this is a preferred method for most staining protocols. The Collagen I layer tends to peel off after fixation, or during staining; this can be avoided by reducing the handling of the coverslips. In many cases, the epithelial spheroids will remain intact, and removing the Collagen I layer can enhance antibody permeabilization; however, with some more mesenchymal contractile cell types (i.e., MiaPaCa2), the spheroids will be lost with the Collagen I layer. In this case, it is best to embed the gel in optimal cutting temperature (OCT) medium and cryosection the fixed Matrigel/Collagen I gel for staining serial sections on slides. Staining of fixed cells on coverslips is performed in a clean 12-well plate and on strips of parafilm.
14. The Triton-X-100 concentration and/or permeabilization timing may need to be modified to optimize staining of different antigens.
15. Depending on the gel thickness and the size of the spheroid when fixed, using the coverslip method may result in the spheroids appearing "squished" or compressed when imaging. To avoid this, either ensure that the Matrigel is well spread or perform this culture assay in a multi-well chamber slide where stained spheroids can be imaged directly in the well. Staining volumes can be found on the chamber slide product sheet (i.e., Ibidi, catalogue 80807). One disadvantage with this method is that mounting media will not harden and the slide will need to be fully imaged within 1 week after processing.

Acknowledgments

This work was supported by grants from the Canadian Institutes of Health Research (CIHR), Canadian Cancer Society-Research (CCS-R), and the Cancer Research Society (CRS). E.M.B. was supported by a doctoral scholarship from the Natural Sciences and Engineering Research Council of Canada (CGS-D NSERC).

References

1. Boyden S (1962) The chemotactic effect of mixtures of antibody and antigen on polymorphonuclear leucocytes. J Exp Med 115: 453–466. https://doi.org/10.1084/jem.115.3.453
2. Greenburg G, Hay ED (1982) Epithelia suspended in collagen gels can lose polarity and express characteristics of migrating mesenchymal cells. J Cell Biol 95(1):333–339

3. Hay ED, Zuk A (1995) Transformations between epithelium and mesenchyme: normal, pathological, and experimentally induced. Am J Kidney Dis 26(4):678–690. https://doi.org/10.1016/0272-6386(95)90610-x
4. Todaro GJ, Lazar GK, Green H (1965) The initiation of cell division in a contact-inhibited mammalian cell line. J Cell Physiol 66(3): 325–333. https://doi.org/10.1002/jcp.1030660310
5. Thiery JP, Acloque H, Huang RY, Nieto MA (2009) Epithelial-mesenchymal transitions in development and disease. Cell 139(5): 871–890. https://doi.org/10.1016/j.cell.2009.11.007
6. Cheung KJ, Padmanaban V, Silvestri V, Schipper K, Cohen JD, Fairchild AN, Gorin MA, Verdone JE, Pienta KJ, Bader JS, Ewald AJ (2016) Polyclonal breast cancer metastases arise from collective dissemination of keratin 14-expressing tumor cell clusters. Proc Natl Acad Sci U S A 113(7): 201508541–201508541. https://doi.org/10.1073/pnas.1508541113
7. Fischer KR, Durrans A, Lee S, Sheng J, Li F, Wong ST, Choi H, El Rayes T, Ryu S, Troeger J, Schwabe RF, Vahdat LT, Altorki NK, Mittal V, Gao D (2015) Epithelial-to-mesenchymal transition is not required for lung metastasis but contributes to chemoresistance. Nature 527(7579):472–476. https://doi.org/10.1038/nature15748
8. Friedl P, Locker J, Sahai E, Segall JE (2012) Classifying collective cancer cell invasion. Nat Cell Biol 14(8):777–783. https://doi.org/10.1038/ncb2548. From Duplicate 2 (Classifying collective cancer cell invasion - Friedl, P; Locker, J; Sahai, E; Segall, J E) And Duplicate 3 (Classifying collective cancer cell invasion - Friedl, P; Locker, J; Sahai, E; Segall, J E) And Duplicate 4 (Classifying collective cancer cell invasion - Friedl, P; Locker, J; Sahai, E; Segall, J E) And Duplicate 5 (Classifying collective cancer cell invasion - Friedl, P; Locker, J; Sahai, E; Segall, J E) Friedl, Peter Locker, Joseph Sahai, Erik Segall, Jeffrey E 15154/Cancer Research UK/United Kingdom CA1000324/CA/NCI NIH HHS/CA104292/CA/NCI NIH HHS/CA76394/CA/NCI NIH HHS/England Nat Cell Biol. 2012;14(8):777–83. https://doi.org/10.1038/ncb2548
9. Yang J, Antin P, Berx G, Blanpain C, Brabletz T, Bronner M, Campbell K, Cano A, Casanova J, Christofori G, Dedhar S, Derynck R, Ford HL, Fuxe J, Garcia de Herreros A, Goodall GJ, Hadjantonakis AK, Huang RJY, Kalcheim C, Kalluri R, Kang Y, Khew-Goodall Y, Levine H, Liu J, Longmore GD, Mani SA, Massague J, Mayor R, McClay D, Mostov KE, Newgreen DF, Nieto MA, Puisieux A, Runyan R, Savagner P, Stanger B, Stemmler MP, Takahashi Y, Takeichi M, Theveneau E, Thiery JP, Thompson EW, Weinberg RA, Williams ED, Xing J, Zhou BP, Sheng G, Association EMTI (2020) Guidelines and definitions for research on epithelial-mesenchymal transition. Nat Rev Mol Cell Biol 21:341–352. https://doi.org/10.1038/s41580-020-0237-9
10. Zheng X, Carstens JL, Kim J, Scheible M, Kaye J, Sugimoto H, Wu CC, LeBleu VS, Kalluri R (2015) Epithelial-to-mesenchymal transition is dispensable for metastasis but induces chemoresistance in pancreatic cancer. Nature 527(7579):525–530. https://doi.org/10.1038/nature16064
11. Frantz C, Stewart KM, Weaver VM (2010) The extracellular matrix at a glance. J Cell Sci 123 (Pt 24):4195–4200. https://doi.org/10.1242/jcs.023820
12. Li ML, Aggeler J, Farson DA, Hatier C, Hassell J, Bissell MJ (1987) Influence of a reconstituted basement membrane and its components on casein gene expression and secretion in mouse mammary epithelial cells. Proc Natl Acad Sci U S A 84(1):136–140
13. Roskelley CD, Desprez PY, Bissell MJ (1994) Extracellular matrix-dependent tissue-specific gene expression in mammary epithelial cells requires both physical and biochemical signal transduction. Proc Natl Acad Sci U S A 91(26): 12378–12382
14. Kleinman HK, Martin GR (2005) Matrigel: basement membrane matrix with biological activity. Semin Cancer Biol 15(5):378–386. https://doi.org/10.1016/j.semcancer.2005.05.004
15. Kubota Y, Kleinman HK, Martin GR, Lawley TJ (1988) Role of laminin and basement membrane in the morphological differentiation of human endothelial cells into capillary-like structures. J Cell Biol 107(4):1589–1598. https://doi.org/10.1083/jcb.107.4.1589
16. Boyd NF, Rommens JM, Vogt K, Lee V, Hopper JL, Yaffe MJ, Paterson AD (2005) Mammographic breast density as an intermediate phenotype for breast cancer. Lancet Oncol 6(10):798–808. https://doi.org/10.1016/S1470-2045(05)70390-9
17. Levental KR, Yu H, Kass L, Lakins JN, Egeblad M, Erler JT, Fong SF, Csiszar K, Giaccia A, Weninger W, Yamauchi M, Gasser DL, Weaver VM (2009) Matrix crosslinking forces tumor progression by enhancing integrin signaling. Cell 139(5):891–906. https://doi.org/10.1016/j.cell.2009.10.027

18. Paszek MJ, Zahir N, Johnson KR, Lakins JN, Rozenberg GI, Gefen A, Reinhart-King CA, Margulies SS, Dembo M, Boettiger D, Hammer DA, Weaver VM (2005) Tensional homeostasis and the malignant phenotype. Cancer Cell 8(3):241–254. https://doi.org/10.1016/j.ccr.2005.08.010
19. Provenzano PP, Eliceiri KW, Campbell JM, Inman DR, White JG, Keely PJ (2006) Collagen reorganization at the tumor-stromal interface facilitates local invasion. BMC Med 4(1):38–38. https://doi.org/10.1186/1741-7015-4-38
20. Nguyen-Ngoc K-VV, Shamir ER, Huebner RJ, Beck JN, Cheung KJ, Ewald AJ (2015) 3D culture assays of murine mammary branching morphogenesis and epithelial invasion. Methods Mol Biol 1189:135–162. https://doi.org/10.1007/978-1-4939-1164-6_10
21. Graves ML, Cipollone JA, Austin P, Bell EM, Nielsen JS, Gilks CB, McNagny KM, Roskelley CD (2016) The cell surface mucin podocalyxin regulates collective breast tumor budding. Breast Cancer Res 18(1):11–11. https://doi.org/10.1186/s13058-015-0670-4
22. Schindelin J, Arganda-Carreras I, Frise E, Kaynig V, Longair M, Pietzsch T, Preibisch S, Rueden C, Saalfeld S, Schmid B, Tinevez JY, White DJ, Hartenstein V, Eliceiri K, Tomancak P, Cardona A (2012) Fiji: an open-source platform for biological-image analysis. Nat Methods 9(7):676–682. https://doi.org/10.1038/nmeth.2019
23. Folkman J, Moscona A (1978) Role of cell shape in growth control. Nature 273(5661):345–349. https://doi.org/10.1038/273345a0
24. Martin M, Veloso A, Wu J, Katrukha EA, Akhmanova A (2018) Control of endothelial cell polarity and sprouting angiogenesis by non-centrosomal microtubules. Elife 7:e33864. https://doi.org/10.7554/eLife.33864
25. To KC, Loh KT, Roskelley CD, Andersen RJ, O'Connor TP (2006) The anti-invasive compound motuporamine C is a robust stimulator of neuronal growth cone collapse. Neuroscience 139(4):1263–1274. https://doi.org/10.1016/j.neuroscience.2006.01.065
26. Bryant DM, Roignot J, Datta A, Overeem AW, Kim M, Yu W, Peng X, Eastburn DJ, Ewald AJ, Werb Z, Mostov KE (2014) A molecular switch for the orientation of epithelial cell polarization. Dev Cell 31(2):171–187. https://doi.org/10.1016/j.devcel.2014.08.027

Chapter 9

CellTrace™ Violet Flow Cytometric Assay to Assess Cell Proliferation

Marina B. Lemieszek, Scott D. Findlay, and Gabrielle M. Siegers

Abstract

CellTrace™ Violet (CTV) is a powerful tool for tracking cell proliferation by permanently binding cellular proteins and rendering the cell fluorescent. After cell division, each daughter cell contains half of the parent cell's fluorescence, enabling quantification of proliferation via flow cytometry. This method enables monitoring of several generations of cell division and tracking of different cell populations in co-culture. Here we describe the use of CellTrace™ Violet in different cell types, and we share important observations we made during protocol optimization.

Key words Cell division, Cell tracking, Dilution of dye, Division tracking, Flow cytometry, Proliferation assay, Proliferation modeling

1 Introduction

CellTrace™ Violet (CTV) is a proliferation dye that fluorescently labels cellular proteins [1] and can be detected using flow cytometry [2]. The quantification of cell proliferation is an essential method to characterize cells in different research areas, such as immunology [3, 4] and cancer research [5, 6]. CTV proliferation quantification with flow cytometry enables single-cell proliferation analysis by dye dilution. This is useful for both homogeneous and heterogeneous populations of cells, the latter being inaccurately quantified by bulk proliferation methods [7]. CTV can also be employed in experiments using imaging flow cytometry, a powerful technique that enables simultaneous visualization of multiple proteins and their localization within a cell [8].

CTV is a succinimidyl ester [8] and is incorporated into cells, permanently binding to cellular proteins through covalent bonds with all free intracellular amine groups. CTV is cleaved by intracellular esterases, becoming fluorescent [1, 8–12]. After cell division, the fluorescence intensity (MFI) of the parent cells is symmetrically

Sherri L. Christian (ed.), *Cancer Cell Biology: Methods and Protocols*,
Methods in Molecular Biology, vol. 2508, https://doi.org/10.1007/978-1-0716-2376-3_9,

divided to each daughter cell [1, 7–9, 12]. Although there are instances of asymmetric cell division, such as in cellular differentiation during development, most cell division should result in equal distribution of dye [8]. Acquisition of cells via flow cytometry and the subsequent analysis of fluorescence histograms enable the inference and quantification of the number of cell divisions [2, 9]. Importantly, it was shown that CTV was not transferred to bystander cells in a culture of Jurkat cells [7]. On the other hand, murine B cells could acquire CTV and other dyes from labeled bystander cells [2]. Also, CTV can be used with other fluorescent dyes or GFP as a method to label and monitor different cell populations in co-culture [13, 14], keeping in mind the aforementioned caveat when using antigen-presenting cells such as B cells. An advantage of labeling cells with CTV compared to genetic manipulation is that CTV preserves cell population heterogeneity and does not require the generation of single-cell-derived clones [14].

CTV is a useful tool to determine proliferation differences after experimental manipulation of gene expression (i.e., overexpression or silencing) in different cell types [4, 6]. Moreover, the estimation of long-term proliferation effects after drug administration can be monitored with CTV due to its capacity to track several generations of cell division [8]. Furthermore, CTV permits analysis of multiple rounds of cell division in both in vitro and in vivo models [2, 10]. CTV is also considered less cytotoxic than other available dyes, such as carboxy-fluorescein succinimidyl ester (CFSE) [9], and showed low toxicity in murine and human lymphocytes [2, 7, 9]. However, CTV inhibited the proliferation of equine lymphocytes [9], likely because CTV labeling was done prior to mitogen stimulation, which conceivably blocked binding of receptors on lymphocytes required for signal transduction and subsequent activation. Thus, it is essential to stimulate proliferation of lymphocytes prior to CTV labeling. It is also important to titrate CTV to determine the best concentration to use for specific cells. For instance, CTV did not affect Nb2-11 cell proliferation when used at 1 μM concentration, while increasing the dye to higher than 2.5 μM caused decreased cell growth and ultimately death [15].

One important technical advantage of CTV fluorescence is that it remains stable after fixation [9], although we noted in our experiments that prolonged fixation can lead to a decrease in signal intensity. Here we describe the CTV staining protocol, including observations we made while optimizing this procedure for breast cancer cell lines and primary γδ T cells [16]. Additionally, we include important considerations for sample acquisition via flow cytometry and subsequent data analyses.

2 Materials

Prepare reagents in a biological safety cabinet. CellTrace™ Violet and fluorescent viability dyes should be protected from light during use and storage.

2.1 CellTrace™ Violet

1. CellTrace™ Violet: 1 vial (ThermoFisher Scientific) stored at or less than −20 °C.
2. Dimethyl sulfoxide (DMSO) for reconstitution, included in abovementioned kit.

2.2 Viability Dyes Compatible[1] with CellTrace™ Violet

1. Propidium iodide (PI) or 7-amino-actinomycin D (7-AAD) for live cell acquisition.
2. Fixable live/dead viability dye, such as Zombie NIR (BioLegend), if live acquisition is not possible.

2.3 Cell Suspension Preparation

1. 1× Phosphate-buffered saline (1× PBS).
2. Culture media containing fetal bovine serum (FBS).
3. Dissociation reagent such as Trypsin–EDTA 0.05% for adherent cells.

2.4 Cell Acquisition on the Flow Cytometer

1. Flow cytometry acquisition tubes—5-ml round-bottom polystyrene tubes.
2. Flow cytometry staining buffer (FACS buffer: 1% FBS and 2 mM EDTA in 1× PBS).
3. Flow cytometer with violet laser (405 nm) and corresponding laser to excite viability dye. CellTrace™ Violet is excited at 405 nm and emits at 450 nm (ex405/em450), which can be detected with the 450/50 detector. PI (ex493/em636) and 7-AAD (ex546/em647) are excited by the blue laser (488 nm); their emissions can be detected with 585/42 (PI) and 670LP (PI and 7-AAD) detectors. If using Zombie NIR (ex633/em746), the red laser (633 nm) is required, and its emission is detected with the 780/60 detector.

2.5 Others

1. Centrifuge and microcentrifuge.
2. 50-ml conical centrifuge tubes, sterile.
3. 1.5-ml microcentrifuge tubes, sterile.
4. Cell incubator at 37 °C.
5. 2% Paraformaldehyde (PFA) in FACS buffer.
6. Three 75-cm^2 flasks.

[1] Compatibility criteria will be discussed in Subheading 3.2.3.

7. Ice.
8. Hemocytometer.
9. FlowJo or other software for analysis of flow cytometry data.

3 Methods

This protocol determines the proliferation of cells at two different time points after CellTrace™ Violet labeling, on days 3 and 6 (*see* **Note 1**).

For immortalized adherent cell lines and for suspension cells, follow method from **step 7**, Subheading 3.2.1.

3.1 Prior to Day 0

1. In a 75-cm^2 flask, grow cells to 70–80% confluence in complete culture medium containing FBS.
2. On day −1, starve cells (*see* **Note 2**) by replacing complete culture medium with serum-free medium after washing with 1× PBS; incubate overnight.

3.2 Day 0

3.2.1 Harvest and Preparation of Single-Cell Suspension

1. Remove serum-free culture medium.
2. Wash the adherent cells with 5 ml 1× PBS (*see* **Note 3**).
3. Remove 1× PBS.
4. Wash the adherent cells again with 5 ml 1× PBS.
5. Remove 1× PBS.
6. Detach cells from flask with dissociation reagent and inactivate according to standard procedure; transfer cells to centrifuge tube.
7. Centrifuge harvested cells at 300 g for 5 min at room temperature.
8. Remove supernatant.
9. Resuspend the cell pellet in 5 ml 1× PBS.
10. Prepare three 10 μl aliquots of the cell suspension for cell counting. Resuspend cells carefully before preparing each aliquot.
11. Add 10 μl trypan blue to each aliquot just before counting (*see* **Note 4**). From the 20 μl total, pipette 10 μl onto a hemocytometer and count. Repeat until all aliquots are counted; average your result. If cell numbers are too high, further dilute aliquot in trypan blue and recount.
12. You will need six million cells per cell line or clone divided into the following groups:
    (a) Unstained control group for flow cytometry setup and acquisition (one million cells *not* stained with CellTrace™ Violet or viability dye).

(b) Viability dye control group for flow cytometry setup and acquisition (one million cells stained with viability dye only).

(c) CellTrace™ Violet control group for flow cytometry setup and acquisition (one million cells stained with Cell-Trace™ Violet only).

(d) CellTrace™ Violet and viability dye group for acquisition by flow cytometry (one million cells stained with both dyes).

(e) CellTrace™ Violet group for cell culture (two million cells stained with CellTrace™ Violet only that will be kept in culture until the next time point; here, this is day 3).

For experiments with single-cell-derived clones from the same cell line, it will be necessary to have unstained and viability dye controls (a and b) from only one clone for flow cytometry setup and acquisition; however, all other groups (c, d, and e) are necessary for each of the single-cell-derived clones.

3.2.2 Cell Staining with CellTrace™ Violet

1. After cell counting, from the 6 ml cell suspension, add one million cells to each labeled 1.5-ml microcentrifuge tube for the unstained control and the viability dye control groups. Put tubes on ice.
2. Centrifuge the tube destined for viability dye staining (group b, **step 12** from Subheading 3.2.1) containing the cell suspension at 300 g for 5 min.
3. Remove the supernatant and resuspend in 100 μl 1× PBS.
4. Put the viability dye tube (group b) at −20 °C to kill the cells while you perform the following steps (*see* **Note 5**).
5. Place four million cells from the initial cell suspension in a new 50-ml centrifuge tube.
6. Centrifuge at 300 g for 5 min at room temperature.
7. Remove the supernatant and resuspend the cell pellet in 4 ml 1× PBS (1 ml PBS per million cells).
8. Turn off the lights in the biological safety cabinet (*see* **Note 6**). Prepare CellTrace™ Violet stock solution immediately before use or thaw aliquot stored at −20 °C (*see* **Note 7**).
9. Add 4 μl of 5 mM stock solution to the 4 ml cell suspension in 1× PBS to a final concentration of 5 μM. CellTrace™ Violet titration is recommended to ensure the best concentration for specific cell lines to ensure strong signals while maintaining viability (*see* Fig. 1 and **Note 8**). Pilot tests should be done to determine the optimal duration of the proliferation assay (*see* Fig. 2 and **Note 9**).

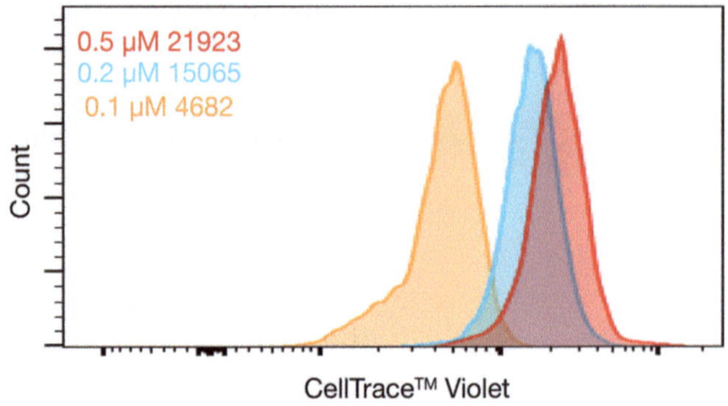

Fig. 1 CellTrace™ Violet titration. T47D breast cancer cells were labeled with 0.5 μM, 0.2 μM, and 0.1 μM CellTrace™ Violet as indicated on day 0 and acquired directly by flow cytometry. Median fluorescence intensity values are shown in the top-left corner of the plot

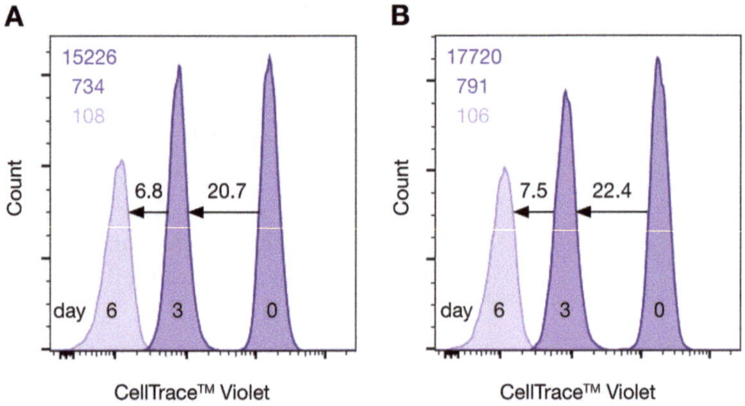

Fig. 2 Dilution of CellTrace™ Violet over 6 days enables comparison of proliferation of wild-type and knockout clones. (**a**) A wildtype clone of the MCF-7 breast cancer cell line was labeled with 0.5 μM CellTrace™ Violet on day 0 and acquired by flow cytometry 3 and 6 days later, as indicated. Median fluorescence intensity (MFI) values are shown in the top-left corner of the plot. Fold changes in MFI are indicated above the arrows. (**b**) As in A, with a clone of MCF-7 in which a gene of interest had been knocked out using CRISPR-Cas9. Fold changes in MFI calculations (MFI^{day0}/MFI^{day3} and MFI^{day3}/MFI^{day6}) suggest that the knockout proliferates at a slightly greater rate than the wild-type clone

10. Mix or vortex gently (*see* Fig. 3 and **Note 10**).
11. Incubate at 37 °C in a cell incubator or water bath for 20 min.
12. Add 20 ml culture media containing at least 1% FBS and incubate for 5 min.
13. Centrifuge at 300 g for 5 min at room temperature.
14. Remove the supernatant.
15. Resuspend the cell pellet in 4 ml culture media (1 ml per million cells).

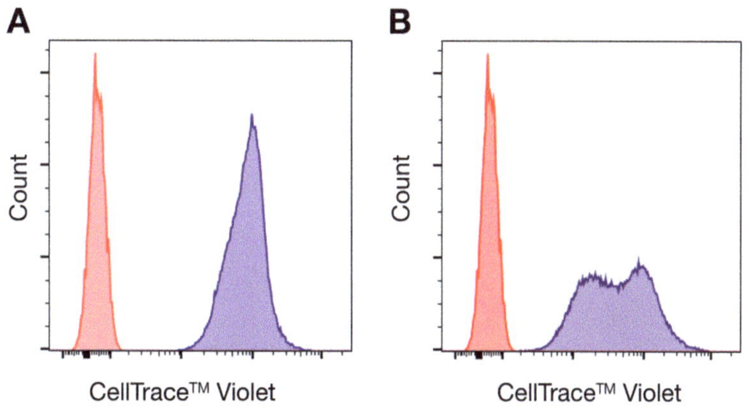

Fig. 3 Thorough mixing is required to ensure homogeneous labeling. (**a**) Example of homogeneous labeling of MCF-7 breast cancer cells stained with 0.5-μM CellTrace™ Violet on day 0 and acquired by flow cytometry. The unstained control is shown in red. (**b**) Example of heterogeneous labeling of MCF-7 cells due to incomplete mixing of the dye with cells during the labeling procedure

16. Add 2 ml (two million cells) to a new 75-cm^2 flask containing the desired volume of complete culture media. Incubate at 37 °C in a cell incubator for 3 days. Do not change the media (*see* **Note 11**).
17. Add 1 ml (one million cells) to a tube labeled CellTrace™ Violet (CTV only group c; *see* Subheading 3.2.1, **step 12**).
18. Add 1 ml (one million cells) to a tube labeled CellTrace™ Violet and viability dye (CTV + viability dye group d; *see* Subheading 3.2.1, **step 12**).
19. Maintain all the tubes on ice.

3.2.3 Cell Staining with Viability Dye for Same-Day Flow Cytometric Acquisition of Fresh Cells

1. Thaw the viability dye only group (group b; *see* Subheading 3.2.1, **step 12**).
2. Centrifuge all samples (containing live and dead cells) at 300 g for 5 min at room temperature. Remove the supernatants.
3. Resuspend the cell pellets in 1 ml 1× PBS.
4. Centrifuge at 300 g for 5 min at room temperature.
5. Resuspend cell pellets in 1 ml FACS buffer and centrifuge at 300 g for 5 min at room temperature.
6. Remove supernatants, and resuspend pellets in 400 μl FACS buffer.
7. Add PI or 7-AAD according to the manufacturer's instructions to all except unstained (group a; *see* Subheading 3.2.1, **step 12**) and CTV only samples (group c; *see* Subheading 3.2.1, **step 12**) just prior to sample acquisition. For acquisition notes, *see* **Note 12**.

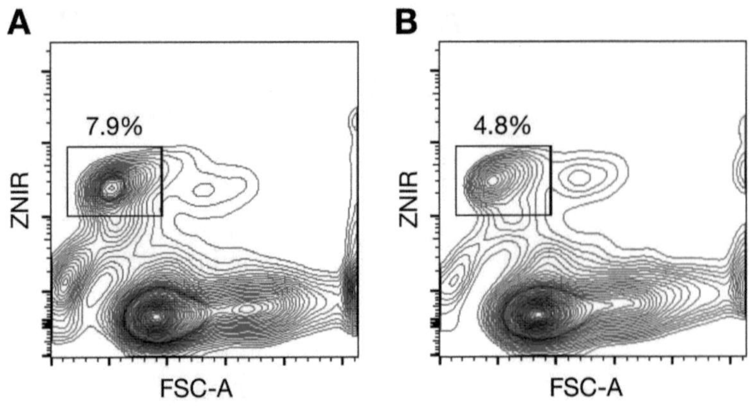

Fig. 4 CellTrace™ Violet labeling prevents efficient binding of amine-reactive fixable live/dead stains. (**a**) MCF-7 breast cancer cells were stained with fixable live/dead Zombie NIR™ (ZNIR) and acquired by flow cytometry. (**b**) MCF-7 cells labeled with 0.5-μM CellTrace™ Violet were stained with ZNIR at the same time as the cells in A and acquired by flow cytometry. The lower percentage of ZNIR+ cells in B suggests incomplete labeling of dead cells with ZNIR due to the lack of free amines in cells already labeled with CellTrace™ Violet

3.2.4 Cell Staining with Viability Dye for Flow Cytometric Acquisition of Fixed Cells

1. Thaw the viability dye only group (group b; *see* Subheading 3.2.1, **step 12**).
2. Centrifuge all tubes (including viability dye only group) at 300 g for 5 min at room temperature. Remove the supernatant.
3. Resuspend the cell pellets in 1 ml 1× PBS.
4. Centrifuge the CTV + viability dye and viability dye only tubes at 300 g for 5 min at room temperature. Remove supernatant.
5. Resuspend and stain cells from CTV + viability dye and viability dye only tubes with fixable live/dead viability dye according to the manufacturer's instructions, though fixable dyes should be used with caution (*see* Fig. 4 and **Note 13**).
6. Centrifuge all samples (including CTV only and unstained tubes) at 300 g for 5 min at room temperature. Remove the supernatants.
7. Resuspend the cell pellets in 1 ml 1× PBS.
8. Centrifuge at 300 g for 5 min at room temperature.
9. Resuspend cell pellets in 1 ml FACS buffer and centrifuge at 300 g for 5 min at room temperature.
10. Dilute PFA to 2% in FACS buffer, add 100 μl to each cell pellet, and resuspend.
11. Incubate on ice for a minimum of 15 min. Fixed cells can be stored at 4 °C for up to 1 week; however, prolonged fixation is not recommended (*see* Fig. 5 and **Note 14**).

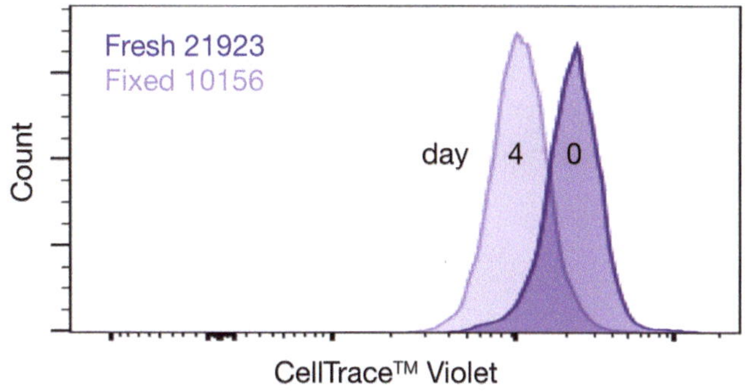

Fig. 5 Prolonged fixation results in signal loss. T47D breast cancer cells were stained with 0.5 μM CellTrace™ Violet on day 0 and acquired by flow cytometry directly. In parallel, a portion of the labeled cells were fixed with FACS buffer containing 2% paraformaldehyde, stored for 4 days at 4 °C, and then acquired on the same flow cytometer with the same settings and compensation values. Median fluorescence intensity values are indicated in the top-left corner of the plot

12. Add the cell suspensions in 2% PFA to flow cytometry acquisition tubes.
13. Add 300 μl FACS buffer before flow cytometry acquisition.

3.3 Day 3: Cell Suspensions and Staining of Cells Labeled with CellTrace™ Violet

1. After 3 days of proliferation, remove the culture media in the biosafety hood with the lights turned off and repeat **steps 2–11** from Subheading 3.2.1 (or **steps 7–11** for suspension cells from Subheading 3.2.1). Prepare three million cells:
   (a) CellTrace™ Violet group for cell culture: Add two million cells to a 75-cm^2 flask containing complete culture media. Incubate at 37 °C for another 3 days.
   (b) One million harvested cells for CellTrace™ Violet and viability dye staining.

   For experiments with single-cell-derived clones, the same CellTrace™ Violet group for cell culture and CellTrace™ Violet and viability dye group for acquisition in flow cytometry (a and b) are required for each clone. Note that these cells have already been labeled with CellTrace™ Violet on day 0 and only require further staining with viability dye.
2. Repeat **steps 2–7** from Subheading 3.2.3 for live cell acquisition or **steps 2–13** from Subheading 3.2.4 if fixing cells (observe that a viability dye only group is not necessary here).

3.4 Day 6: Cell Suspension and Staining of Cells Labeled with CellTrace™ Violet

1. After 3 days of proliferation, remove the culture media in the biosafety hood with the lights turned off and repeat **steps 2–11** from Subheading 3.2.1 (or **steps 7–11** for suspension cells from Subheading 3.2.1).
2. One million cells are required: CellTrace™ Violet and viability dye group for flow cytometric acquisition (one million cells stained with both dyes; note that these cells were already labeled with CellTrace™ Violet on day 0 and only require further staining with viability dye).

   For single-cell-derived clones, CellTrace™ Violet and viability dye group from each clone are required for acquisition via flow cytometry.
3. Repeat **steps 2–7** from Subheading 3.2.3 for live cell acquisition or **steps 2–13** from Subheading 3.2.4 if fixing cells (observe that a viability dye only group is not necessary here).
4. For data analysis tips, *see* **Note 15**.

4 Notes

1. These time points will vary depending on the intensity of initial labeling, the seeding density, the proliferation rate of the cells, and the intended experimental objectives. We chose days 3 and 6 because our breast cancer cell lines require splitting and fresh medium every 3 days.
2. When assessing proliferation of different clones derived from a cell line, such as those in Fig. 2, overnight starvation is used to synchronize the cell lines for easier comparison. If comparing proliferation of cells in the presence or absence of an exogenous stimulus, as was done in Fig. 7, then starvation is not necessary, as groups (treatment and control) were taken from the same population of cells in culture just prior to treatment and as such would already be considered synchronized.
3. Do not directly add PBS to the adherent cells but rather to the flask's lateral wall to avoid premature cell detachment.
4. Trypan blue should only be added to cells just prior to counting, as it is toxic to cells.
5. The duration of freezing should be tested for each cell line to ensure enough cells die to provide positively labeled cells for setting the appropriate voltage and performing compensation on the flow cytometer.
6. Working with the lights off in the biological safety cabinet is done to prevent loss of fluorescence due to exposure to light.
7. To prepare 5 mM stock solution, add 20 μl DMSO to one vial of CellTrace™ Violet reagent and mix well. Aliquot small

volumes (to avoid repeated freeze/thaw cycles) and store at −20 °C for future experiments.

8. While the manufacturer recommends final concentrations of 1–10 μM, we were able to use even lower concentrations with good results.
9. The assay should not be carried out beyond the point at which the dye is diluted so as to no longer be distinguishable from an unstained control. This will vary with initial CellTrace™ Violet labeling concentration, with greater initial signal intensity enabling longer duration of the assay.
10. Mixing is essential to ensure homogeneous labeling (*see* Fig. 3).
11. Parallel flasks or wells of CTV-labeled cells can be prepared and stimulated with reagents of interest (antibodies, small molecule inhibitors, therapeutic drugs, experimental compounds) in order to compare effects of these substances on cell proliferation. Mitogenic stimulation directly after CTV labeling may not be successful (*see* Ref. [9]; thus, we recommend mitogenic stimulation of primary cells some time prior to CTV labeling. The interval between mitogenic stimulation and CTV labeling should be determined by experimentation. If proliferation after stimulation is unsuccessful, then the interval should be lengthened.
12. Voltages and compensation on the flow cytometer should be set up on day 0 using samples prepared and acquired fresh or fixed. The experiment should be saved in the flow cytometer's software, and additional samples prepared on subsequent days should be acquired within the same experiment with the same settings. Try to maintain consistent timing between sample preparation and acquisition time on all sample acquisition days. If possible, acquire 100,000 live singlet events per sample. In the case of limiting material, or for immortalized cell lines that divide synchronously, as few as 20,000 live singlet events may be sufficient, but coefficients of variance for different event counts should be determined in a pilot experiment (*see* Fig. 6).
13. Use of fixable viability dye that relies on binding to free amines in the cell is not ideal, as CTV labeling blocks binding sites resulting in an underestimate of dead cell content in the sample.
14. If fixation is unavoidable, flow cytometric acquisition on the same day is recommended for better results. Above all, consistency in the time between sample preparation and acquisition is paramount since CTV signal decreases over time once the cells have been fixed (*see* Fig. 5). Samples can be centrifuged after an overnight fixation, and fixation buffer can be replaced with standard FACS buffer for further storage. Samples should be acquired within 1 week of harvest.

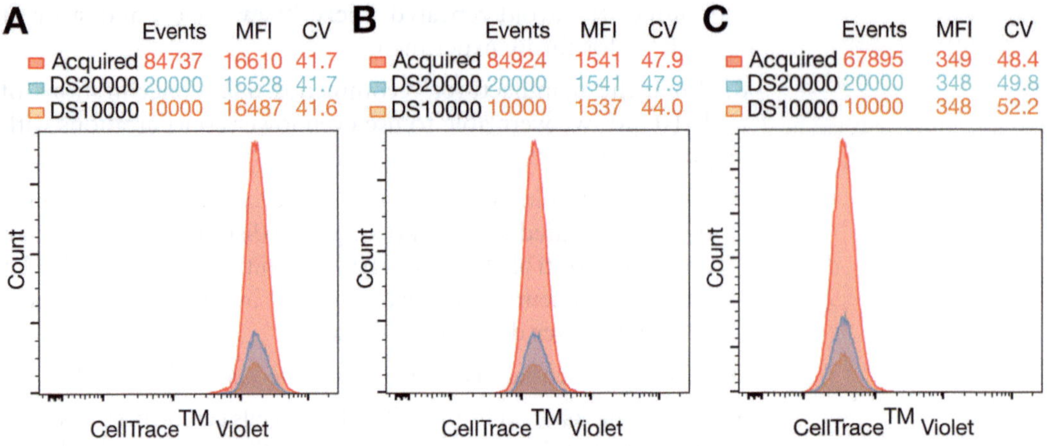

Fig. 6 Coefficient of variation (CV) comparison for determination of event numbers to acquire. T47D cells were labeled with 0.5 μM CellTrace™ Violet and acquired on (**a**) day 0, (**b**) day 3, and (**c**) day 6. The number of acquired events, median fluorescence intensities (MFIs), and coefficients of variation (CV) are listed above the plots. Downsampling to 20,000 (DS20000) and 10,000 events (DS10000) was performed in FlowJo, and MFIs and CVs were compared. As the CellTrace™ Violet signal and/or number of events decreases, the CV increases. DS20000 values are very similar to those recorded with the full event count, suggesting that acquisition of 20,000 live singlets suffices when evaluating proliferation of immortalized cell lines that proliferate in a synchronous manner

15. For cells that proliferate synchronously, such as those shown in Figs. 2 and 6, median fluorescence intensity (MFI) values should be calculated. Fold changes in MFI can then be calculated by dividing the CellTrace™ Violet MFI for day 0 by the MFIs for subsequent acquisition days. Results from biological replicates can be compared by plotting these fold changes over time, averaging the fold changes, and performing statistical analyses. For evaluation of primary cell proliferation, in which many cell divisions can be observed (*see* Fig. 7), proliferation modeling software, such as that available in FlowJo versions 10.5.3 or 10.7.1 or later, is recommended. Versions of FlowJo between 10.5.3 and 10.7.1 are not recommended for this modeling, as a bug in the software prevents proper calculation of percent cells divided; this has been fixed in version 10.7.1. Additional output parameters beyond those shown in Fig. 7—such as proliferation index, division index, percent divided, expansion index, and replication index—can also be measured using this modeling tool.

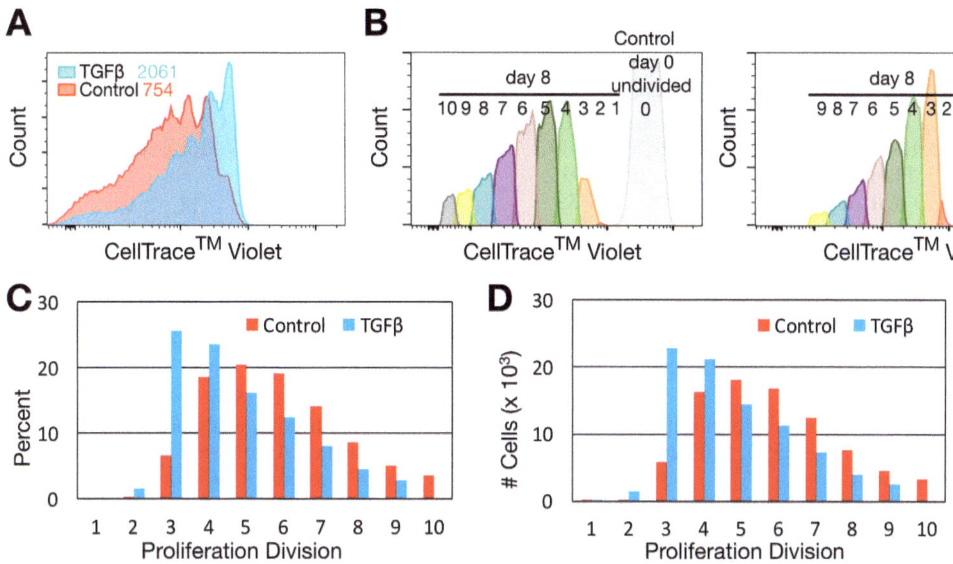

Fig. 7 Proliferation modeling. (**a**) Primary human γδ T cells that had been cultured for 11 days were labeled with 1 μM CellTrace™ Violet, treated with vehicle control ("Control," red) or 1.7 ng/ml TGF-β (blue) on day 0 and acquired by flow cytometry 8 days later. Median fluorescence intensity values are indicated. (**b**) The proliferation modeling tool in FlowJo 10.7.1 was used to fit control (left panel) and TGF-β (right panel) histogram curves, and determine proliferation divisions. These are numbered in the plots, with the day 0 histogram indicated in gray. (**c**) Graph of the percent of total γδ T cells in each sample that were in each proliferation division. (**d**) Cell counts from each proliferation division. These data suggest that TGF-β treatment inhibits proliferation of γδ T cells expanded using our culture conditions [16]

References

1. Begum J, Day W, Henderson C, Purewal S, Cerveira J, Summers H, Rees P, Davies D, Filby A (2013) A method for evaluating the use of fluorescent dyes to track proliferation in cell lines by dye dilution. Cytometry A 83(12):1085–1095. https://doi.org/10.1002/cyto.a.22403
2. Quah BJ, Parish CR (2012) New and improved methods for measuring lymphocyte proliferation in vitro and in vivo using CFSE-like fluorescent dyes. J Immunol Methods 379(1–2): 1–14. https://doi.org/10.1016/j.jim.2012.02.012
3. Tempany JC, Zhou JH, Hodgkin PD, Bryant VL (2018) Superior properties of CellTrace Yellow as a division tracking dye for human and murine lymphocytes. Immunol Cell Biol 96(2):149–159. https://doi.org/10.1111/imcb.1020
4. Winter S, Martin E, Boutboul D, Lenoir C, Boudjemaa S, Petit A, Picard C, Fischer A, Leverger G, Latour S (2018) Loss of RASGRP1 in humans impairs T-cell expansion leading to Epstein-Barr virus susceptibility. EMBO Mol Med 10(2):188–199. https://doi.org/10.15252/emmm.201708292
5. Perego M, Maurer M, Wang JX, Shaffer S, Muller AC, Parapatics K, Li L, Hristova D, Shin S, Keeney F, Liu S, Xu X, Raj A, Jensen JK, Bennett KL, Wagner SN, Somasundaram R, Herlyn M (2018) A slow-cycling subpopulation of melanoma cells with highly invasive properties. Oncogene 37(3): 302–312. https://doi.org/10.1038/onc.2017.341
6. Zhou P, Chen X, Li M, Sun X, Tan J, Wang X, Chu Y, Zhang Y, Cheng T, Zhou J, Wang G, Yuan W (2019) Overexpression of PRDM5 promotes acute myeloid leukemia cell proliferation and migration by activating the JNK pathway. Cancer Med 8(8):3905–3917. https://doi.org/10.1002/cam4.2261
7. Filby A, Begum J, Jalal M, Day W (2015) Appraising the suitability of succinimidyl and lipophilic fluorescent dyes to track proliferation in non-quiescent cells by dye dilution. Methods 82:29–37. https://doi.org/10.1016/j.ymeth.2015.02.016

8. Filby A, Day W, Purewal S, Martinez-Martin N (2016) The analysis of cell cycle, proliferation, and asymmetric cell division by imaging flow cytometry. In: Bartneva NS, Vorobjev IA (eds) Imaging flow cytometry methods and protocols. Methods in molecular biology, vol 1389, 1st edn. Humana Press, New York, pp 71–95. https://doi.org/10.1007/978-1-4939-3302-0
9. Lacy KJ, Parlane NA, Riley CB, Gee EK, Roberts JM, McIlwraith CW (2020) CellTrace Violet inhibits equine lymphocyte proliferation. Vet Immunol Immunopathol 223: 110037. https://doi.org/10.1016/j.vetimm.2020.110037
10. CellTrace™ Cell Proliferation Kits user guide (2017) vol MAN0002595, E.00 edn. ThermoFisher Scientific
11. Tario JD Jr, Conway AN, Muirhead KA, Wallace PK (2018) Monitoring cell proliferation by dye dilution: considerations for probe selection. Methods Mol Biol 1678:249–299. https://doi.org/10.1007/978-1-4939-7346-0_12
12. McKinnon KM (2018) Flow cytometry: an overview. Curr Protoc Immunol 120:5 1 1–5 1 11. https://doi.org/10.1002/cpim.40
13. Haramoto Y, Onuma Y, Mawaribuchi S, Nakajima Y, Aiki Y, Higuchi K, Shimizu M, Tateno H, Hirabayashi J, Ito Y (2020) A technique for removing tumourigenic pluripotent stem cells using rBC2LCN lectin. Regen Ther 14:306–314. https://doi.org/10.1016/j.reth.2020.03.017
14. Ivanov DP, Parker TL, Walker DA, Alexander C, Ashford MB, Gellert PR, Garnett MC (2015) In vitro co-culture model of medulloblastoma and human neural stem cells for drug delivery assessment. J Biotechnol 205: 3–13. https://doi.org/10.1016/j.jbiotec.2015.01.002
15. Chung S, Kim SH, Seo Y, Kim SK, Lee JY (2017) Quantitative analysis of cell proliferation by a dye dilution assay: application to cell lines and cocultures. Cytometry A 91(7): 704–712. https://doi.org/10.1002/cyto.a.23105
16. Siegers GM, Ribot EJ, Keating A, Foster PJ (2013) Extensive expansion of primary human gamma delta T cells generates cytotoxic effector memory cells that can be labeled with Feraheme for cellular MRI. Cancer Immunol Immunother 62(3):571–583. https://doi.org/10.1007/s00262-012-1353-y

Chapter 10

Enzyme-Linked Immunosorbent Assay (ELISA)

Mahdis Sadat Tabatabaei and Marya Ahmed

Abstract

Enzyme-linked immunosorbent assay (ELISA) is one of the most specific and straightforward assays for detecting biomolecules in research and clinics. With advances in analytical methods, ELISA assay has been constantly optimized to improve its sensitivity, and different types of ELISA are now available to detect various biomarkers. This chapter provides an overall summary of the basic principle of ELISA, discusses different components of ELISA assay, and clearly outline protocols for different types of ELISA assays, including direct, indirect, sandwich, competitive, and nanoparticle-based ELISA.

Key words Direct ELISA, Indirect ELISA, Nanoparticle-based ELISA, Sandwich ELISA, Competitive ELISA, Biotin–streptavidin interactions, Assay protocols

1 Introduction

Enzyme-linked immunosorbent assay (ELISA) is the leading detection method in clinical laboratories around the globe for the detection of disease-associated biomarkers. Commercial ELISA kits are readily available to diagnose a variety of pathological conditions, including neurological disorders, cancers, and inflammatory diseases. They can detect the slight changes in expression of several disease-associated biomarkers, including beta-amyloid [1] and tau protein [2] for Alzheimer's diagnosis and BLTA, CD137, CTLA-4, and TNFRSF14 for cancer diagnosis. The basic principle of ELISA is derived from the natural immune system of the body and is based on the interactions between antibodies produced by the immune system and antigens of the disease [3]. ELISA was first introduced as a radioimmunoassay (RIA) by Hoffman et al. in 1973 and has evolved since then [4]. A traditional ELISA takes place on a solid plastic support called ELISA plate, where antibodies are employed to capture antigens, and then a substrate is used to generate a measurable signal with the help of an enzyme. The intensity of the generated signal is proportional to the concentration of the antigen [5]. Hence, ELISA can be used to determine both the

Sherri L. Christian (ed.), *Cancer Cell Biology: Methods and Protocols*,
Methods in Molecular Biology, vol. 2508, https://doi.org/10.1007/978-1-0716-2376-3_10,

presence of the antigen and its concentration. This is made possible by the step-by-step addition of components of ELISA to the assay plate followed by an incubation period. The main components of ELISA are primary antibody, secondary antibody, enzyme-conjugated antibody, antigen, and substrate. Several buffers are used during ELISA for dilution of the components, washing the excess chemicals, blocking vacancies on the plate, and stopping the substrate reaction and will be discussed in this chapter.

Over the years, different types of ELISA were developed to improve the specificity, reduce the background noise, and improve the assay's sensitivity. These methods can be categorized into five main types: direct, indirect, sandwich, competitive, and nanoparticle-based ELISA. All ELISA types follow the same basic principle of antigens binding with antibodies but are different in order and number of the steps [6]. Overall, an ELISA assay can be divided into the following steps: (1) coating of proteins on ELISA plate, (2) blocking of free binding sites of the plate, (3) attachment of antibodies with antigens, (4) washing the plate to remove excess reagents, (5) attachment of enzyme-conjugated antibody with the antigen or antibody, (6) substrate addition, (7) addition of stopping reagent to stop the substrate reaction, and (8) signal measurement. Obtained signals can be analyzed to calculate the limit of detection (LOD) and sensitivity of the assay [7]. Depending on the type of antibody and antigen used in the ELISA assay, the concentrations of the proteins, incubation times, and composition of the buffers can be optimized to yield a high sensitivity of the assay.

1.1 Primary Antibodies

Antibodies or immunoglobulins (Ig) are proteins produced by the immune system to identify and neutralize foreign molecules. Antibodies are Y-shaped proteins with two fragmented antigen-binding (Fab) regions that specifically interact with antigens and a fragment crystalline (Fc) arm that interacts with other molecules (such as white blood cells) of the immune system. Each antigen has a complementary antibody with a strong and specific affinity toward its antigen [8]. ELISA assay takes advantage of this specificity and is developed as one of the most specific testing methods available for the clinical diagnosis of pathological conditions. Antibodies have several isotypes such as IgG, IgM, IgE, IgD, and IgA that are different from each other in terms of the immune response they generate and the parts of the body they can be found in [9]. Any of the isotypes can be employed in ELISA but requires optimized incubation times and concentrations. IgG is the most abundant antibody with a relatively simple structure and is chosen as the sample antibody for the ELISA protocols explained in this chapter [10].

1.1.1 IgG

IgG is the most abundant antibody in the bloodstream and has excellent specificity toward antigens. Hence, it is the primary choice of antibody in immunological tests. IgG has four subclasses: IgG1, IgG2, IgG3, and IgG4. The subclasses are slightly different in structure, and some subclasses are more abundant than others. IgG retains in the bloodstream for an extended time after infection. The detection of IgG in body fluids indicates vaccination or prior infection.

1.1.2 IgM

IgM is the largest antibody in size. It is the first antibody that is produced when an infection occurs. IgG gradually substitutes IgM for a more effective immune response as the disease proceeds. The relative percentage of IgG and IgM in the blood can be translated to determine the stage and type of the infection. The large size of IgM contributes to the difficulty in movement. Therefore, a longer incubation time is needed when coating IgM on an ELISA plate [11].

1.1.3 IgA

IgA is produced in parts of the body that are directly in contact with the outside environment. For instance, it can be found in tears, saliva, and breast milk. IgA prevents foreign pathogens from entering the internal organs. IgA has two subclasses. The level of IgA is measured via ELISA for the detection of bacterial infections [12].

1.1.4 IgE

IgE is the least abundant antibody in the body but produces the strongest inflammatory response. The production of IgE is related to allergic reactions. It is expressed in reactions such as allergies, asthma, sinusitis, food allergies, and against parasites. Since the concentration of expressed IgE is very high, it is not measured quantitatively [13].

1.2 Antigens

Antigen, also termed as antibody generator (Ag), is a protein produced by a foreign pathogen, and its presence in the body will induce the immune response resulting in the generation of antibodies [14]. Antigen–antibody interactions are the essential factor that makes the basis of ELISA. Therefore, the antibodies and antigens should be chosen carefully to have a specific and strong affinity for each other. Each antibody has two Fab fragments for its matching antigen, a factor that needs to be considered in the calculations and analysis of ELISA [10]. Antigens are classified based on their origins. They can either be foreign particles or can be generated inside the body in response to infection. Some examples of foreign antigens are proteins or their fragments produced by viruses or bacteria. Usually, the antigens that originate from inside the body do not trigger the immune system.

1.3 Secondary Antibodies

Secondary antibodies are used for indirect detection of the antigen. Secondary antibodies bind with the primary antibody after the binding of the primary antibody with the antigen. Secondary antibodies must be specific to the primary antibody. A primary antibody is injected into a different host to obtain a secondary antibody. For example, if a mouse IgG is being used as the primary antibody, mouse IgG gets injected into another host such as goat to obtain a secondary antibody specific to the mouse IgG, and this antibody is called goat anti-mouse IgG [15]. Mouse, goat, rabbit, and horse are common hosts for the production of secondary antibodies, but other animals can be used as well [10].

1.4 Enzyme-Conjugated Antibodies

Enzymes are biological catalysts that accelerate biochemical reactions but do not undergo any changes themselves. The role of enzymes in ELISA is to catalyze a chemical reaction that produces fluorescent, luminescent, or UV-absorbing products. A variety of enzymes can be used in ELISA. Horseradish peroxidase (HRP) and alkaline phosphatase (ALP) are the most commonly used enzymes for ELISA applications [16]. In ELISA, enzymes are used in conjugation with antibodies. Enzymes and antibodies can be conjugated before addition to the ELISA plate or conjugated on the ELISA plate. The former is the most popular approach and is typically done by employing bioconjugation methods such as maleimide chemistry [17], carbodiimide chemistry, and the periodate method [18]. A commonly used non-covalent approach for enzyme antibody conjugation is biotin–streptavidin chemistry. Biotin streptavidin interactions are the strongest non-covalent interactions in nature and are widely employed to label enzymes, fluorophores, and nanoparticles with proteins [19]. In ELISA, a biotinylated antibody is immobilized on the ELISA plate, followed by the addition of streptavidin-conjugated enzyme for enzymatic labeling of the biotinylated antibody. In this chapter, biotin–streptavidin interactions will be utilized for the enzymatic labeling of antibodies in ELISA protocols.

1.4.1 Horseradish Peroxidase (HRP)

HRP is a strong catalyst for the oxidation of substrates. The strong catalytic activity of HRP makes it the primary choice for ELISA. In addition, HRP catalyzes the oxidation reaction of several different substrates. Substrates such as 3,3′,5,5′-tetramethylbenzidine (TMB), 2,2′-azino-bis (3-ethylbenzothiazoline-6-sulfonic acid) (ABTS), and luminol can be oxidized in the presence of HRP (*see* Fig. 1) [20, 21].

1.4.2 Alkaline Phosphatase (ALP)

ALP is an enzyme that is usually active at alkaline pH and can be found in the human body. It has good heat and pH stability. 5-Bromo-4-chloro-3-indolyl phosphate and nitro blue tetrazolium (BCIP/NBT), para-nitrophenylphosphate (pNPP), and Fast Red are substrates that are used with ALP in ELISA.

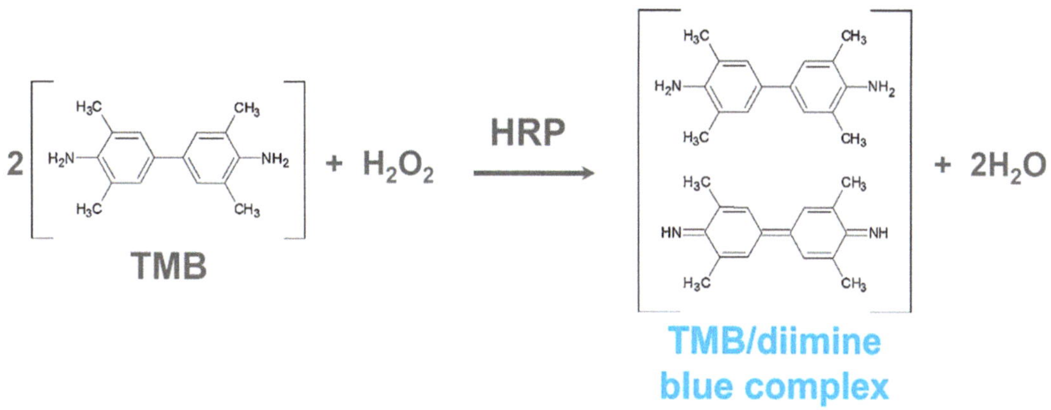

Fig. 1 Oxidation reaction of 3,3′,5,5′-tetramethylbenzidine (TMB) with hydrogen peroxide, catalyzed by horseradish peroxidase (HRP). (Reproduced with permission from Ref. [21])

1.5 Substrates

Substrates are chemicals that react in the presence of enzymes to produce products with photometrical, fluorescence, or luminescence features. The concentration of a substrate can be quantified by measuring its UV absorbance, fluorescence emission, or luminescence [22]. The quantified optical values are in a linear relationship with the concentration of the substrate and are indirectly correlated with the concentration of the analyte immobilized on the ELISA plate. Upon the addition of substrate to the plate, a chemical reaction starts in the presence of the enzyme that can be detected by the color change and is stopped by adding a stop solution to the reaction [23].

1.5.1 3,3′,5,5′-Tetramethylbenzidine (TMB)

TMB is used as a substrate in the presence of an HRP enzyme that catalyzes the rapid oxidation of substrate into its oxidized form (diimine) (*see* Fig. 4) [20, 24]. TMB is commercially available in many forms, and each form yields a different detection limit. The TMB oxidation reaction can be stopped by adding phosphoric or sulfuric acid, and the absorbance values are typically read at 450 nm (*see* Fig. 2).

1.5.2 2,2′-Azinobis (3-Ethylbenzothiazoline-6-Sulfonic Acid) Diammonium Salt (ABTS)

ABTS is another substrate specific for HRP but is less sensitive than TMB. Oxidation of ABTS in the presence of HRP in ELISA yields a green color. The absorbance of ABTS after the addition of the stop buffer is measured at 734 nm.

1.5.3 p-Nitrophenylphosphate, Disodium Salt (pNPP)

pNPP is the commonly used substrate in the presence of the ALP enzyme. It is commercially available in both solution and tablet forms. pNPP yields a yellow color after the oxidation reaction, and its absorbance is measured at 405 nm.

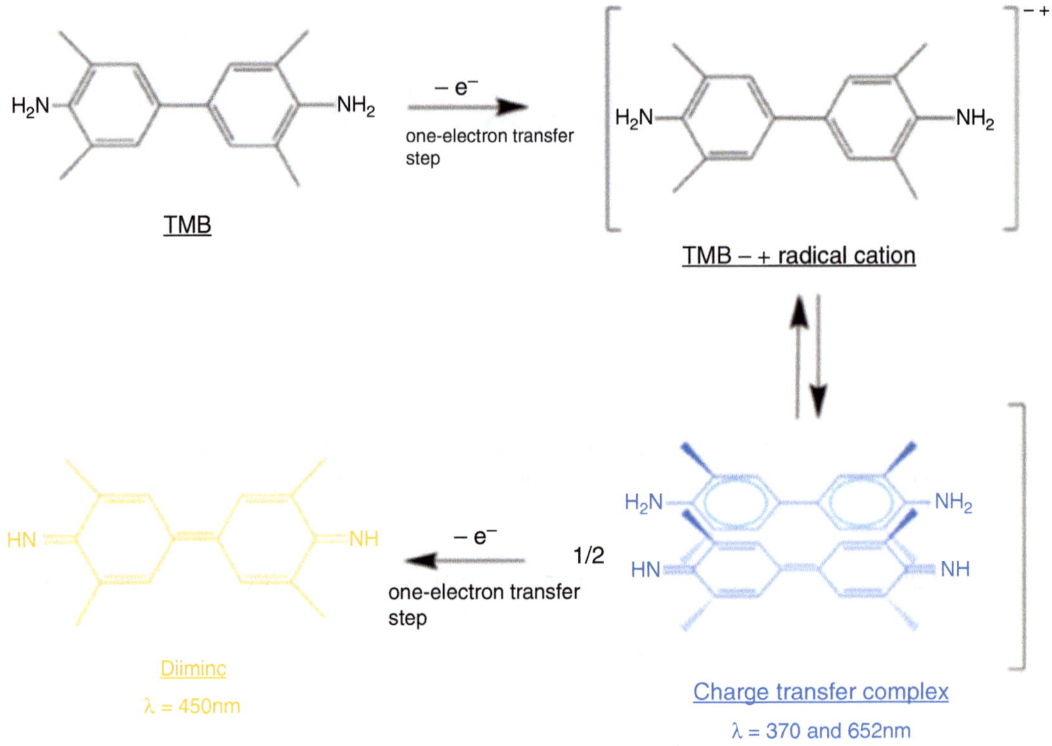

Fig. 2 3,3′,5,5′-Tetramethylbenzidine (TMB) in oxidized form (blue) and after stopping the oxidation reaction (yellow) [24]

1.5.4 Luminescent and Fluorescent Probes

Luminol/H_2O_2 is an example of a typical luminescent substrate. The reduction of luminol/H_2O_2 is catalyzed in the presence of horseradish peroxidase (HRP), and a phenolic compound is added for enhancement and prolongation of the luminescence signal intensity. Similarly, 4-methylumbelliferyl phosphate substrate converts to 4-methylumbelliferone in the presence of alkaline phosphate enzyme. 4-Methylumbelliferone is a fluorescent product with an excitation wavelength of 360 nm, and fluorescent emission is measured at 440 nm.

1.6 Stop Buffer

The reaction stop buffer contains chemicals that interfere with the reaction of the substrate catalyzed by the enzyme. An advantage of using a stop buffer is to compare different individual ELISA assays with each other. The composition of the stop solution depends on the substrate. For example, the addition of sulfuric acid to TMB halts the oxidation reaction and results in a color change from blue to yellow that is recorded at 450 nm. For TMB, the stop solution is comprised of 2-M phosphoric and 0.16-M sulfuric acid. For ATMB, the stop solution is 1% SDS. For pNPP, the stop solution is 1-M NaOH.

1.7 Coating Buffer

Coating buffer is used to dilute the antibody or antigen before its addition to the ELISA plate. Coating buffer should not contain any other proteins to maintain specific binding of desired antigens or antibodies with ELISA plate. The role of coating buffer is to stabilize the proteins (antigen and antibody) and facilitate their attachment to the ELISA plate [25]. A mixture of sodium carbonate and sodium bicarbonate in diH_2O between pH 7.4 and 9.6 is used to dilute and stabilize antibodies and antigens when coating them on an ELISA plate. The high pH of the buffer helps with the solubility of the proteins and deprotonates to facilitate their binding on the positively charged plate. The phosphate buffer concentration and its pH depend on the assay conditions and should be carefully optimized.

1.8 Blocking Buffer

After coating the surface of the ELISA plate, a blocking buffer is employed to block any free binding sites on the ELISA plate. Blocking buffer is composed of a mixture of proteins and salts to block the vacant spots and moderate the ELISA plate's pH. Casein, gelatine, and bovine serum albumin (BSA) are commonly used proteins for blocking the ELISA plate. These proteins are dissolved in buffers such as phosphate-buffered saline (PBS) and Tris-buffered saline (TBS) to make an effective blocking buffer. The use of detergents such as Tween 20 is also recommended for ELISA assays that require more potent blocking agents.

1.9 Washing Buffer

ELISA plates are washed with washing buffer to remove the excess proteins. Washing buffers typically consist of detergents dissolved in a buffer. For a more intense washing, increasing the percentage of the detergent or increasing the number of washing steps can be helpful. For washing the ELISA, plate phosphate-buffered saline or Tween-buffered saline containing Tween 20 are used.

1.10 Types of ELISA

1.10.1 Direct ELISA

Direct ELISA is more straightforward but less specific than other types of ELISA. In this ELISA, an analyte of interest is directly coated onto the ELISA plate, followed by adding an enzyme-conjugated antibody specific to the coated analyte. The substrate is then added that produces a detectable signal in the presence of an enzyme (*see* Fig. 3a) [26]. The addition of the stop buffer stops the reaction, and then, the signal is measured.

1.10.2 Indirect ELISA

In this ELISA, a secondary antibody is used as a linker to indirectly conjugate the first antibody to the enzyme (*see* Fig. 3b) [26]. An advantage of this method over direct ELISA is its increased sensitivity. In indirect ELISA, only the secondary antibody needs to be biotinylated.

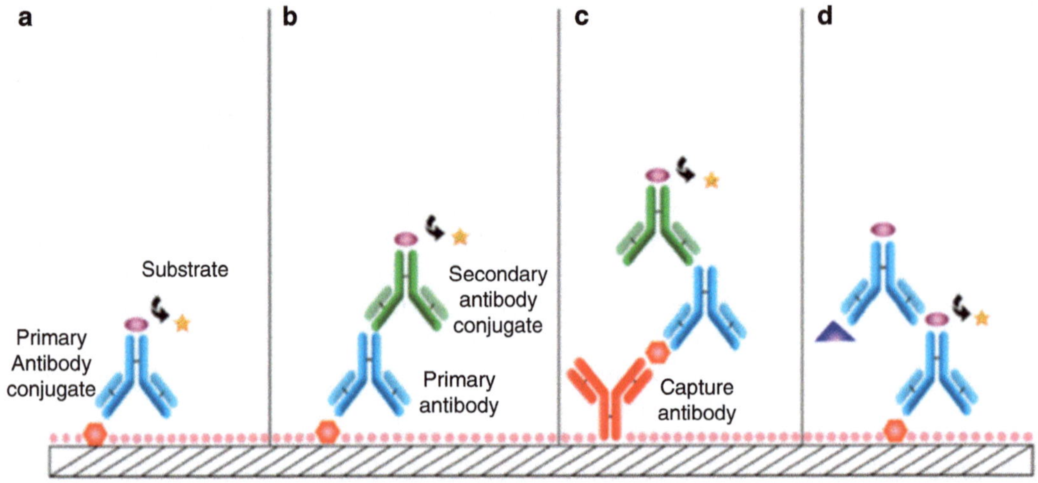

Fig. 3 Types of enzyme-linked immunosorbent assay (ELISA). (**a**) Direct ELISA, (**b**) indirect ELISA, (**c**) sandwich ELISA, (**d**) competitive ELISA. (Reproduced with permission from Ref. [26])

1.10.3 Sandwich EISA

In sandwich ELISA, the antigen is sandwiched between two antibodies. The antibody directly immobilized onto the ELISA plate, termed as capture antibody, which binds with the antigen. The secondary antibody is then added, followed by adding enzyme-conjugated or detection antibody [27]. The addition of substrate initiates the reaction, and the product formed is analyzed by measuring the absorbance of the signal, as is discussed above (*see* Fig. 3c) [26]. The main advantage of sandwich ELISA over other types is reduced background signal, high sensitivity, and applicability for detecting antigens in biological fluids. For example, antigens are typically present in other biomolecules, such as in serum proteins. Therefore, direct coating of antigens on the ELISA plate might result in the non-specific attachment of serum proteins, resulting in a high background signal and "false-positive" results [28]. In sandwich ELISA, pre-coating with primary antibodies allows specific antigen binding on the ELISA plate and allows selective detection of a molecule of interest in the given sample.

1.10.4 Competitive ELISA

In competitive ELISA, the antigen of interest competes with a reference antigen to bind with the antibody [29]. Any of the previously explained types of ELISA can be performed competitively. For example, in a competitive sandwich ELISA, the plate is first coated with the capture antibody. The reference antigen will then bind to the capture antibody, followed by adding the target antigen and the detection antibody simultaneously. Any detection antibody that is not conjugated to the target antigen will be immobilized on the plate through binding to the reference antigen. Target antigens and their attached antibodies will be washed away, and the enzyme and substrate will be added to detect the immobilized antibodies (*see* Fig. 3d) [26]. In this type of ELISA, a higher signal indicates a lower concentration of the target antigen.

1.10.5 Nanoparticle-Based ELISA

Several modifications to the abovementioned original ELISA format have been explored throughout the years to improve the sensitivity of ELISA-based assay. For instance, nanoparticles have been used to replace the enzyme and the substrate in ELISA, and luminescent [30] and fluorescent markers have been used instead of a UV-absorbing substrate [31]. To date, many nanoparticles such as magnetic and metal oxide [32, 33] nanoparticles have been shown to possess enzymatic activities. The advantage of nanoparticle-based ELISA over traditional enzyme-based ELISA is the thermal and chemical stability, ease of synthesis, and low cost of the former. Gold nanoparticles have been explored as signal-generating markers because of their intrinsic surface plasmon resonance (SPR) feature [34]. In ELISA, nanoparticles have also been used as a scaffold to connect the enzymes and antibodies on a three-dimensional surface. For instance, spherical gold nanoparticles can accommodate several enzymes and enzyme-conjugated antibodies, and increased concentration of enzymes on nanoparticles surface leads to enhancement of ELISA signal (*see* Fig. 4) [35]. The detailed protocols of different ELISA types are provided as follows:

2 Materials

1. 96-well ELISA plate (*see* **Notes 1** and **2**).
2. Pipettors, from 2 to 1000 L (*see* **Note 3**).
3. Pipettor tips, from 2 to 1000 L (*see* **Note 4**).
4. ELISA plate seal (*see* **Note 5**).
5. Solvent reservoir (*see* **Note 6**).
6. Centrifuge tubes.
7. Plate shaker.
8. Microplate reader.
9. Plate washer (*see* **Note** 7).
10. Biotinylated human IgG (*see* **Notes 8** and **9**).
11. Mouse anti-human IgG (*see* **Notes 9** and **10**).
12. Antigen (*see* **Note 11**).
13. Streptavidin–HRP conjugate.
14. Gold nanoparticles functionalized with human IgG–HRP conjugates.
15. TMB solution: Dissolve 100 mg of TMB in 10-mL dimethyl sulfoxide (DMSO), divide the solution into 1-mL aliquots,

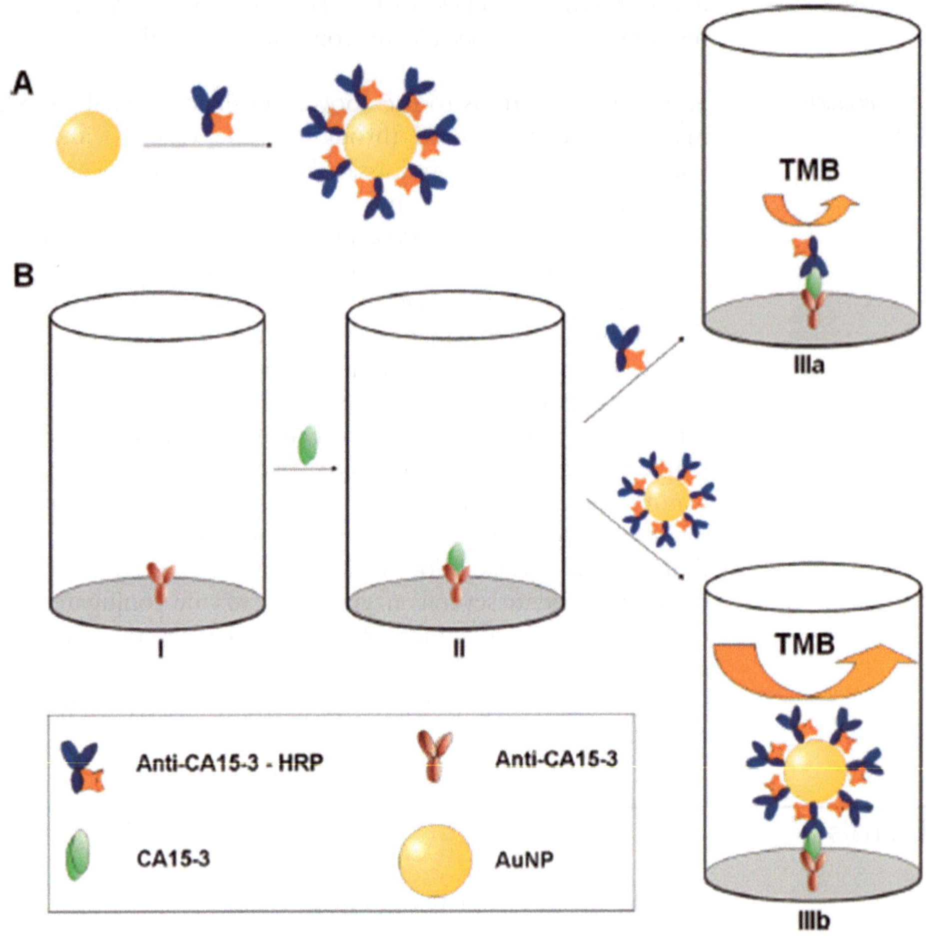

Fig. 4 Gold nanoparticle–assisted ELISA. (**A**) Conjugation of antibodies with gold nanoparticles and (**B**) evaluation of antibody coated nanoparticles in ELISAassay. (Reproduced with permission from Ref. [35])

cover the vials with thin foil, and store at 4 °C. Make a solution of 0.07-M sodium acetate and 0.03-M acetic acid by dissolving 577 mg of sodium acetate and 178 mg of acetic acid to 60 mL of diH_2O. Use 10-N HCl to bring the pH to 5.2. Add 1 mL of TMB solution (TMB is commercially available from Cedarlane. TMB contains 3,3′,5,5′-tetramethylbenzidine in a mildly acidic buffer and is supplied as ready-to-use solution). Filter the solution using a 0.45-m filter membrane. Add 30-mL hydrogen peroxide. Bring the total volume to 100 mL with diH_2O. Store at 4 °C.

16. Coating buffer: 0.2-M sodium bicarbonate, 0.2-M sodium carbonate, pH 9.4. Dissolve 6.7-g sodium bicarbonate and 2.1-g sodium carbonate (anhydrous) in 800-mL diH_2O. Use HCl to adjust the pH to 9.4. Bring the total volume to 1 L by adding diH_2O. Store at 4 °C.

17. Phosphate-buffered saline (PBS): 137-mM NaCl, 10-mM phosphate, 2.7-mM KCl, pH 7.4. Make PBS by adding 8-g NaCl, 1.44-g Na_2HPO_4, 0.24-g KH_2PO_4, and 0.2-g KCl to 800-mL diH_2O. Adjust the pH to 7.4 by adding HCl or NaOH. Bring the volume to 1 L by adding diH_2O.
18. Blocking buffer: 1% BSA in PBS.
19. Washing buffer: 0.1% Tween 20 in PBS.
20. Reaction stop buffer: 0.16-M sulfuric acid.

3 Methods

As mentioned above, ELISA is performed to identify the presence and concentration of the biomarker in a solution [36]. For quantifying the concentration of the biomarker, a standard curve is needed to correlate the concentration of the biomarker with the produced signal. To obtain a calibration curve, an ELISA should be performed on different known concentrations of the biomarkers. A control containing no biomarkers should always be used to monitor the background signal. The protocols in this section explain how to prepare standard curves for different types of ELISA. The standard curve of a typical ELISA is depicted in Fig. 5.

3.1 Direct ELISA

1. Prepare seven different concentrations of the target antigen in coating buffer, from 0.01 to 10,000 ng/mL, and a negative control using Table 1. Name the highest concentration "Cal 1" and the second highest "Cal 2," and continue the same pattern for the rest of the concentrations. Name the negative control "Cal 8". (*see* **Notes 12** and **13**).
2. Add 100 L of Cal 1 to wells A1, A2, and A3. Continue adding three replicates of each concentration to wells B–H. Negative control (Cal 8) would be added to wells H1, H2, and H3 (*see* Fig. 6).
3. Seal the plate and put it on a plate shaker at room temperature (rt). Set the speed to 300 rpm and let the plate shake for 1 h.
4. Take the seal off, discard the remaining liquid in the wells (*see* **Note 14**).
5. Add 200 L of blocking buffer to all the wells that were previously added a solution to in **step 2** (*see* **Note 15**).
6. Repeat **step 3**.
7. Take the seal off. Discard the remaining liquid in the wells. Add 300 L of washing buffer to all the blocked wells. Discard the buffer from the wells. Wash the plate with washing buffer two more times (*see* **Notes 15** and **16**).

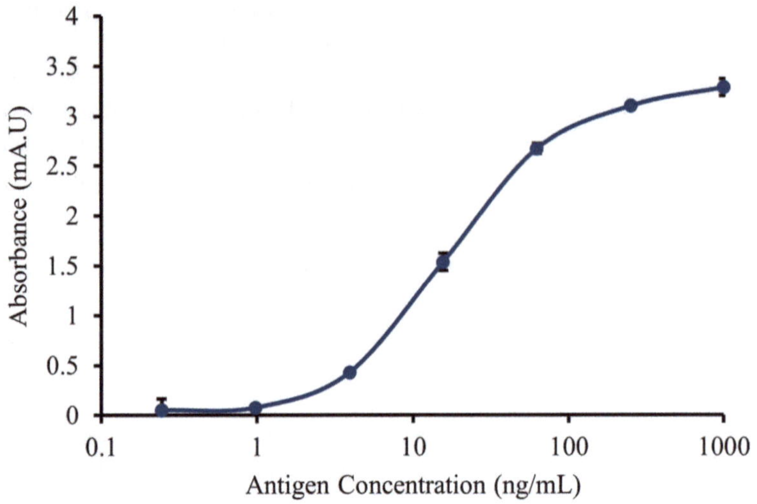

Fig. 5 An example of enzyme-linked immunosorbent assay (ELISA) standard curve

Table 1
Guide for dilution of antigens

Sample name	Source name	Source volume (L)	Diluent volume (L)	Final volume (L)	Final concentration (ng/mL)
Cal 1	Stock	–	–	500	10,000
Cal 2	Cal 1	50	450	500	1000
Cal 3	Cal 2	50	450	500	100
Cal 4	Cal 3	50	450	500	10
Cal 5	Cal 4	50	450	500	1
Cal 6	Cal 5	50	450	500	0.1
Cal 7	Cal 6	50	450	500	0.01
Cal 8	–	–	500	500	0

8. Prepare 3 mL of the biotinylated antibody (IgG) with the concentration of 1 g/mL in blocking buffer (*see* **Note 17**).
9. Add 100 L of the biotinylated antibody to all the washed wells (*see* **Note 15**).
10. Repeat **step 3**.
11. Repeat **step 7**.
12. Prepare 3 mL of streptavidin–HRP with the concentration of 200 ng/mL in blocking buffer.
13. Add 100 L of streptavidin–HRP to all the washed wells.

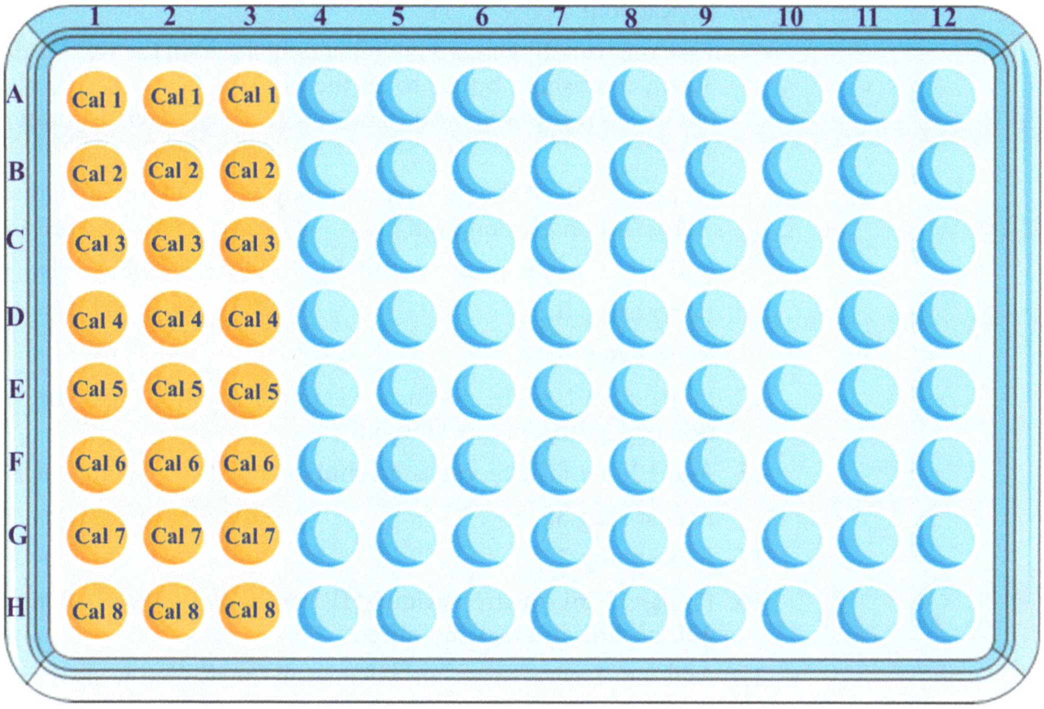

Fig. 6 Visual presentation of 96-well enzyme-linked immunosorbent assay (ELISA) plate map explained in this protocol. (Figure produced with the help of Servier Medical Art)

14. Repeat **step 3**.
15. Repeat **step** 7 (*see* **Note 18**).
16. Add 100 L of the TMB solution to each washed well.
17. Put the plate in the dark and wait for 20 min (see **Note 19**).
18. Add 100 L of the stop buffer to the wells that contain TMB (see **Note 20**).
19. Put the plate in a microplate reader and measure the absorbance value of the wells at 450 nm with reference at 620 nm.

3.2 Indirect ELISA

1. Prepare seven different concentrations of the target antigen in coating buffer, from 0.01 to 10,000 ng/mL, and a negative control using Table 1.
2. Add 100 L of Cal 1 to wells A1, A2, and A3. Continue adding three replicates of each concentration to wells B–H. Negative control (Cal 8) would be added to wells H1, H2, and H3.
3. Seal the plate and put it on a plate shaker at room temperature (rt). Set the speed to 300 rpm and let the plate shake for 1 h.
4. Take the seal off, and discard the remaining liquid in the wells.
5. Add 200 L of blocking buffer to all the wells that were previously added a solution to in **step 2**.

6. Repeat **step 3**.
7. Take the seal off. Discard the remaining liquid in the wells. Add 300 L of washing buffer to all the blocked wells. Discard the buffer from the wells. Add and discard the buffer two more times.
8. Prepare 3 mL of the primary antibody with the concentration of 1 g/mL in blocking buffer.
9. Add 100 L of the primary antibody to all the washed wells.
10. Repeat **step 3**.
11. Repeat **step 7**.
12. Prepare 3 mL of the biotinylated antibody with the concentration of 1 g/mL in blocking buffer.
13. Add 100 L of the biotinylated antibody to all the washed wells.
14. Repeat **step 3**.
15. Repeat **step 7**.
16. Prepare 3 mL of streptavidin–HRP with the concentration of 200 ng/mL in blocking buffer.
17. Add 100 L of streptavidin–HRP to all the washed wells.
18. Repeat **step 3**.
19. Repeat **step 7**.
20. Add 100 L of the TMB solution to each washed well.
21. Put the plate in a dark place and wait for 20 min.
22. Add 100 L of the stop buffer to the wells that contain TMB.
23. Put the plate in a plate reader and measure the absorbance value of the wells at 450 nm and 620 nm for reference.

3.3 Sandwich ELISA

1. Prepare 3 mL of capture antibody with the concentration of 1 g/mL in coating buffer.
2. Add 100 L of capture antibody in triplicates to each row of the ELISA plate.
3. Seal the plate and put it on a plate shaker at room temperature (rt). Set the speed to 300 rpm and let the plate shake for 1 h.
4. Take the seal off, and discard the remaining liquid in the wells.
5. Add 200 L of blocking buffer to all the wells containing capture antibodies.
6. Repeat **step 3**.
7. Take the seal off. Discard the remaining liquid in the wells. Add 300 L of washing buffer to all the blocked wells. Discard the buffer from the wells. Add and discard the buffer two more times.

8. Prepare seven different concentrations of the target antigen in blocking buffer, from 0.01 to 10,000 ng/mL, and a negative control using Table 1.
9. Add 100 L of Cal 1 to wells A1, A2, and A3. Continue adding three replicates of each concentration to wells B–H. Negative control (Cal 8) would be added to wells H1, H2, and H3.
10. Repeat **step 3**.
11. Repeat **step** 7.
12. Prepare 3 mL of the biotinylated detection antibody with the concentration of 1 g/mL in blocking buffer.
13. Add 100 L of the detection antibody to all the washed wells.
14. Repeat **step 3**.
15. Repeat **step** 7.
16. Add 100 L of streptavidin–HRP to all the washed wells.
17. Repeat **step 3**.
18. Repeat **step** 7.
19. Add 100 L of the TMB solution to each washed well.
20. Put the plate in a dark place and wait for 20 min.
21. Add 100 L of the stop buffer to the wells that contain TMB.
22. Put the plate in a plate reader and measure the absorbance value of the wells at 450 nm and 620 nm for reference.

3.4 Competitive ELISA

1. Prepare 3 mL of reference antigen with the concentration of 10 g/mL in coating buffer.
2. Add 100 L of reference antigen in triplicates to each row of the ELISA plate.
3. Seal the plate and put it on a plate shaker at room temperature (rt). Set the speed to 300 rpm and let the plate shake for 1 h.
4. Take the seal off, and discard the remaining liquid in the wells.
5. Add 200 L of blocking buffer to all the wells containing reference antigen.
6. Repeat **step 3**.
7. Take the seal off. Discard the remaining liquid in the wells. Add 300 L of washing buffer to all the blocked wells. Discard the buffer from the wells. Add and discard the buffer two more times.
8. Prepare seven different concentrations of the target antigen in blocking buffer, from 0.01 to 10,000 ng/mL, and a negative control using Table 1.
9. Prepare 3 mL of the biotinylated antibody with the concentration of 1 g/mL in blocking buffer.

10. Add 100 L of the antibody to all the washed wells.
11. Add 100 L of Cal 1 to wells A1, A2, and A3. Continue adding three replicates of each concentration to wells B–H. Negative control (Cal 8) would be added to wells H1, H2, and H3.
12. Repeat **step 3**.
13. Repeat **step** 7.
14. Add 100 L of streptavidin–HRP to all the washed wells.
15. Repeat **step 3**.
16. Repeat **step** 7.
17. Add 100 L of the TMB solution to each washed well.
18. Put the plate in a dark place and wait for 20 min.
19. Add 100 L of the stop buffer to the wells that contain TMB.
20. Put the plate in a plate reader and measure the absorbance value of the wells at 450 nm and 620 nm for reference.

3.5 Gold Nanoparticle–Based ELISA

1. Prepare 3 mL of capture antibody with the concentration of 1 g/mL in coating buffer.
2. Add 100 L of capture antibody in triplicates to each row of the ELISA plate.
3. Seal the plate and put it on a plate shaker at room temperature (rt). Set the speed to 300 rpm and let the plate shake for 1 h.
4. Take the seal off, and discard the remaining liquid in the wells.
5. Add 200 L of blocking buffer to all the wells that you previously added a solution to in **step 2**.
6. Repeat **step 3**.
7. Take the seal off. Discard the remaining liquid in the wells. Add 300 L of washing buffer to all the blocked wells. Discard the buffer from the wells. Add and discard the buffer two more times.
8. Prepare seven different concentrations of the target antigen in blocking buffer, from 0.01 to 10,000 ng/mL, and a negative control using Table 1.
9. Add 100 L of Cal 1 to wells A1, A2, and A3. Continue adding three replicates of each concentration to wells B–H. Negative control (Cal 8) would be added to wells H1, H2, and H3.
10. Repeat **step 3**.
11. Repeat **step** 7.
12. Prepare 5 mL of 10-g/mL gold nanoparticles functionalized with HRP-conjugated detection antibodies.
13. Add 200 L of functionalized gold nanoparticles to the washed wells.

14. Repeat **step 3**.
15. Take the seal off. Discard the remaining liquid in the wells. Add 300 L of washing buffer to all the blocked wells. Discard the buffer from the wells. Add and discard the buffer seven more times.
16. Add 100 L of the TMB solution to each washed well.
17. Put the plate in a dark place and wait for 20 min.
18. Add 100 L of the stop buffer to the wells that contain TMB.
19. Put the plate in a plate reader and measure the absorbance value of the wells at 450 nm and 620 nm for reference.

3.6 Analysis

Run the unknown samples along with the standards. Make a standard curve by plotting the absorbance values (Y-axis) against known concentrations (X-axis) of analytes in Microsoft Excel. Add the trendline and find the best fit and linear regression equation. Solve the equation for X. Put the absorbance values of unknowns in the equation. Solve the equation for the concentration of unknown samples.

4 Notes

1. ELISA plates with removable strips are recommended. Removable strips are handy when performing different assays on the same plate. For example, when optimizing the washing buffer, the strips can be separated and washed with different buffers, but the rest of the steps, such as incubation, can be done using only one plate and with the same conditions.
2. ELISA plates are typically made from polystyrene (PS) and have been shown to have a high binding affinity toward antibodies and antigens.
3. Using a multichannel pipettor is unnecessary, but it will improve the standard deviations and reduce the overall assay duration. Working with multichannel pipettors requires practice and precision, but it can be a great asset once mastered. For ELISA, a multichannel pipettor with a variable volume from 100 to 300 L is suggested.
4. The pipettor tips must match the pipettors. They should not be loose on the pipettor and easy to discard as well.
5. Seals are sticky plastic or metal sheets that fit on top of an ELISA plate and block dust and contaminations. Metallic heat-resistant seals are available for assays that require incubations in temperatures other than room temperature.

6. Solvent reservoirs are only required when using a multichannel pipettor. Solvents will be poured into the reservoirs for all pipettor channels to take the solvent at the same time.
7. Using a plate washer is not necessary and will only reduce assay duration.
8. Biotinylated antibodies are commercially available, but simple carbodiimide chemistry can also do the biotinylation.
9. This antibody can be replaced with any desired antibody with a high binding affinity toward the antigen of interest.
10. This antibody will be used as the secondary antibody in indirect ELISA (*see* Subheading 3.2).
11. The type of antigen should be chosen by the user based on the target of ELISA.
12. The diluent of the target antigen for preparation of standard curve samples depends on the diluent of the unknown sample to be tested. Many analytes are in serum. In this case, direct and indirect ELISA are not recommended because coating the serum sample on the plate can cause a high background signal resulting from binding serum proteins to the plate.
13. The proposed concentrations and dilutions of standards cover a wide range and, depending on the resulted signal, can be further optimized. It is possible that an oversaturated absorbance signal appears for the highest concentrations. On the other hand, the lowest concentrations might have an absorbance of equal or lower than the negative control. In addition, it is possible to include more concentrations in the standard curve. This can be done by using a different dilution factor. For instance, in this protocol, tenfold dilution series have been used, but twofold dilution series would result in more points in the standard curve. It is recommended that the standard curve has at least five different concentrations.
14. Discarding the remaining liquid in the wells should be done very carefully. It should happen suddenly and forcefully to make sure the liquid from one well does not pour into another well and cause contamination.
15. Using a multichannel pipette is recommended for this step, as it helps with consistency and lowers standard deviations between the replicates.
16. This step can also be done by an automated plate washer. Using a plate washer will not make any difference to the results but will shorten the overall assay time.
17. If the absorbance values of the different concentrations are similar, lower the concentration of the antibody. If the absorbance value of the highest concentration is lower than 1 mA. U, increase the antibody concentration.

18. If the negative control shows a signal, perform more than three washes at this step to reduce the background signal.
19. This incubation time can be lowered to 5 min if the difference in color intensity of different concentrations is distinguishable by the naked eye and the UV absorbance value of the highest concentration is higher than 3 mA.U.
20. Care must be taken when using the substrate since its high sensitivity can pick up any contamination and result in a false signal.

References

1. Bruggink KA, Jongbloed W, Biemans EALM et al (2013) Amyloid-β oligomer detection by ELISA in cerebrospinal fluid and brain tissue. Anal Biochem 433:112–120
2. Harrington CR, Edwards PC, Wischik CM (1990) Competitive ELISA for the measurement of tau protein in Alzheimer's disease. J Immunol Methods 134:261–271
3. Darwish IA (2006) Immunoassay methods and their applications in pharmaceutical analysis: basic methodology and recent advances. Int J Biomed Sci 2:217–235
4. Hoffman DR (1973) Estimation of serum IgE by an enzyme-linked immunosorbent assay (ELISA). J Allergy Clin Immunol 51:303–307
5. Engvall E, Perlmann P (1971) Enzyme-linked immunosorbent assay (ELISA). Quantitative assay of immunoglobulin G. Immunochemistry 8(9):871–874. https://doi.org/10.1016/0019-2791(71)90454-x. PMID: 5135623.
6. Banerjee A, Hargreaves C (2007) Chapter 2: Basic principles of ELISA. In: A resuscitation room guide. Oxford University Press, Oxford/New York
7. Danielak D, Banach G, Walaszczyk J et al (2020) A novel open source tool for ELISA result analysis. J Pharm Biomed Anal 189:113415
8. Cruse JM, Lewis RE (1999) Atlas of immunology. Springer Berlin Heidelberg, Berlin, Heidelberg
9. French MAH (ed) (1986) Immunoglobulins in health and disease. Springer Netherlands, Dordrecht
10. Janeway CA Jr, Travers P, Walport M et al (2001) Immunobiology: The Immune System in Health and Disease. 5th edition. New York: Garland Science; 2001. Chapter 1, Basic Concepts in Immunology.
11. Florent G, Wiseman A (1990) An IgM specific ELISA for the serodiagnosis of viral bovine respiratory infections. Comp Immunol Microbiol Infect Dis 13:203–208
12. Sahu PS, Parija SC, Sahu PK (2008) Tear IgA-ELISA: a novel and sensitive method for diagnosis of ophthalmic cysticercosis. Acta Trop 106:168–174
13. He S, Li X, Gao J et al (2017) Development of sandwich ELISA for testing bovine β-lactoglobulin allergenic residues by specific polyclonal antibody against human IgE binding epitopes. Food Chem 227:33–40
14. DeLisi C (1976) Antigen antibody interactions. Springer Berlin Heidelberg, Berlin, Heidelberg
15. Steward MW (1984) Antibodies. Springer Netherlands, Dordrecht
16. Hosseini S, Vázquez-Villegas P, Rito-Palomares M et al (2018) Enzyme-linked immunosorbent assay (ELISA). Springer Singapore, Singapore
17. Methods I, Jim E, Sciences P et al (1988) The use of N-[3-(4-diazophenyl)ethyl] maleimide as a coupling agent in the preparation of enzyme-antibody conjugates. J Immunol Methods 110:47–53
18. Jeanson A, Cloes JM, Bouchet M et al (1988) Comparison of conjugation procedures for the preparation of monoclonal antibody-enzyme conjugates. J Immunol Methods 111:261–270
19. Bu D, Zhuang H, Zhou X et al (2014) Biotin-streptavidin enzyme-linked immunosorbent assay for detecting Tetrabromobisphenol A in electronic waste. Talanta 120:40–46
20. Zhang Y, Ni S, Chong C et al (2021) Biocatalysts at atom level: from coordination structure to medical applications. Appl Mater Today 23:101029
21. Busa LSA, Maeki M, Ishida A et al (2016) Simple and sensitive colorimetric assay system for horseradish peroxidase using microfluidic paper-based devices. Sensors Actuators B Chem 236:433–441

22. Crowther JR, Angarita L, Anderson J (1990) Evaluation of the use of chromogenic and fluorogenic substrates in solid-phase enzyme linked immunosorbent assays (ELISA). Biologicals 18:331–336
23. Gibbs J, Vessels M, Rothenberg M (2017) Selecting the detection system – colorimetric, fluorescent, luminescent methods for ELISA assays. ELISA Tech Bull – No 5, Corning Life Sci 1–14
24. Harpaz D, Eltzov E, Ng TSE et al (2020) Enhanced colorimetric signal for accurate signal detection in paper-based biosensors. Diagnostics 10:1–15
25. Craig JC, Parkinson D, Goatley L et al (1989) Assay sensitivity and differentiation of monoclonal antibody specificity in ELISA with different coating buffers. J Biol Stand 17: 281–289
26. Cerda-Kipper AS, Montiel BE, Hosseini S (2019) Immunoassays radioimmunoassays and enzyme-linked immunosorbent assay. Encyclopedia of analytical science (pp. 55–75). https://doi.org/10.1016/B978-0-12-409547-2.14510-X
27. Lequin RM (2005) Enzyme immunoassay (EIA)/enzyme-linked immunosorbent assay (ELISA). Clin Chem 51:2415–2418
28. Drijvers JM, Awan IM, Perugino CA et al (2017) The enzyme-linked immunosorbent assay: the application of ELISA in clinical research. Editor(s): Morteza Jalali, Francesca Y.L. Saldanha, Mehdi Jalali, Basic Science Methods for Clinical Researchers, Academic Press, 2017, Pages 119–133.
29. González-Martínez MÁ, Puchades R, Maquieira Á (2018) Immunoanalytical Technique: Enzyme-Linked Immunosorbent Assay (ELISA), Editor(s): Da-Wen Sun, Modern Techniques for Food Authentication (Second Edition), Academic Press, Pages 617–657
30. Duan CF, Yu YQ, Cui H (2008) Gold nanoparticle-based immunoassay by using non-stripping chemiluminescence detection. Analyst 133:1250–1255
31. Zhao Q, Lu D, Zhang G et al (2021) Recent improvements in enzyme-linked immunosorbent assays based on nanomaterials. Talanta 223:121722
32. Guo L, Qian P, Yang M (2017) Determination of immunoglobulin G by a Hemin-Manganese (IV) oxide-labeled enzyme-linked immunosorbent assay. Anal Lett 50:1803–1811
33. Deng D, Hao Y, Xue J et al (2019) A colorimetric enzyme-linked immunosorbent assay with CuO nanoparticles as signal labels based on the growth of gold nanoparticles in situ. Nanomaterials 9:4
34. Zhu FF colorimetric E method based on D hybridization reaction and non-crosslinking gold nanoparticles aggregation for the detection of amantadine, Peng J, Huang Z et al (2018) Specific colorimetric ELISA method based on DNA hybridization reaction and non–crosslinking gold nanoparticles aggregation for the detection of amantadine. Food Chem 257:382–387
35. Ambrosi A, Airò F, Merkoçi A (2010) Enhanced gold nanoparticle based ELISA for a breast cancer biomarker. Anal Chem 82: 1151–1156
36. Jianwen H (2013) Practical guide to ELISA development. The Immunoassay Handbook 381–393. https://doi.org/10.1016/B978-0-08-097037-0.00025-7.

Chapter 11

Flow Cytometry Analysis to Detect Lapatinib-Induced Modulation of Constitutive and IFN-γ-Induced HLA Class I Expression in HER2-Positive Breast Cancer Cells

Louisa L. Wiede and Sheila Drover

Abstract

Drug-induced modulation of HLA molecules on cancer cell lines can easily be detected using flow cytometry and HLA-specific antibodies to ascertain the number of positive cells and their expression levels. Loss or downregulation of HLA-I molecules on cancer cells, a well-documented immune escape mechanism, may occur via activation and integration of numerous signalling pathways that are operative in cancer. Whereas IFN-γ, produced during an adaptive anti-tumor immune response upregulates HLA expression, activation of the human epidermal growth factor receptor 2 (HER2) pathway and its downstream signalling pathways are reported to decrease HLA-I. Here we describe the flow cytometry procedure used to determine whether lapatinib, known to negate HER2 signalling, increased HLA-I expression on HER2+ cell lines, in the presence and absence of IFN-γ. Contrary to our prediction, the flow cytometry data clearly show lapatinib-mediated downregulation of both constitutive and IFN-γ-induced HLA class I expression. These results, for which we do not yet have an explanation, may have important implications for our understanding of lapatinib resistance in metastatic HER2+ cancer.

Key words Human leukocyte antigen class I (HLA-I), Breast cancer, Human epidermal growth factor receptor 2 (HER2), Lapatinib, IFN-γ, Flow cytometry

1 Introduction

Human leukocyte antigen (HLA) class I molecules, expressed on nearly all nucleated cells, are obligatory for CD8+T-cell recognition of peptides derived from viruses and tumor antigens [1]. During an adaptive anti-tumor immune response, production of IFN-γ upregulates HLA class I on tumor cells, leading to enhanced antigen presentation and activation of tumor-specific CD8+ cytolytic T cells [2–4]. Unfortunately, a frequent occurrence in cancers, including breast cancer enriched for the human epidermal growth factor receptor 2 (HER2+), is loss or downregulation of HLA class I molecules [5, 6] leading to tumor aggressiveness and metastases

Sherri L. Christian (ed.), *Cancer Cell Biology: Methods and Protocols*,
Methods in Molecular Biology, vol. 2508, https://doi.org/10.1007/978-1-0716-2376-3_11,

[7–10]. Of potential significance to immunotherapy are reports that HER2 knockdown may lead to upregulation of HLA class I on HER2+ cancer cells [11].

Lapatinib, a reversible tyrosine kinase inhibitor, used to treat HER2+ metastatic breast cancer attenuates HER2 signalling [12–14]. It acts by targeting the intracellular domain of HER2, competing for the ATP-binding domain of protein kinases, thereby interrupting receptor phosphorylation and subsequent activation of downstream signalling pathways such as MAPK and PI3K/Akt [15–17]. Both pathways are implicated in the regulation of constitutive and IFN-γ-induced HLA expression [18, 19]. Thus, the upregulatory effects of IFN-γ on tumor cells may be thwarted by the downstream events of HER2 signalling.

Flow cytometry, using either fluorochrome-labelled primary antibodies or fluorochrome-labelled secondary antibodies, is an excellent and well-described method [reviewed in 20] to determine quantitative changes in numbers and expression levels of biomarkers present on cancer cells. Our laboratory has used flow cytometry to show upregulated HLA on IFN-γ-treated synoviocytes and hormone-induced modulation of HLA on breast cancer cells [21, 22]. To evaluate the effect of lapatinib on constitutive and IFN-γ-induced HLA class I expression on HER2+ tumor cells, we treated BT-474 and SKBR3 cells with pre-determined concentrations of lapatinib in the absence and presence of IFN-γ for 24 h. This was followed by flow cytometry, using a well-described and validated mouse monoclonal antibody, W6/32 [23], followed by PE-labelled goat anti-mouse antibody to determine changes in HLA expression.

2 Materials

2.1 Cell Culture

1. HER2+ breast cancer cell lines: SKBR3 (ATCC® HTB-30) and BT-474 (ATCC® HTB-20).
2. Complete Medium (CM): Iscove's Modified Dulbecco's Medium (IMDM), 10% inactivated fetal bovine serum (iFBS) and 2 mM L-glutamine and 1% antibiotic-antimycotic mixture.
3. Humidified incubator, set at 37 °C with 5% CO_2 and 95% atmospheric pressure.
4. Phase contrast microscope.
5. Pipettes (various sizes) and pipette aid.
6. 10 cm^2 tissue culture plates.
7. 1× phosphate buffered saline (PBS), pH 7.2.
8. 0.25% Trypsin-EDTA.
9. 15 ml conical tubes.

10. Cell counter.
11. 6-well culture plates.
12. Micropipettors and micropipettes.
13. Dimethyl sulfoxide (DMSO).
14. Lapatinib, stock solution 10 mM in DMSO, stored at −20 °C.
15. IFN-γ 10,000 units/ml in CM, stored at −70 °C.
16. Flow cytometer.

2.2 Flow Cytometry Assay

1. 5 ml round bottom tubes, commonly called FACS tubes.
2. Wash Buffer (WB): 1× PBS, containing 0.2% iFBS and 0.02% NaN_3, stored at 4 °C.
3. 1% Paraformaldehyde (PFA) in 1× PBS, stored in brown bottle at 4 °C.
4. Mouse anti-HLA class I antibody, clone w6/32, supernatant diluted 1/25 in WB.
5. Mouse IgG2a, negative control, diluted to 10 μg/ml in WB.
6. R-phycoerythrin-labelled goat anti-mouse IgG Fcγ-specific, (PE-GAM), optimally diluted in WB.

3 Method

BT-474 and SKBR3 are adherent breast cancer cells, cultured and maintained in CM in a humidified 37 °C incubator, 5% CO_2, and 95% atmospheric pressure (*see* **Note 1**). All manipulations are performed using sterile technique in a Class II Biosafety Cabinet. On the day of the experiment, cells should be at 75–85% confluency, as determined by phase contrast microscopy. The method is described in detail below and outlined in Fig. 1.

3.1 Cell Preparation: Day 1

1. Harvest each cell line individually.
2. Aspirate CM and any detached or dead cells.
3. Add 10 ml sterile 1× PBS, swirl gently, and then aspirate.
4. Add 3–4 ml pre-warmed 0.25% trypsin-EDTA, and incubate at 37 °C for 3–5 min. Check detachment progress after 3 min by phase contrast microscopy (*see* **Note 2**).
5. Once detachment is complete, immediately add 3–4 ml CM, gently mix, and pipette the cells into a 15 ml conical tube.
6. Centrifuge tubes at 500 g/5 min/10 °C.
7. Decant supernatant, and dislodge cell button by tapping the bottom of the tube; add 6 ml CM and centrifuge at 500 g/ 5 min/10 °C.

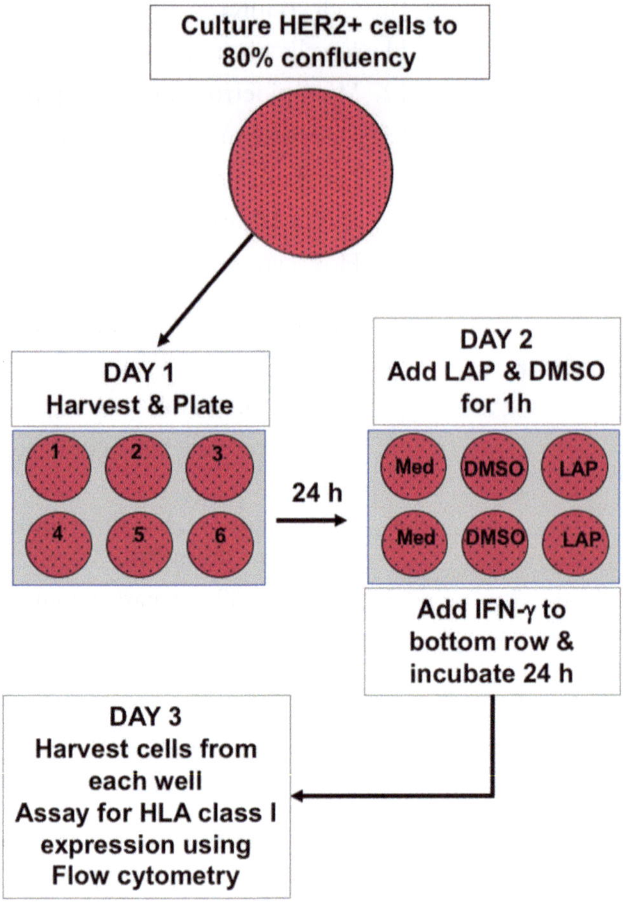

Fig. 1 Graphical depiction of workflow from Day 1 to Day 3

8. Decant supernatant, dislodge cell button by gentle tapping, and resuspend cells in 6 ml cold CM.
9. Remove 0.5 ml for a cell count, and keep cells on ice to prevent adherence to the surface of the tubes.
10. Count cells using a cell counter, and calculate the number of cells/ml (*see* **Note 3**).
11. Dilute SKBR3 cells to 1.75×10^5/ml and dilute BT-474 to 2.5×10^5/ml in CM.
12. Label two 6-well plates, one for SKBR3 and one for BT-474. Dispense 2 ml of appropriate and well-mixed cells into each of the wells. The total number of cells seeded/well is 3.5×10^5 and 5×10^5 for SKBR3 and BT-474, respectively.
13. Incubate plates for 18–24 h in 37 °C incubator in 5% CO_2.

3.2 Treatment of Cells with Lapatinib and IFN-γ: Day 2

The concentrations of lapatinib (*see* **Note 4**) and IFN-γ used in this protocol were predetermined and not described herein.

1. Predilute stock lapatinib, 10 mM to 50 μM in CM (1 μl stock + 199 μl CM); dilute DMSO, 1 μl stock + 199 μl CM.
2. Aspirate CM from all wells using a vacuum attached to a sterile Pasteur pipette, and replace with 2 ml CM.
3. For SKBR3: Wells 1 and 3 are untreated; to wells 2 and 4, add 40 μl 50 μM lapatinib for a final concentration of 0.1 μM and to wells 5 and 6, add 40 μl of the diluted DMSO.
4. For BT-474, treatment is the same as **step 3**, except add 20 μl lapatinib for a final concentration of 0.5 μM/ml, and add 20 μl of diluted DMSO.
5. Incubate plates at 37 °C/1 h.
6. Add 20 μl IFN-γ, 10,000 units/ml to wells 4–6 of both plates; leave wells 1–3 untreated.
7. Incubate plates for 24 h.

3.3 Preparation of Treated Cells for Flow Cytometry: Day 3

1. Remove plates from the incubator, and examine the cells in each well using a phase contrast microscope – note any differences in growth between untreated and treated wells (*see* **Note 5**).
2. Label a 5 ml tube for each of the six wells.
3. Aspirate CM (using the vacuum and Pasteur pipette method) from each well.
4. Add 2 ml sterile PBS to each well and aspirate.
5. Add 0.5 ml pre-warmed 0.25% trypsin-EDTA and incubate at 37 °C for 3–5 min. Check detachment progress after 3 min by phase contrast microscopy.
6. Once detachment is complete, immediately add 2 ml CM to each well. Using a new pipette for each well, gently mix the cells, and pipette them into the appropriately labelled 5 ml tube.
7. Centrifuge tubes at 500 g/5 min/10 °C.
8. Decant supernatant, and dislodge cell button by tapping the bottom of the tube; add 3 ml cold CM and centrifuge at 500 g/5 min/10 °C.
9. Decant supernatant, dislodge cell button by gentle tapping, and resuspend cells in 2 ml cold CM and hold on ice while proceeding to the flow cytometry assay.

3.4 Flow Cytometry to Determine HLA Class I Expression

1. Add 2 ml WB to cells from **step 9** and centrifuge at 500 g/ 8 °C/5 min.

2. Decant supernatants and resuspend in 250 μl WB – the cell count will be roughly 1.0×10^5/ml–2.5×10^6/ml (*see* **Note 6**).
3. Add 100 μl cells to appropriate tubes, followed by 25 μl of appropriate antibody; mix well.
4. Place tubes either in refrigerator or in an ice bath for 30 min.
5. Wash by adding 2.5 ml of WB to each tube, and centrifuge at 500/5 min at 10 °C; decant supernatant by tipping into a beaker containing 10% bleach solution.
6. Repeat the washing twice. After last wash, tip out fluid and drain on a piece of absorbent paper.
7. Add 25 μl of optimally diluted PE-GAM, and incubate for 30 min in the cold and dark using tin foil to wrap around the tube rack.
8. Wash × 2 as in **steps 5** and **6**.
9. Add 200 μl of 1% paraformaldehyde in PBS to each tube (*see* **Note** 7).
10. Cover tubes with tin foil to reduce quenching of the fluorochrome and refrigerate at 4 °C. Cells should be analyzed using a flow cytometer within 24–48 h.
11. Acquire 10,000 events using FSC and SSC parameters; exclude dead cells and debris by gating the healthy cell population (*see* **Note 8**). Determine the level of fluorescence using the FL-2 channel for the PE fluorochrome (Fig. 2).

3.5 Interpretation of Data and Statistical Analysis

1. The flow cytometry data are analyzed using a software program such as Kaluza and include the mean channel fluorescence intensity (MFI) and percentage (%) positive cells as compared to the negative control (Fig. 3).
2. HLA class I expression is determined using the following formula:

$$(\text{MFI w6/32}) - (\text{MFI mIgG2a})$$

3. Results are considered positive if MFI is at least twice the background in three independent experiments.
4. To determine the effect of lapatinib on HLA class I expression, calculate the following ratio:

$$\frac{(\text{MFI w6/32}_{\text{LAPATINIB}}) - \text{MFI mIgG2a}_{\text{LAPATINIB}})}{(\text{MFI w6/32}_{\text{DMSO}}) - \text{MFI mIgG2a}_{\text{DMSO}})}$$

The ratios are interpreted as follows: 1, no effect; >1, upregulation; <1, downregulation. One-way ANOVA and Tukey HSD test may be used to analyze the results for three independent experiments.

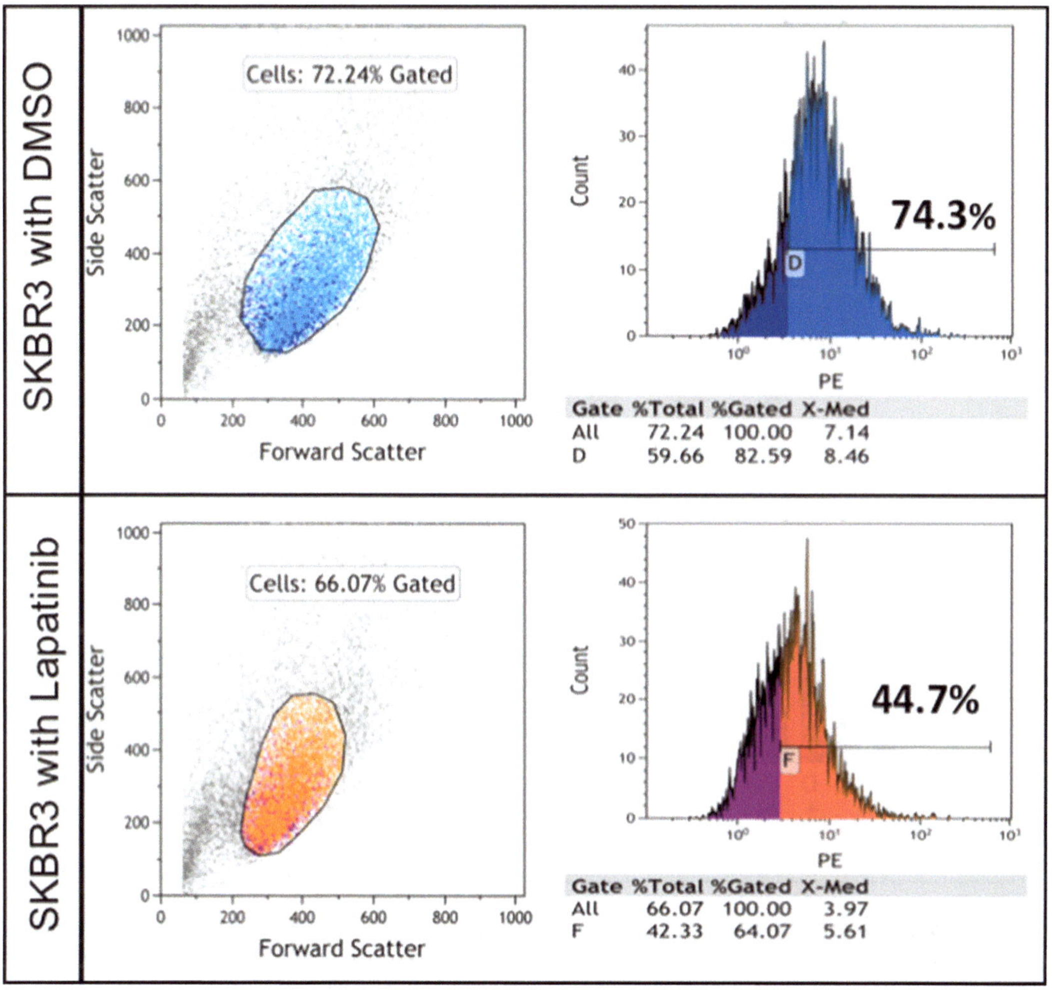

Fig. 2 Representative example of cellular events collected and gated using the forward scatter and side scatter (FSC and SSC) settings of the Cytoflex flow cytometer. Top: Representative dot plot (left) and histogram (right) of SKBR3 treated with DMSO. Bottom: Representative dot plot (left) and histogram (right) of SKBR3, treated with lapatinib. Dot plots and histograms are representative of one experiment

4 Notes

1. The breast cancer cells lines, stored in liquid nitrogen, should be thawed 1–2 weeks prior to experiments. Seed 10 cm^2 TC dishes with 5×10^5 (SKBR3) and 1×10^6 (BT-474) as the latter grows more slowly. After 3–4 days the CM should be removed and replaced with fresh CM. At the end of the week, the monolayer should be almost confluent, at which time cells should be harvested and an aliquot frozen and reseeded for experiments. It is important that the cells not be overgrown as this may induce genetic changes in the cells.

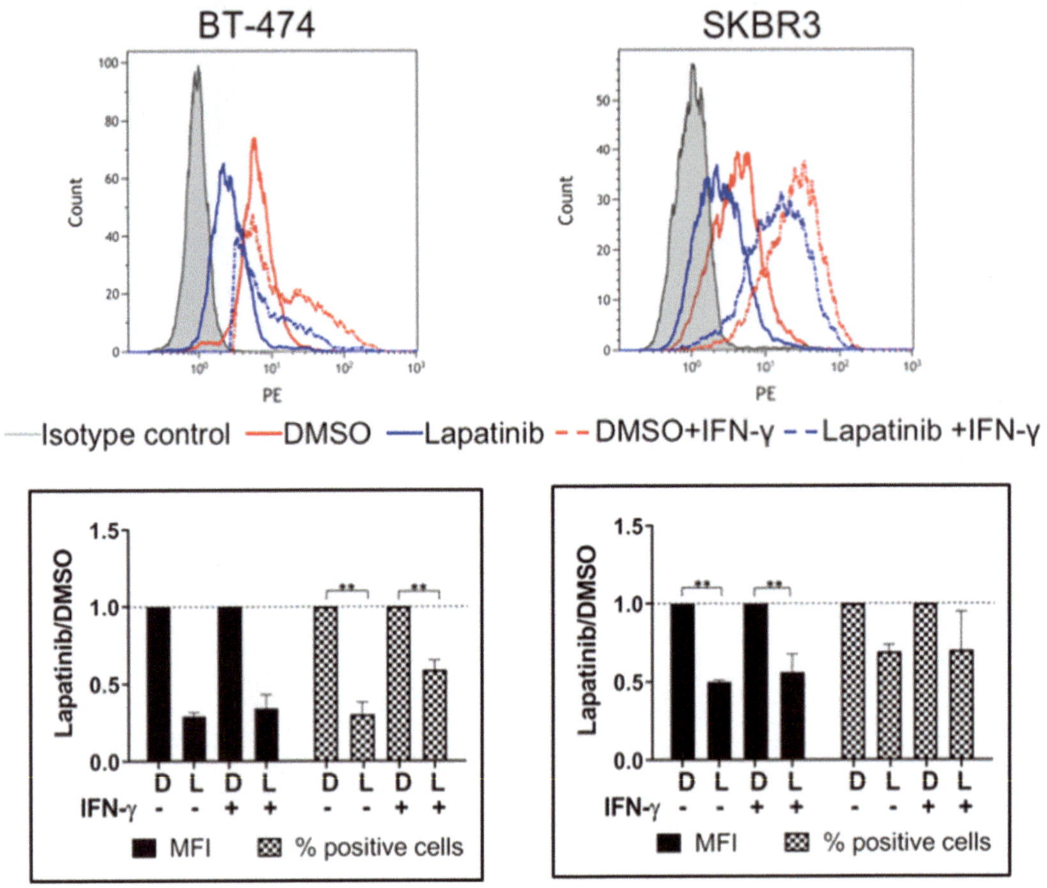

Fig. 3 Flow cytometric analysis of HLA class I expression on BT-474 and SKBR3, treated with lapatinib or DMSO (vehicle control), ± IFN-γ (100 U/mL) for 24 h. Top: Histogram overlays are representative of one experiment. Bottom: Bar graphs represent the mean MFI or percentage (%) of positive cells of treated (L ± IFN-γ) over untreated (D ± IFN-γ) with error bars ± SEM of three independent experiments. Statistical analysis was done by one-way ANOVA and Tukey HSD test (**$p < 0.01$)

2. The cells should now have a round morphology and have lifted from the plastic and float in the trypsin-EDTA solution. BT-474 takes a little longer to detach.
3. If an automatic counter is not available, use a Neubauer counting chamber and phase contrast microscopy to determine the number of viable cells. Usually, cells are >90% viable as any dead cells are removed by the pre-washing with PBS (*see* **step 3**). If a phase contrast microscope is not available, cells can be mixed 1:1 with trypan blue, which stains dead cells.
4. Prior to using any drug, it is important to establish toxicity by treating the cells with a range of concentrations including a physiologically relevant concentration, over the time period that will be used for the experiment. Then, determine the viable cell count, either by harvesting and counting the cells

or using a commercially available kit that measures cell proliferation. The results for each concentration should be compared to cells growing in equivalent vehicle controls (diluent used to prepare the drug).

5. If there is drug-induced toxicity, the cell monolayer will be patchy with some cells lifted from the plastic, as compared to the vehicle control or CM. In case of toxicity, it is best to do two PBS washes prior to trypsinization as the dead cells will mostly be removed by suction. Cells can also be counted using trypan blue (*see* **Note 3** above).
6. These numbers are extrapolated based on the knowledge that cells grow more slowly due to anti-proliferative effects of lapatinib and IFN-γ. Thus, if the total number of cells seeded (3.5×10^5 and 5×10^5 for SKBR3 and BT-474, respectively) remains after harvest, resuspending in 0.4 ml WB would yield about 1.4×10^5/ml–2.0×10^6/ml. However, if there is toxicity (*see* **Note 5** above), a viable count should be performed, as a minimum of 1×10^5 cells are desirable for the flow cytometry assay.
7. The 1% PFA is diluted from a stock 4% solution, prepared as follows: dissolve 4 mg of paraformaldehyde in 90 ml of PBS using a hot plate and stir bar in a chemical fume hood – when the solution reaches 60 °C, add a few drops of 10 N NaOH to clarify it. Cool and adjust to 100 ml, dispense in 20 ml aliquots, and store at −20 °C. For the 1% solution, add 20 ml thawed 4% PFA to 60 ml 1× PBS. Mix and store in a brown glass bottle in the fridge.
8. A cell density plot is created using FSC, which detects cell size, and SSC, which detects cell granularity. A gate is then created around cells to be collected and to exclude dead cells and cellular debris, located in the left bottom corner of the FCS versus SCC density plot (*see* Fig. 2). The concentration of lapatinib used to treat cells was optimally determined to inactivate HER2 signalling without inducing much toxicity. In case of too much toxicity, it is best to stain with a live/dead fluorophore such as Zombie-Aqua, following the manufacturer's instructions.

Acknowledgments

This work was funded by the Canadian Breast Cancer Foundation (BC-RG-13-1 grant #310985) and by a Dean's Fellowship, Faculty of Medicine, Memorial University of Newfoundland.

References

1. Bukur J, Jasinski S, Seliger B (2012) The role of classical and non-classical HLA class I antigens in human tumors. Semin Cancer Biol 22(4): 350–358
2. Seliger B, Ruiz-Cabello F, Garrido F (2008) IFN inducibility of major histocompatibility antigens in tumors. Adv Cancer Res 101: 249–276
3. Ostroumov D, Fekete-Drimusz N, Saborowski M, Kühnel F, Woller N (2018) CD4 and CD8 T lymphocyte interplay in controlling tumor growth. Cell Mol Life Sci 75(4):689–713
4. de Groot AF, Blok EJ, Charehbili A, Engels CC, Smit VTHBM, Dekker-Ensink NG, Putter H, Meershoek-Klein Kranenbarg E, van de Velde CJH, Liefers GJ, Nortier JWR, Kuppen PJK, van der Burg SH, Kroep JR (2019) Strong CD8+ lymphocyte infiltration in combination with expression of HLA class I is associated with better tumor control in breast cancer patients treated with neoadjuvant chemotherapy. Breast Cancer Res Treat 175(3):605–615
5. Rodríguez JA (2017) HLA-mediated tumor escape mechanisms that may impair immunotherapy clinical outcomes via T-cell activation. Oncol Lett 14(4):4415–4427
6. Campoli M, Chang CC, Oldford SA, Edgecombe AD, Drover S, Ferrone S (2004) HLA antigen changes in malignant tumors of mammary epithelial origin: molecular mechanisms and clinical implications. Breast Dis 20: 105–125
7. Payne KK, Manjili MH (2012) Adaptive immune responses associated with breast cancer relapse. Arch Immunol Ther Exp 60(5): 345–350
8. Saio M, Teicher M, Campbell G, Feiner H, Delgado Y, Frey AB (2004) Immunocytochemical demonstration of down regulation of HLA class-I molecule expression in human metastatic breast carcinoma. Clin Exp Metastasis 21(3):243–249
9. Algarra I, Garrido F, Garcia-Lora AM (2021) MHC heterogeneity and response of metastases to immunotherapy. Cancer Metastasis Rev. https://doi.org/10.1007/s10555-021-09964-4
10. Garrido F, Ruiz-Cabello F (1991) MHC expression on human tumors--its relevance for local tumor growth and metastasis. Semin Cancer Biol 2(1):3–10
11. Choudhury A, Charo J, Parapuram SK, Hunt RC, Hunt DM, Seliger B, Kiessling R (2004) Small interfering RNA (siRNA) inhibits the expression of the Her2/neu gene, upregulates HLA class I and induces apoptosis of Her2/neu positive tumor cell lines. Int J Cancer 108(1):71–77
12. Escrivá-de-Romaní S, Arumí M, Bellet M, Saura C (2018) HER2-positive breast cancer: current and new therapeutic strategies. Breast 39:80–88
13. Rimawi MF, Schiff R, Osborne CK (2015) Targeting HER2 for the treatment of breast cancer. Annu Rev Med 66:111–128
14. Duchnowska R, Loibl S, Jassem J (2018) Tyrosine kinase inhibitors for brain metastases in HER2-positive breast cancer. Cancer Treat Rev 67:71–77
15. Iqbal N, Iqbal N (2014) Human epidermal growth factor receptor 2 (HER2) in cancers: overexpression and therapeutic implications. Mol Biol Int 2014:852748. https://doi.org/10.1155/2014/852748
16. D'Amato V, Raimondo L, Formisano L, Giuliano M, De Placido S, Rosa R, Bianco R (2015) Mechanisms of lapatinib resistance in HER2-driven breast cancer. Cancer Treat Rev 41(10):877–883
17. Scaltriti M, Verma C, Guzman M, Jimenez J, Parra JL, Pedersen K, Smith DJ, Landolfi S, Ramon y Cajal S, Arribas J, Baselga J (2009) Lapatinib, a HER2 tyrosine kinase inhibitor, induces stabilization and accumulation of HER2 and potentiates trastuzumab-dependent cell cytotoxicity. Oncogene 28(6): 803–814
18. Mimura K, Shiraishi K, Mueller A, Izawa S, Kua LF, So J, Yong WP, Fujii H, Seliger B, Kiessling R, Kono K (2013) The MAPK pathway is a predominant regulator of HLA-A expression in esophageal and gastric cancer. J Immunol 191(12):6261–6272
19. Chandrasekaran S, Sasaki M, Scharer CD, Kissick HT, Patterson DG, Magliocca KR, Seykora JT, Sapkota B, Gutman DA, Cooper LA, Lesinski GB, Waller EK, Thomas SN, Kotenko SV, Boss JM, Moreno CS, Swerlick RA, Pollack BP (2019) Phosphoinositide 3-kinase signaling can modulate MHC class I and II expression. Mol Cancer Res 17(12):2395–2409
20. McKinnon KM (2018) Flow cytometry: an overview. Curr Protoc Immunol 120: 5.1.1–5.1.11

21. Spurrell DR, Oldford SA, Frost T, Larsen B, Codner D, Edgecombe A, Drover S (2004) Discordant expression of HLA class II-associated co-chaperones and HLA-DRB alleles in cultured fibroblast-like synoviocytes. Hum Immunol 65(12):1516–1529
22. Mostafa AA, Codner D, Hirasawa K, Komatsu Y, Young MN, Steimle V, Drover S (2014) Activation of ERα signaling differentially modulates IFN-γ induced HLA-class II expression in breast cancer cells. PLoS One 9(1):e87377
23. Barnstable CJ, Bodmer WF, Brown G, Galfre G, Milstein C, Williams AF, Ziegler A (1978) Production of monoclonal antibodies to group A erythrocytes, HLA and other human cell surface antigens-new tools for genetic analysis. Cell 14(1):9–20

Chapter 12

An End-to-End Workflow for Interrogating Tumor-Infiltrating Myeloid Cells Using Mass Cytometry

Maunish Barvalia and Kenneth W. Harder

Abstract

Myeloid cells are a highly heterogeneous group of innate immune cells which include a diverse collection of cell types and cell states. Distinct subsets can impact tumor progression differently, with conventional type 1 DCs important in protective anti-tumor immune responses, while immunosuppressive tumor-associated macrophages and myeloid-derived suppressive cells (MDSCs) play tumor-promoting roles. Deep phenotyping of myeloid cells using single-cell technologies such as mass cytometry provides the unprecedented opportunity to comprehensively characterize the underlying heterogeneity of myeloid cells. Here we provide a detailed end-to-end workflow including both experimental and computational protocols enabling deep phenotyping of tumor-infiltrating myeloid cells using mass cytometry. A protocol that facilitates interrogation of phosphoproteins in circulating and tumor-infiltrating myeloid cells has been provided together with detailed scripts for Phenograph analysis of tumor-infiltrating myeloid cells.

Key words Mass cytometry, CyTOF, Myeloid cells, Deep phenotyping, Phenograph, Monocytes, Macrophages, Dendritic cells, Myeloid-derived suppressor cells

1 Introduction

Myeloid cells include mononuclear cells such as monocytes, dendritic cells (DCs), tissue macrophages (macs), and polymorphonuclear cells such as neutrophils, eosinophils, basophils, and mast cells [1]. Myeloid cells are important effectors of the innate immune system and can not only initiate but also amplify adaptive immune responses via antigen presentation and cytokine secretion [2]. Myeloid cells are present in every organ and in the circulation and constantly survey the entire organism for any foreign entities or tissue damage [1, 3]. Due to their central role in maintaining homeostasis, myeloid cell deregulation has profound impacts on the health and well-being of an individual and has been associated with a wide variety of ailments including infectious disease, cancer, autoimmunity, heart disease, obesity, neurodegeneration, aging, and allergy [4, 5].

Sherri L. Christian (ed.), *Cancer Cell Biology: Methods and Protocols*,
Methods in Molecular Biology, vol. 2508, https://doi.org/10.1007/978-1-0716-2376-3_12,

Traditionally, myeloid cells have been classified into distinct cell states based on their cell surface protein expression, ontogeny, and the differential dependence on lineage-defining transcription factors and growth factor signaling [2]. However, recent deep phenotyping studies using single-cell technologies such as mass cytometry (CyTOF) and single-cell RNA-sequencing (scRNA-seq) have identified marked heterogeneity and further highlighted the complex ontogeny of the various myeloid cell subsets [1, 2, 6]. In CyTOF, antibodies are conjugated with distinct, stable, transition element isotopes and are used to mark target intracellular or cell surface epitopes. Cells bound with antibody-isotope conjugates are sprayed as single-cell droplets using a nebulizer into an inductively coupled argon plasma. The plasma vaporizes each cell and induces the ionization of its atomic constituents which are filtered by a 100 Dalton quadrupole, thereby removing common biological elements and enriching transition element reporter ions. The resulting elemental ions are sampled by a time-of-flight (TOF) mass spectrometer and quantified. Lastly, the signal for each transition elemental isotope reporter is integrated for a given cell on a per cell basis allowing for the simultaneous measurement of 50 proteins on or in each cell for 1000 cells per second of acquisition [7, 8]. Machine learning approaches, such as unsupervised clustering and dimensionality reduction, are critical for effective CyTOF data analysis [9–11]. CyTOF has been used to analyze myeloid cell subsets and phenotypes in a wide variety of contexts. For example, by simultaneously measuring 34 parameters in bone marrow immune cells, CyTOF revealed the continuous tree-like structure of healthy human hematopoiesis. Additionally, intracellular signalling profiling in response to ex vivo stimuli and inhibitor treatments identified cell subset specific signalling phenotypes [8].

In the presence of persistent stimulants of myeloid cells such as colony-stimulating factors (CSF1, CSF2, and CSF3) and cytokines such as IL-1β and IL-6, monocytes and neutrophils enter a pathogenic activation state rendering them potently immunosuppressive [12]. These cells are called myeloid-derived suppressor cells (MDSCs) and, depending on their origin from granulocytic or monocytic myeloid lineages, are classified as granulocytic/polymorphonuclear MDSCs (G- MDSCs/PMN-MDSCs) or monocytic MDSCs (M-MDSCs), respectively. MDSCs play an important role in diverse settings such as cancer, autoimmunity, as well as chronic and acute infections including COVID-19 [13]. Hence, there is tremendous complexity in the myeloid system, with diverse discrete cell types and a multitude of cell states at homeostasis, and during pathologic contexts including cancer.

Most immune cell types play crucial roles in cancer biology, in which different myeloid subsets can play distinct tumor-promoting or tumor-inhibitory roles [14, 15]. For example, a subset of DCs, known as type 1 conventional DCs (cDC1s), are pivotal for generating anti-tumor immune responses, due to their ability to transport antigens to the tumor-draining lymph nodes, cross present tumor-derived antigens, and mediate potent $CD8^+$ T cell priming [16]. By contrast, systemic expansion and intratumoral accumulation of G-MDSCs, M-MDSCs, and tumor-associated macrophage (TAM) subsets have been shown to be immunosuppressive and tumor-promoting across multiple cancer types [12, 17]. Lastly, various rare myeloid cell subsets have been relatively ignored in cancer and metastasis studies. For example, a subset of monocytes termed patrolling monocytes (pMos) has been shown to limit lung metastasis, but their role in primary tumors and other organ metastasis is unknown [16, 18]. Hence, single-cell technologies can provide an unprecedented opportunity to comprehensively capture the underlying heterogeneity of myeloid cells and improve our understanding of the roles of the different cell types in tumor immunity.

Cancer immunotherapy has revolutionized cancer treatment, with current US Food and Drug Administration (FDA) approval of 7 immune checkpoint inhibitors across 19 different cancer types [16]. Additionally, more than 2000 immuno-oncology agents are currently being tested or in use, with around 1000 in human clinical trials [19]. Despite the prolific pace of clinical trials, detailed mechanistic understanding of many of these therapies is currently lacking. Single-cell technologies have provided hints regarding the systemic and tumor stroma immunological changes that occur after immune checkpoint blockade, especially with anti-CTLA4 and anti PD-L1/PD-1 therapy [20–22]. Along with elucidating the important role of T cells in immune checkpoint blockade, single-cell profiling has revealed there is significant remodeling of the myeloid tumor stroma compartment as well, such as an expansion of activated $iNOS^+$ proinflammatory macrophages in tumors from immune checkpoint inhibitors-treated mice [20, 21]. In conjunction with the profound changes in the tumor microenvironment, tumors cause marked changes in the immune landscape in circulation and in primary and secondary lymphoid organs (also known as the immune macroenvironment) [23]. Tumor rejection following immunotherapy requires immune activation in the periphery. High-dimensional single-cell profiling of the peripheral blood can be a remarkably powerful tool to apply to clinical trials, to simultaneously infer the immune responses associated with a given therapy. For example, CyTOF profiling of peripheral blood from patients with stage IV melanoma before and after anti-PD-1 immunotherapy showed that the frequency of $CD14^+CD16^-HLA\text{-}DR^{hi}$ monocytes before therapy could effectively stratify responders and non-responders [22]. Peripheral immune biomarkers can greatly assist the diagnosis and prognosis of cancer and response to therapy [16, 22, 24]. Along with profiling surface markers, analysis of intracellular signalling protein phosphorylation profiles, as well as the expression of transcription factors, cytokines, and immunometabolism-relevant enzymes, provides an opportunity to elucidate the functional heterogeneity and activation profiles of various myeloid cells in the context of disease [8]. Indeed, CyTOF analysis of signaling patterns in acute myeloid leukemia (AML) identified a hematopoietic stem and progenitor cell-like signaling signature in AML cells which outperformed surface marker analysis in predicting survival [10]. Investigating signaling patterns using CyTOF has been useful in diverse settings including glioblastoma, recovery from surgery, and COVID-19 [25–27].

Here, we describe an end-to-end protocol for interrogating murine myeloid cells using mass cytometry. By using buffers which can preserve phosphorylation signals, we show that multiple phosphoproteins can be detected in multiple immune cell subsets at single-cell resolution (Fig. 1). Using CyTOF and scRNA-seq, we recently showed that Lyn tyrosine kinase deficiency in mice causes a selective expansion of pMos and modulates the surface marker expression, signaling landscape, and gene expression program of monocytes [28]. Here, Phenograph analysis of the same CyTOF dataset (antibody panel in Table 1) provided additional insights. For example, the superior k-nearest neighbor approach of Phenograph revealed contaminating CD11b^{hi} cells and DCs in the monocyte gate. A significant expansion of CD274^{hi} pMos was observed in *Lyn*$^{-/-}$ mice (CD274 is also known as PD-L1). The CD274^{hi} pMos exhibited elevated expression of CD172a (also known as SIRPα) and CD115 (also known as CSF1R or MCSFR). Examination of baseline signaling profiles revealed that CD274^{hi} pMos had increased pAkt, pS6, and pSTAT1 levels compared to other pMos. After identification of the different cell states, all the surface marker expression and phosphoprotein levels were quantified. Upregulation of various Fc receptors including CD16/32, CD64, and FcγRIV was detected across the various monocyte cell states in Lyn-deficient monocytes. Investigating phosphoprotein signals across the monocyte subsets also revealed decreased pTyr and pSTAT5 and a corresponding increase in pSTAT3 levels across various *Lyn*$^{-/-}$ pMos cell states (Fig. 2).

Although many alternative strategies other than the one described here have been developed for profiling phosphoprotein signatures of circulating immune cells, profiling solid tissues, including solid tumors, is technically challenging. Along with identification of appropriate fixatives, additional mechanical and

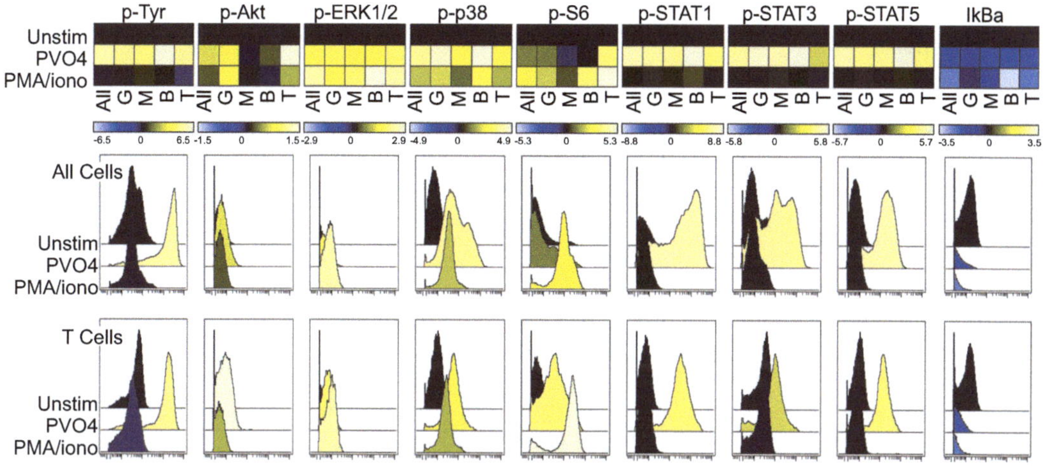

Fig. 1 Testing antibodies for probing intracellular proteins using CyTOF. Blood leukocytes from wild-type (wt) C57BL/6 mice ($n = 3$, pooled) were stimulated with PMA/ionomycin or pervanadate (PV04) and analyzed using phospho-CyTOF. The heatmaps depict the arcsinhx transformed signal intensities of the indicated phosphoproteins and IkBα, across multiple immune cell types, where G indicates granulocytes, M monocytes, B B cells, and T T cells. Blue and yellow indicate low and high expression, respectively. Data from an independent experiment is shown

enzymatic digestion steps are required for generation of a single-cell suspension. Here, we demonstrate the utility of CyTOF for profiling tumor-infiltrating myeloid cells in a murine model of breast cancer. Using CyTOF (antibody panel in Table 2), we could identify tumor-infiltrating monocyte subsets (cMos and pMos), macrophages, DC subsets, neutrophils/G-MDSCs, B cells, T cell subsets, and NK cells. Next, "gated" (subset) myeloid cells were subjected to further phenotyping using Phenograph. This unsupervised clustering approach identified 19 myeloid cell clusters, which were annotated as distinct cell states based on their surface marker expression profiles. Using this approach, we could annotate the myeloid cells in an unbiased fashion. For example, among macrophages we identified Ly6C$^+$ macs, CD11c$^-$MHCII$^+$ macs, CD11c$^+$MHCII$^-$ macs, and CD11c$^+$MHCII$^+$ macs. Among Neuts/G-MDSCs, we identified a cluster of cells, which express CD274. Despite being a minor myeloid population in the tumor proper, plasmacytoid DCs (pDCs) exhibited substantially higher phosphoprotein signals as compared to the other myeloid subsets. Lastly, among macrophages differential signaling landscapes were associated with different cell states, with CD11c$^+$ macs having lower levels of multiple phosphoproteins compared to the other macrophage subsets (Fig. 3).

Table 1
Lyn CyTOF panel

Sr. no.	Metal	Marker	Secondary	Clone	Antibody vendor	Catalog #
1	115Ln	CD11b		M1/70	Biolegend	101249
2	141Pr	MHCII		M5/144.15.2	Biolegend	107637
3	142Nd	B220		RA36B2	UBC Antibody Lab (AbLab.ca)	21-0030-01
4	143Nd					
5	144Nd	pTyr		3144003A	Fluidigm	p-Tyr100
6	145Nd	cKit		ACK2	Thermo Fisher (eBioSciences)	14-1172-82
7	146Nd	CD16/32		93	Biolegend	101335
8	147Sm					
9	148Nd					
10	149Sm					
11	150Nd	pStat5 [Y694]		47	Fluidigm	3150005A
12	151Eu	SIRPα (CD172)		P84	Biolegend	144001
13	152Sm	pAkt [S473]		D9E	Fluidigm	3152005A
14	153Eu	pStat1 [Y701]		58D6	Fluidigm	3153003A
15	154Sm					
16	155Gd					
17	156Gd	p-p38 [T180/Y182]		D3F9	Fluidigm	3156002A
18	157Gd					
19	158Gd	pStat3 [Y705]		4	Fluidigm	3158005A
20	159Tb	CD43		eBioR2/60	Thermo Fisher (eBioSciences)	14-0431-82
21	160Gd	CD115		AFS98	Biolegend	135521
22	161Dy	CD274		MIH5	Thermo Fisher (eBioSciences)	14-5982-82
23	162Dy	Sca1		D7	Biolegend	108135
24	163Dy					
25	164Dy	IkBa		L35A5	Fluidigm	3164004A
26	165Ho					
27	166Dy	CD11c		N418	UBC Antibody Lab (AbLab.ca)	21-0051-01

(continued)

Table 1
(continued)

Sr. no.	Metal	Marker	Secondary	Clone	Antibody vendor	Catalog #
28	167Er	TCRB		H57–597	Biolegend	109235
29	168Er	Ly6G		1A8	UBC Antibody Lab (AbLab.ca)	21-0066-01
30	169Tm	Ly6C		HK1.4	Biolegend	128039
31	170Er	CD62L		MEL-14	Biolegend	104443
32	171Yb	pERK 1/2 [T202/Y204]		D13.14.4E	Fluidigm	
33	172Yb	CD8		53-6.7	UBC Antibody Lab (AbLab.ca)	21-0012-01
34	173Yb	CCR7		4B12	Biolegend	120101
35	174Yb	CD64		X54-5/7.1	Biolegend	139301
36	175Lu	pS6 [S235/S236]		N7-548	Fluidigm	3175009A
37	176Yb	FcgRIV-APC	Anti-APC	9E9/APC 003	Biolegend/Fluidigm	149506/3176007B
38	209Bi	Ter119		Ter-119	UBC Antibody Lab (AbLab.ca)	21-0031-01

2 Materials

2.1 Blood Collection and Fixation/RBC Lysis

1. 1-Step Fix/Lyse Buffer (eBioscience, REF# 00-5333) (10×): Make to 1× by diluting with distilled water (dH_2O); reverse osmosis treated water can also be used.
2. 15 mL Falcon tubes.

2.2 NOP12 Tumor Cell Culture

1. NOP12 cells [29].
2. Plain RPMI.
3. NOP12 cell culture media: RPMI-1640, 10% (v/v) HI (heat inactivated) FBS (fetal bovine serum), 100 U/mL penicillin-G and 100 μg/mL streptomycin, 55 μM β-mercaptoethanol, 0.5 mL ITS (insulin/transferrin/selenium).
4. T75 tissue culture flasks.
5. 0.05% Trypsin-EDTA, phenol red.
6. MMTV-*neu*$^{OTI/OTII}$ mice.

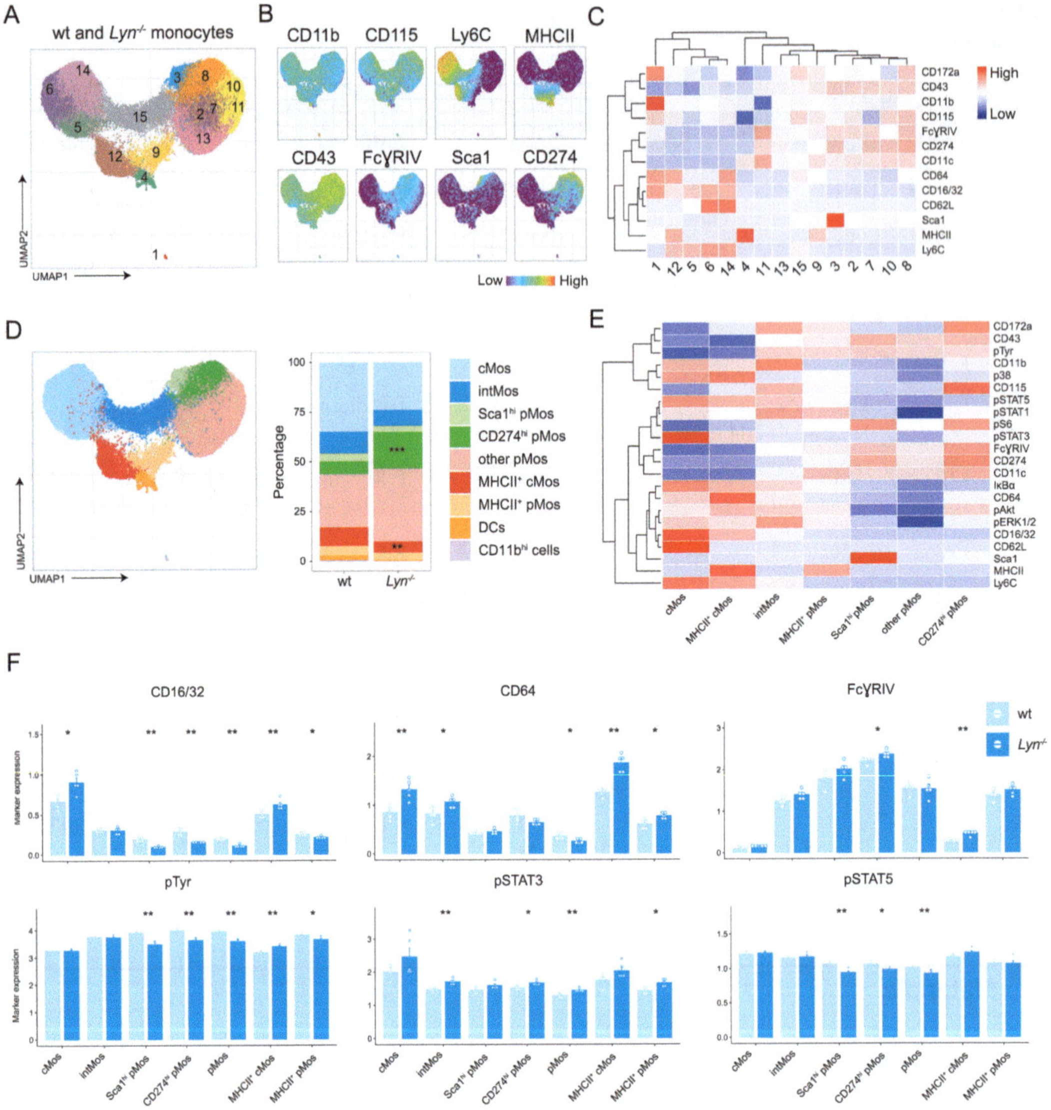

Fig. 2 Phenograph analysis of wt and $Lyn^{-/-}$ monocytes (**a**) wt ($n = 5$) and $Lyn^{-/-}$ ($n = 5$) monocytes previously profiled using mass cytometry by us [28] were reanalyzed using Phenograph for clustering and UMAP for dimensionality reduction. The monocytes are annotated by their Phenograph clusters. (**b**) Plots are colored according to marker intensity of the indicated markers. Dark blue and red indicate low and high expression, respectively. (**c**) Heatmap depicting the Z-score normalized expression level for the indicated markers across the Phenograph clusters. (**d**) Phenograph clusters are annotated on the UMAP plot as indicated cell subsets based on surface marker expression patterns (left), and stack bar graph depicts the cell subset frequency in wt and $Lyn^{-/-}$ monocytes. (**e**) Heatmap depicting the Z-score normalized expression level for the indicated markers across the monocyte subsets. (**c**, **e**) Blue and red indicate low and high, respectively. (**f**) Expression levels of the indicated markers in the wt and $Lyn^{-/-}$ monocyte subsets. Statistical significance was assessed using the unpaired Student t-test, $^{*}p \leq 0.05$, $^{**}p \leq 0.01$, $^{***}p \leq 0.001$

Table 2
NOP12 tumor CyTOF panel

Sr. no.	Metal	Marker	Secondary	Clone	Antibody vendor	Catalog #
1	115Ln	CD11b		M1/70	Biolegend	101249
2	141Pr	MHCII		M5/144.15.2	Biolegend	107637
3	142Nd	B220		RA36B2	UBC Antibody Lab (AbLab.ca)	21-0030-01
4	143Nd					
5	144Nd	pTyr		3144003A	Fluidigm	p-Tyr100
6	145Nd	cKit		ACK2	Thermo Fisher (eBioSciences)	14-1172-82
7	146Nd	CD16/32		93	Biolegend	101335
8	147Sm	CD45 (pan)		I3/2	UBC Antibody Lab (AbLab.ca)	13-0047-01
9	148Nd					
10	149Sm					
11	150Nd	pStat5 [Y694]		47	Fluidigm	3150005A
12	151Eu	SIRPα (CD172)		P84	Biolegend	144001
13	152Sm	pAkt [S473]		D9E	Fluidigm	3152005A
14	153Eu	pStat1 [Y701]		58D6	Fluidigm	3153003A
15	154Sm					
16	155Gd	NK1.1		PK136	UBC Antibody Lab (AbLab.ca)	21-0056-01
17	156Gd	p-p38 [T180/ Y182]		D3F9	Fluidigm	3156002A
18	157Gd					
19	158Gd	pStat3 [Y705]		4	Fluidigm	3158005A
20	159Tb	CD43		eBioR2/60	Thermo Fisher (eBioSciences)	14-0431-82
21	160Gd	CD115		AFS98	Biolegend	135521
22	161Dy	CD274		MIH5	Thermo Fisher (eBioSciences)	14-5982-82
23	162Dy	Sca1		D7	Biolegend	108135
24	163Dy					
25	164Dy	IkBa		L35A5	Fluidigm	3164004A
26	165Ho					

(continued)

Table 2 (continued)

Sr. no.	Metal	Marker	Secondary	Clone	Antibody vendor	Catalog #
27	166Dy	CD11c		N418	UBC Antibody Lab (AbLab.ca)	21-0051-01
28	167Er	TCRB		H57-597	Biolegend	109235
29	168Er	Ly6G		1A8	UBC Antibody Lab (AbLab.ca)	21-0066-01
30	169Tm	Ly6C		HK1.4	Biolegend	128039
31	170Er	CD62L		MEL-14	Biolegend	104443
32	171Yb	pERK 1/2 [T202/Y204]		D13.14.4E	Fluidigm	
33	172Yb	CD8		53-6.7	UBC Antibody Lab (AbLab.ca)	21-0012-01
34	173Yb	CCR7		4B12	Biolegend	120101
35	174Yb	CD64		X54-5/7.1	Biolegend	139301
36	175Lu	pS6 [S235/S236]		N7-548	Fluidigm	3175009A
37	176Yb	FcgRIV-APC	Anti-APC	9E9/APC 003	Biolegend/Fluidigm	149506/ 3176007B
38	209Bi	Ter119		Ter-119	UBC Antibody Lab (AbLab.ca)	21-0031-01

2.3 Tumor Processing

1. 10 mL Safety-lok™ syringe.
2. gentleMACS™ C tubes.
3. gentleMACS™ dissociator.
4. Enzyme digestion cocktail: 125 U/ml collagenase type XI, 60 U/mL hyaluronidase type I-S, 60 U/mL DNase I type II-S, and 450 U/mL collagenase type I, in 10 mL RPMI with 5% (v/v) HI-FBS.
5. FACS buffer: 1× PBS, 2.5 mM EDTA, 0.05% NaN_3, 2% HI-ABS (adult bovine serum).
6. Bright-Line™ hemocytometer.
7. Sorvall Legend RT Refrigerated Benchtop centrifuge.

2.4 CyTOF Staining

1. Fluidigm Cell-ID 20-Plex Pd Barcoding Kit.
2. MaxPar cell staining buffer (MCSB, CAT# 201068 Fluidigm).
3. MaxPar Fix and Perm buffer.

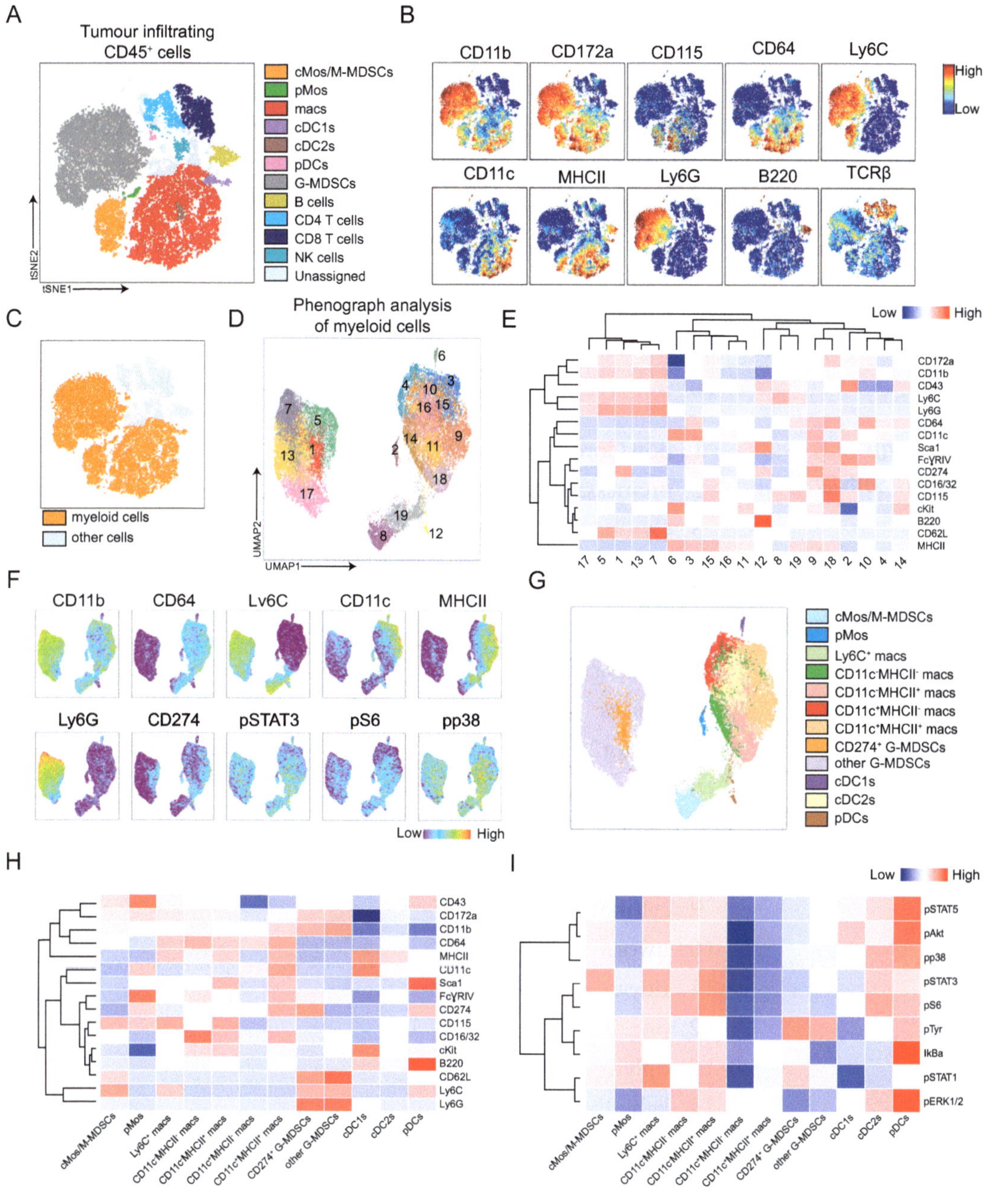

Fig. 3 Phenograph analysis of NOP12 tumor-infiltrating myeloid cells. (**a**) Tumors from NOP12 tumor-bearing mice ($n = 4$) were processed into a single-cell suspension and were stained with a 31 marker CyTOF panel. $CD45^{+}$ cells were gated, and the t-distributed stochastic neighbor embedding algorithm (t-SNE) was used to visualize the CyTOF data (viSNE) with immune cell types annotated based on their canonical marker expression profiles. (**b**) viSNE plots are colored according to the expression of the indicated markers. (**c**) Myeloid cells are highlighted among all the tumor-infiltrating immune cells. (**d**) Myeloid cells were subsetted, and deep phenotyping of these cells was done using Phenograph for clustering and UMAP for dimensionality reduction. The myeloid cells are annotated by their Phenograph clusters. (**e**) Heatmap depicting the Z-score normalized expression levels for the indicated markers across the Phenograph clusters. (**f**) The UMAP plots are colored based on the signal intensity of the indicated proteins. (**g**) Phenograph clusters are annotated on the UMAP plot as indicated cell subsets based on their surface marker expression patterns. (**h**, **i**) Heatmap depicting the Z-score normalized signal intensities for the indicated cell surface and intracellular phosphoproteins, respectively. For (**e**, **h**, and **i**) Blue and red indicate low and high signals, respectively

4. MaxPar water.
5. Methanol (MeOH, certified ACS) (pre-chilled on ice).
6. Cell-ID Intercalator-Ir.
7. 4EQ beads.
8. Cell ID-Cisplatin 195Pt (5 mM stock): Use at 5 uM (dilute 1: 1000 in PBS).
9. Falcon® 5 mL round bottom polystyrene tubes (FACS tubes, for staining).
10. 5 mL round bottom polystyrene tubes with cell-strainer cap.

2.5 CyTOF Data Analysis Software

1. flowCore R package.
2. premessa R package.
3. Cytobank (www.premium.cytobank.org).

2.6 Myeloid Cell Deep Phenotyping R Packages

1. flowCore.
2. MASS.
3. RColorBrewer.
4. Gplots.
5. ggplot2.
6. cowplot.
7. reshape2.
8. ggridges.
9. cytofkit.
10. uwot.
11. tidyr.
12. plyr.
13. dplyr.
14. kableExtra.
15. ggpubr.
16. miscTools.

3 Methods

3.1 Growing NOP12 Tumor Cells

All cell culture should be performed using sterile technique in a tissue culture hood.

1. Remove frozen NOP12 cells (~1–1.5 * 10^6 cells in a cryovial) from liquid N_2, and thaw using standard cell culture techniques (*see* **Notes 1** and **2**).

2. Resuspend the thawed cells in 9 mL of NOP12 cell culture media.
3. Centrifuge cells at 485 g for 5 min at 4 °C in a bench top centrifuge.
4. Resuspend the cells in 1 mL of NOP12 media, and transfer the cells into a T75 flask containing 10–12 mL of NOP12 media.
5. Grow cells at 37 °C at 5% CO_2 in an incubator for 2–3 days until they reach 80% confluency.
6. After cells have reached 80% confluency, rinse the adherent side of the flask with 5 mL plain (unsupplemented) RPMI.
7. Trypsinize with 3 mL 0.05% Trpsin-EDTA for 5 min at 37 °C, tapping flask to remove the adherent cells.
8. Neutralize the trypsin with NOP12 media by addition of 5 mL NOP 12 medium to side of the flask which has cells.
9. Collect into a sterile 15 mL tube, and centrifuge cells at 485 g (1500 rpm) for 5 min at 4 °C.
10. Discard the supernatant and resuspend cells in 1 mL of NOP12 media, and add back-calculated amount based on split (split/expand the cells at 1/5 or 1/10 for 2 days) to fresh media in the same flask (can use the same flask four to five times).

3.2 Transplantable Tumor Model

1. Dissociate NOP12 cells using the same method as described in **steps 6–9** in heading 3.1.
2. Wash the cells twice with 10 mL of sterile ice-cold PBS by centrifuging cells at 485 g for 5 min at 4 °C.
3. Resuspend cells at 10^6 cells/100 uL of sterile PBS, and inject subcutaneously into syngeneic MMTV-$neu^{OTI/OTII}$ mice in the left flank.
4. Assess tumor volume when they become palpable using a caliper and the spherical approximation formula: 4/3π * radial length * radial width * radial height (rl × rw × rh); the mice are euthanized in 4–6 weeks when the tumors reach ~1 cm^3 in volume.

3.3 Blood and Tumor Processing at Endpoint for CyTOF Profiling

1. Humanely euthanize mice via overdose of either CO_2 or isoflurane according to animal ethics protocol.
2. Collect blood via cardiac puncture, and rapidly add the blood to 12 mL of 1-step Fix/Lyse solution (Invitrogen CAT#00-5333).
3. Incubate the blood in 1-step Fix/Lyse for 15–60 min at room temperature (RT), ensuring that all the samples are incubated with the fixative for same amount of time.
4. After the collection of the blood, perfuse the mouse with 20 mL of 1-step Fix/Lyse solution via the left ventricle.

5. Dissect the subcutaneous tumor, and place it into 10 mL of Fix/Lyse solution in gentleMACS™ C tubes.
6. Chop the tumor into smaller pieces, and mechanically dissociate twice using the m_impTumor_02 protocol on the gentleMACS®. The tumor cell suspension should be kept in fixative solution for 15–60 min after collection ensuring that all the samples are incubated with the fixative for same amount of time.
7. After fixation centrifuge the samples at 800 g for 5 min at 4 °C, discard the fixative, and resuspend cells in 10 mL of plain (unsupplemented) RPMI.
8. Repeat wash as in **step** 7.
9. Digest the tumor samples using the enzyme digestion cocktail at 37 °C for 40 min.
10. After digestion, wash the samples with cold RPMI +5% HI-FBS, and pass through a 70 um cell strainer and wash again with FACS buffer.
11. Pass homogenate through 70 uM filter into 50 ml falcon tube, rinse plunger, and sieve with FACS buffer.
12. Centrifuge at 800 g, 4 °C, for 5 min, and resuspend in 5 ml FACS buffer.
13. Count cells.
14. Add three million cells to 3 ml FACS buffer in FACS tubes and centrifuge.
15. Count the cells using a hemocytometer and proceed to CyTOF cell staining.

3.4 CyTOF Staining

1. *Barcoding*
   1. Add 1 mL of MCSB to each Falcon® 5 mL round bottom polystyrene tube (*see* **Note 3**).
   2. Add up to 3 M cells to each tube.
   3. Centrifuge (800 g, 5 min, 4 °C or RT) and pour off supernatant and blot on paper towel to remove excess liquid on tube rim (*see* **Notes 4** and **5**).
   4. Wash with 1 mL 1× Fix/Perm buffer at RT.
   5. Centrifuge (800 g, 5 min, 4 °C or RT) and pour off supernatant and blot on paper towel.
   6. Repeat **steps 4** and **5** for a total of two washes.
   7. Resuspend each sample in 800 uL of 1× Perm Buffer from the Fluidigm Cell-ID 20-Plex Pd Barcoding Kit.
   8. Resuspend each barcode in 100 uL of 1× Perm Buffer, and add barcode to each sample.

9. Incubate at RT for 30 min.
10. Centrifuge (800 g, 5 min, 4 °C or RT) and pour off supernatant and blot on paper towel.
11. Wash with 2 mL of MCSB.
12. Centrifuge (800 g, 5 min, 4 °C or RT) and pour off supernatant and blot on paper towel.
13. Repeat **steps 11** and **12** for a total of two washes.
14. Combine all barcoded samples into 1 tube.
15. Centrifuge (800 g, 5 min, 4 °C or RT) and pour off supernatant and blot on paper towel.
16. Proceed to surface staining.

2. *CyTOF Surface Staining*
   1. Add 5% (v/v) rat serum in MCSB to cells. Final volume of cells should be 50 uL/sample. For barcoded samples scale up the volume, every 3 M cells = 1 samples worth of stain.
   2. Incubate on ice for at least 15 min. Add 50 uL of antibodies (diluted in MCSB; scale up accordingly as in **step 1**).
   3. Incubate on ice for 45 min. Wash with 1 mL MCSB and centrifuge (800 g, 5 min, 4 °C or RT), and pour off supernatant and blot on paper towel (for barcoded samples add up to 4 mL of wash).
   4. Wash with 1 mL MCSB and centrifuge, and pour off as in **step 5** (for barcoded samples add up to 4 mL of wash).
   5. Top up tubes to 50 uL in each tube with MCSB (scale up accordingly as in **step 1**).
   6. Add 50 uL of secondary antibodies diluted in MCSB (scale up accordingly as in **step 1**).
   7. Incubate on ice for 30 min.
   8. Wash with 1 mL MCSB and centrifuge, and pour off as in **step 5** (for barcoded samples, add up to 4 mL of wash).
   9. Wash with 1 mL MCSB and centrifuge, and pour off as in **step 5** (for barcoded samples, add up to 4 mL of wash).
   10. Proceed to intracellular phosphoprotein antibody staining.

3. *Phosphoprotein Staining (Intracellular Staining)*
   1. Add 1 mL (up to 4.5 mL for barcoded samples) of cold MeOH to each tube (*see* **Note 6**).
   2. Incubate on ice for 15 min.
   3. Centrifuge (800 g, 5 min, 4 °C or RT) and pour off supernatant and blot on paper towel.
   4. Wash with 2 ml of MCSB. Centrifuge (800 g, 5 min, 4 °C or RT) and pour off supernatant and blot on paper towel.

5. Repeat **steps 4** and **5** for a total of two washes (refer to **Note 7**; if there are chunks in sample, pass cells through cell strainer before last wash).
6. Top up tubes to 50 uL in each tube with MCSB (for barcoded samples scale up the volume accordingly).
7. Add 50 uL of phosphoprotein antibody cocktail diluted in MCSB (scale up accordingly as in **step 7**).
8. Incubate at RT for 30 min. Wash with 2 mL of MCSB (for barcoded samples, add up to 4 mL of MCSB).
9. Centrifuge (800 g, 5 min, 4 °C or RT) and pour off supernatant and blot on paper towel.
10. Repeat **steps 10** and **11** for a total for two washes. Proceed to nuclear staining (*see* **Note 8**).

4. *Nuclear Staining*
   1. Add 1 mL of MaxPar Fix and Perm buffer with 125 nM Cell-ID intercalator (= 1/4000 dilution = 0.25 uL per 1 mL buffer; for barcoded samples, add up to 4 mL).
   2. Incubate at RT for 1 h (can also go overnight at 4 °C).
   3. Centrifuge (900 g, 5 min, 4 °C), and pour off and blot on paper towel.
   4. Wash with 2 mL MCSB and centrifuge (900 g, 5 min, 4 °C), and pour off and blot on paper towel.
   5. Wash with 3 mL MaxPar water and centrifuge (900 g, 5 min, 4 °C), and pour off and blot on paper towel (refer to **Note 9**).
   6. Resuspend cells in MaxPar water + 4EQ beads (count cells and resuspend at a maximum concentration of 500,000 cells/mL with 35 uL of beads per 500 uL of MaxPar water).
   7. Just before putting cells on machine, sieve cells into a new 5 mL round bottom polystyrene tubes with cell-strainer cap by passing cell suspensions through the filter in the lid.
   8. Run cells on the CyTOF instrument based on the recommendations provided by the core facility (*see* **Note 10**).

3.5 CyTOF Data Pre-processing

Depending on the CyTOF run size, it is likely that a given sample file would be split into multiple FCS files for size considerations. To that end, the .fcs files generated post CyTOF acquisition need to be concatenated prior to downstream analysis. The premessa package in R is a convenient tool for CyTOF data pre-processing, allowing concatenation, normalization, and debarcoding.

1. Load the premessa package from https://github.com/ParkerICI/premessa.
2. Load the flowCore [30] package from Bioconductor.

3. Use the concatenate_fcs_files function, and save the output as a single merged file.
4. Use the normalize_GUI() function to normalize the CyTOF data by selecting the Fluidigm Beads (140,151,153,165,175) option and choosing "current files" as a baseline for normalization.

Use the debarcoder_GUI() function to debarcode the samples into individual fcs files by modifying the barcode key to only include the barcodes used in the experiment, and choose optimal minimum barcode separation and maximum Mahalanobis distance (default = 30; *see* **Notes 11** and **12**).

3.6 Major Immune Cell Type Annotation

Each of the samples will contain cellular debris, and for tumor tissue, there are additional stromal cells; hence "gating" on immune cells ($CD45^+$) should be performed first. This can be done using FlowJo, software routinely used for flow cytometry data, or using Cytobank, a cloud computing-based, easy-to-use software, specifically designed for high-dimensional CyTOF data analysis (*see* **Notes 13** and **14**).

1. Gate on singlets based on the Intercalator stain (NC1 and NC2) and event length.
2. Gate on $CD45^+$ cells.
3. Perform t-SNE analysis.
4. Annotate major immune cell types based on their surface marker expression patterns by drawing gates around the populations in the Gating editor.
5. Create a gate based on the tSNE plot which encompasses all the myeloid cell types using the polygon gate button.
6. Apply the gates and return to the original illustration.
7. Select the "clone experiment" action button and choose the above-drawn myeloid gate as the population to split the individual sample. fcs files.
8. Download the myeloid subsetted individual sample .fcs files.

3.7 Myeloid Lineage Cell Deep Phenotyping

The Phenograph workflow presented here has been developed by the Cancer Omics Group: (https://biosurf.org/cytof_data_scientist.html) who provide an elegant end-to-end pipeline for CyTOF data analysis using R [31]. The protocol presented below is a modification of that strategy for myeloid cell deep phenotyping (*see* **Note 15**).

1. Load flowCore, MASS, RColorBrewer, gplots, ggplot2, cowplot, reshape2, ggridges, cytofkit, uwot, tidyr, plyr, dplyr, kableExtra, ggpubr, and miscTools R packages.

2. Create a list of the myeloid subsetted individual sample .fcs files using the list.files function.
3. Create a list of the pre-gating channels such as 4EQBeads, NC1, and NC2 (pregating_channels).
4. Create a list of the lineage channels, which are all surface markers and useful for immunophenotyping (lineage_channels).
5. Create a list of clustering markers which are relevant for myeloid cell annotation and hence used in dimensionality reduction and unsupervised clustering (clustering_channels).
6. Create a list of phospho markers (also include all the intracellular enzyme/protein markers in this list, phospho_channels).
7. Read the myeloid .fcs files using the read.FCS function.
8. Extract the expression matrix, and make the column names (marker names) machine-readable using the information in the parameter data slot.
9. Arcsinhx transform the marker expression values using a cofactor of 5.
10. Clean up the fcs_file names using g.
11. set equal number of events from all the files, and cluster the cells using the cytof_cluster function from the cytofkit package [32] with method = "Rphenograph."
12. Perform dimensionality reduction using the umap function from the uwot package, with n_neighbours = 15, min_dist = 0.2, and metric = 'euclidean'.
13. Generate the UMAP plot using the ggplot2 package.
14. Visualize the marker expression across the various myeloid clusters as a heatmap or based on the marker expression for relevant markers across the UMAP plot.
15. Based on the marker expression patterns, annotate the myeloid clusters into specific immune cell types/populations.
16. Group the relevant individual sample files (here: wt1:wt5 and ko1:ko5) into sample categories (wt and ko).
17. Compare the frequency (proportion) of the various clusters across the individual sample files, and visualize data as a stacked bar graph.
18. Compare the proportion of the identified cell types/populations across the two sample categories.
19. Perform statistical analyses using the ggpubr package to compare the frequency of the various populations across the two sample categories.

20. Visualize the marker expression patterns (surface markers and phospho markers) across the annotated cell populations as a heatmap.
21. Compare the expression of each of the markers in the CyTOF panel across the various annotated populations between both the sample categories.

4 Notes

1. For growing NOP 12 cells, it is recommended to freeze multiple aliquots in liquid nitrogen at the start of the project, to ensure that all the cells used in the various experiments are from the same passage.
2. NOP cell lines express Neu linked to $OVA^{OTI/OTII}$ peptide sequences and a dominant-negative p53 transgene, MMTV-$neu^{OTI/OTII}$.
3. Take an aliquot of buffers that you need using disposable plastic pipettes. Do not use glass pipettes because glassware can be contaminated with metals adversely impacting the CyTOF. Also do not pipette directly from stock bottles – take an aliquot.
4. Perform all steps of the staining on ice (unless otherwise indicated) like you would for regular FACS experiments.
5. While CyTOF staining, pour off supernatant by inverting the tubes in one smooth motion, and then blot upside-down tubes on a stack of paper towels to remove as much of the supernatant as possible without disturbing the pellet.
6. For the phosphoprotein perm step in the CyTOF staining protocol, cells should already be cold before you add MeOH. If they are not cold, put the cells in ice for ~10 min before adding MeOH.
7. During CyTOF staining, occasionally chunks can be seen after methanol treatment. In case that happens, pass the cells suspension through a 70 μm cell strainer prior to intracellular antibody staining. See the link for more details: http://cytoforum.stanford.edu/viewtopic.php?f=4&t=1332.
8. The CyTOF barcoding protocol assumes the cells are already fixed and so you can skip the fixation step using Fix I Buffer as the cells are already fixed using one-step Fix/Lyse. If cells are not already fixed, first perform Cisplatin viability staining, and then proceed to fixation as per the standard Fluidigm protocol.
   1. Add 3 M cells per FACS tube.
   2. Centrifuge (1500 rpm, 5 min, 4 °C) and pour off supernatant and vortex rack to resuspend cells.

3. Wash with 3 ml PBS and centrifuge and pour off as in #2.
4. Add 300 ul of 5 uM cisplatin in PBS (= 10^7 cells/ml).
5. Incubate at RT for 5 min.
6. Add 2 ml of MCBS and centrifuge and pour off as in #2.
7. Wash with 1 ml MCSB & centrifuge and pour off as in #2.

9. If the stained cells are not going to be acquired immediately, cells can be left pelleted in MAXPAR water at 4°C for up to two days.
10. Preferably use the super-sampler (http://www.victorianairships.com/) on CyTOF1 and CyTOF2 as the recovery is significantly better compared to using the loops.
11. For optimal debarcoding of the samples, choose the minimum barcode separation value just before the event yield dramatically drops in the barcode separation plots. It is best to only adjust the minimum barcode separation distance, and only when the debarcoded samples have a high incidence of cross-sample or cell-debris aggregates, should the Mahalanobis distance be changed. See the Cell-ID 20-Plex Pd Barcoding Kit User Guide for more details.
12. After debarcoding and prior to reading all the files, read a single fcs and ensure that the marker names are machine readable using the g function.
13. CyTOF data analysis can be performed in FlowJo as well, especially the new versions of FlowJo such as v10.5+ which can perform dimensionality reduction and clustering natively.
14. Subsetting of myeloid cells can also be performed by serially gating out non-myeloid cells including T cells, B cells, NK cells, ILCs, etc.
15. The detailed R markdown files of the myeloid cell deep phenotyping approach and the *Lyn*$^{+/+}$ and *Lyn*$^{-/-}$ monocyte .fcs files can be found at: https://github.com/maunishb/Barvalia_2021_MIMB_analysis.

Acknowledgments

We would like to thank Dr. Morgan Roberts for generating the original data presented in Figs. 1 and 2 and Israel Matos for generating the original data presented in Fig. 3. We would like to extend our gratitude to Jessica Silva for her thoughtful critique of this chapter.

References

1. Bassler K, Schulte-Schrepping J, Warnat-Herresthal S, Aschenbrenner AC, Schultze JL (2019) The myeloid cell compartment-cell by cell. Annu Rev Immunol 37:269–293. https://doi.org/10.1146/annurev-immunol-042718-041728
2. Becher B, Schlitzer A, Chen J, Mair F, Sumatoh HR, Teng KW, Low D, Ruedl C, Riccardi-Castagnoli P, Poidinger M, Greter M, Ginhoux F, Newell EW (2014) High-dimensional analysis of the murine myeloid cell system. Nat Immunol 15(12): 1181–1189. https://doi.org/10.1038/ni.3006
3. Geissmann F, Manz MG, Jung S, Sieweke MH, Merad M, Ley K (2010) Development of monocytes, macrophages, and dendritic cells. Science 327(5966):656–661. https://doi.org/10.1126/science.1178331
4. Guilliams M, Mildner A, Yona S (2018) Developmental and functional heterogeneity of monocytes. Immunity 49(4):595–613. https://doi.org/10.1016/j.immuni.2018.10.005
5. Guilliams M, Thierry GR, Bonnardel J, Bajenoff M (2020) Establishment and maintenance of the macrophage niche. Immunity 52(3): 434–451. https://doi.org/10.1016/j.immuni.2020.02.015
6. Villani AC, Satija R, Reynolds G, Sarkizova S, Shekhar K, Fletcher J, Griesbeck M, Butler A, Zheng S, Lazo S, Jardine L, Dixon D, Stephenson E, Nilsson E, Grundberg I, McDonald D, Filby A, Li W, De Jager PL, Rozenblatt-Rosen O, Lane AA, Haniffa M, Regev A, Hacohen N (2017) Single-cell RNA-seq reveals new types of human blood dendritic cells, monocytes, and progenitors. Science 356(6335). https://doi.org/10.1126/science.aah4573
7. Bendall SC, Nolan GP, Roederer M, Chattopadhyay PK (2012) A deep profiler's guide to cytometry. Trends Immunol 33(7):323–332. https://doi.org/10.1016/j.it.2012.02.010
8. Bendall SC, Simonds EF, Qiu P, Amir el AD, Krutzik PO, Finck R, Bruggner RV, Melamed R, Trejo A, Ornatsky OI, Balderas RS, Plevritis SK, Sachs K, Pe'er D, Tanner SD, Nolan GP (2011) Single-cell mass cytometry of differential immune and drug responses across a human hematopoietic continuum. Science 332(6030):687–696. https://doi.org/10.1126/science.1198704
9. Amir el AD, Davis KL, Tadmor MD, Simonds EF, Levine JH, Bendall SC, Shenfeld DK, Krishnaswamy S, Nolan GP, Pe'er D (2013) viSNE enables visualization of high dimensional single-cell data and reveals phenotypic heterogeneity of leukemia. Nat Biotechnol 31(6):545–552. https://doi.org/10.1038/nbt.2594
10. Levine JH, Simonds EF, Bendall SC, Davis KL, Amir el AD, Tadmor MD, Litvin O, Fienberg HG, Jager A, Zunder ER, Finck R, Gedman AL, Radtke I, Downing JR, Pe'er D, Nolan GP (2015) Data-driven phenotypic dissection of AML reveals progenitor-like cells that correlate with prognosis. Cell 162(1):184–197. https://doi.org/10.1016/j.cell.2015.05.047
11. Weber LM, Robinson MD (2016) Comparison of clustering methods for high-dimensional single-cell flow and mass cytometry data. Cytometry A 89(12):1084–1096. https://doi.org/10.1002/cyto.a.23030
12. Veglia F, Sanseviero E, Gabrilovich DI (2021) Myeloid-derived suppressor cells in the era of increasing myeloid cell diversity. Nat Rev Immunol. https://doi.org/10.1038/s41577-020-00490-y
13. Bronte V, Brandau S, Chen SH, Colombo MP, Frey AB, Greten TF, Mandruzzato S, Murray PJ, Ochoa A, Ostrand-Rosenberg S, Rodriguez PC, Sica A, Umansky V, Vonderheide RH, Gabrilovich DI (2016) Recommendations for myeloid-derived suppressor cell nomenclature and characterization standards. Nat Commun 7:12150. https://doi.org/10.1038/ncomms12150
14. Hanahan D, Coussens LM (2012) Accessories to the crime: functions of cells recruited to the tumor microenvironment. Cancer Cell 21(3): 309–322. https://doi.org/10.1016/j.ccr.2012.02.022
15. Gonzalez H, Hagerling C, Werb Z (2018) Roles of the immune system in cancer: from tumor initiation to metastatic progression. Genes Dev 32(19–20):1267–1284. https://doi.org/10.1101/gad.314617.118
16. Hiam-Galvez KJ, Allen BM, Spitzer MH (2021) Systemic immunity in cancer. Nat Rev Cancer. https://doi.org/10.1038/s41568-021-00347-z
17. Kiss M, Caro AA, Raes G, Laoui D (2020) Systemic reprogramming of monocytes in cancer. Front Oncol 10:1399. https://doi.org/10.3389/fonc.2020.01399
18. Hanna RN, Cekic C, Sag D, Tacke R, Thomas GD, Nowyhed H, Herrley E, Rasquinha N, McArdle S, Wu R, Peluso E, Metzger D, Ichinose H, Shaked I, Chodaczek G, Biswas SK, Hedrick CC (2015) Patrolling monocytes

control tumor metastasis to the lung. Science 350(6263):985–990. https://doi.org/10.1126/science.aac9407

19. Galon J, Bruni D (2019) Approaches to treat immune hot, altered and cold tumours with combination immunotherapies. Nat Rev Drug Discov 18(3):197–218. https://doi.org/10.1038/s41573-018-0007-y
20. Wei SC, Levine JH, Cogdill AP, Zhao Y, Anang NAS, Andrews MC, Sharma P, Wang J, Wargo JA, Pe'er D, Allison JP (2017) Distinct cellular mechanisms underlie anti-CTLA-4 and anti-PD-1 checkpoint blockade. Cell 170(6):1120–1133 e1117. https://doi.org/10.1016/j.cell.2017.07.024
21. Gubin MM, Esaulova E, Ward JP, Malkova ON, Runci D, Wong P, Noguchi T, Arthur CD, Meng W, Alspach E, Medrano RFV, Fronick C, Fehlings M, Newell EW, Fulton RS, Sheehan KCF, Oh ST, Schreiber RD, Artyomov MN (2018) High-dimensional analysis delineates myeloid and lymphoid compartment remodeling during successful immune-checkpoint cancer therapy. Cell 175(5):1443. https://doi.org/10.1016/j.cell.2018.11.003
22. Krieg C, Nowicka M, Guglietta S, Schindler S, Hartmann FJ, Weber LM, Dummer R, Robinson MD, Levesque MP, Becher B (2018) High-dimensional single-cell analysis predicts response to anti-PD-1 immunotherapy. Nat Med 24(2):144–153. https://doi.org/10.1038/nm.4466
23. Allen BM, Hiam KJ, Burnett CE, Venida A, DeBarge R, Tenvooren I, Marquez DM, Cho NW, Carmi Y, Spitzer MH (2020) Systemic dysfunction and plasticity of the immune macroenvironment in cancer models. Nat Med 26(7):1125–1134. https://doi.org/10.1038/s41591-020-0892-6
24. Nixon AB, Schalper KA, Jacobs I, Potluri S, Wang IM, Fleener C (2019) Peripheral immune-based biomarkers in cancer immunotherapy: can we realize their predictive potential? J Immunother Cancer 7(1):325. https://doi.org/10.1186/s40425-019-0799-2
25. Leelatian N, Sinnaeve J, Mistry AM, Barone SM, Brockman AA, Diggins KE, Greenplate AR, Weaver KD, Thompson RC, Chambless LB, Mobley BC, Ihrie RA, Irish JM (2020) Unsupervised machine learning reveals risk stratifying glioblastoma tumor cells. elife 9. https://doi.org/10.7554/eLife.56879
26. Gaudilliere B, Fragiadakis GK, Bruggner RV, Nicolau M, Finck R, Tingle M, Silva J, Ganio EA, Yeh CG, Maloney WJ, Huddleston JI, Goodman SB, Davis MM, Bendall SC, Fantl WJ, Angst MS, Nolan GP (2014) Clinical recovery from surgery correlates with single-cell immune signatures. Sci Transl Med 6(255):255ra131. https://doi.org/10.1126/scitranslmed.3009701
27. Arunachalam PS, Wimmers F, Mok CKP, Perera R, Scott M, Hagan T, Sigal N, Feng Y, Bristow L, Tak-Yin Tsang O, Wagh D, Coller J, Pellegrini KL, Kazmin D, Alaaeddine G, Leung WS, Chan JMC, Chik TSH, Choi CYC, Huerta C, Paine McCullough M, Lv H, Anderson E, Edupuganti S, Upadhyay AA, Bosinger SE, Maecker HT, Khatri P, Rouphael N, Peiris M, Pulendran B (2020) Systems biological assessment of immunity to mild versus severe COVID-19 infection in humans. Science 369(6508):1210–1220. https://doi.org/10.1126/science.abc6261
28. Roberts ME, Barvalia M, Silva J, Cederberg RA, Chu W, Wong A, Tai DC, Chen S, Matos I, Priatel JJ, Cullis PR, Harder KW (2020) Deep phenotyping by mass cytometry and single-cell RNA-sequencing reveals LYN-regulated signaling profiles underlying monocyte subset heterogeneity and lifespan. Circ Res 126(10):e61–e79. https://doi.org/10.1161/CIRCRESAHA.119.315708
29. Wall EM, Milne K, Martin ML, Watson PH, Theiss P, Nelson BH (2007) Spontaneous mammary tumors differ widely in their inherent sensitivity to adoptively transferred T cells. Cancer Res 67(13):6442–6450. https://doi.org/10.1158/0008-5472.CAN-07-0622
30. Hahne F, LeMeur N, Brinkman RR, Ellis B, Haaland P, Sarkar D, Spidlen J, Strain E, Gentleman R (2009) flowCore: a Bioconductor package for high throughput flow cytometry. BMC Bioinform 10:106. https://doi.org/10.1186/1471-2105-10-106
31. Pedersen CB, Olsen LR (2019) Analysis of mass cytometry data. Methods Mol Biol 1989:267–279. https://doi.org/10.1007/978-1-4939-9454-0_17
32. Chen H, Lau MC, Wong MT, Newell EW, Poidinger M, Chen J (2016) Cytofkit: a bioconductor package for an integrated mass cytometry data analysis pipeline. PLoS Comput Biol 12(9):e1005112. https://doi.org/10.1371/journal.pcbi.1005112

Chapter 13

Manual Immunofluorescence of Formalin-Fixed Paraffin-Embedded Human Tumor Tissues

Sarah Nersesian and Jeanette E. Boudreau

Abstract

Immunofluorescence (IF) of tumor tissues has become a key tool in the study of cancer. With a wide variety of formalin-fixed paraffin-embedded (FFPE) preserved tissues available, there are possibilities to assess large cohorts using archived tissue which may have archived associated clinical outcomes. Although best practice guidelines for the assessment of tissues have been published, a standardized method for immunofluorescence of FFPE tumor tissues is elusive. Here we provide a protocol for using classical secondary fluorescent antibodies that bind directly to the primary antibody of interest. This protocol can easily be adapted to use several primary antibodies, of different species, with unique secondary fluorophores that correspond to each species of origin. It can also be adapted for cyclic amplification-based immunofluorescence of FFPE tissues. We aim to provide a beginner-friendly and highly accessible method for immunofluorescence of FFPE-embedded tissues, hoping to enable more laboratories to take on this highly informative technique and empower them to begin IF analysis in their own tissues of interest.

Key words Immunofluorescence, FFPE tissue staining, Immunofluorescence staining, Immunofluorescence methods, Tissue staining methods

1 Introduction

Whether investigating a marker to validate biological function, appraising experimental outcomes, or exploring putative or established biomarkers, immunofluorescence (IF) of tumor tissues is an important tool for the study of cancer. Using formalin-fixed paraffin-embedded (FFPE) preserved tissues to explore these questions opens the possibility to assess large cohorts using archived tissue and for which clinical outcomes may be collated or for introducing immunofluorescence analysis into existing workflows. IHC analysis is therefore a powerful tool which, properly developed and validated, can provide important insights into the biology of cancer.

Although best practice guidelines for the assessment of tissues have been published, a standardized method for IF of FFPE tumor tissues is elusive [1]. Tumor preservation method, length of

Sherri L. Christian (ed.), *Cancer Cell Biology: Methods and Protocols*,
Methods in Molecular Biology, vol. 2508, https://doi.org/10.1007/978-1-0716-2376-3_13,

preservation time, thickness of tissue sections, methods of post-deparaffinization fixation and antigen retrieval, antibodies, and fluorophores all vary, necessitating careful optimization of IF staining and tissue analysis. This variation, and requirement for expertise, creates barriers to entry for many studies on FFPE tissues in laboratories not specialized in IF. Here, we present a primer for conducting manual IF, including cost-conscious alternatives to standard equipment, with a goal of mobilizing IF-based analysis of FFPE in more laboratories.

When planning an IF experiment with FFPE tissues, it will be important to consider that both the tissue studied and the reagents used can impact success. Each antibody should be carefully validated and titrated, and additional steps for quenching of autofluorescence may be necessary to achieve interpretable staining. We recommend a thorough literature review and use of databases such as https://antibodyregistry.org and https://www.antibodypedia.com/ to select antibodies, with consideration to specificity and indications [2]. Generally, monoclonal antibodies have greater sensitivity, but polyclonal antibodies have greater specificity [3]; the markers and features of interest should inform the investigator's choice of antibody formulation [4]. While databases, published reports, and manufacturer's recommendations are informative, each protocol should be locally optimized and validated [4].

Sensitivity and specificity can be determined with a series of optimization experiments. We recommend using cell lines or samples known to be positive or negative for the marker of interest to test and optimize antibodies. Cell lines can be pelleted, fixed, and embedded to create similar preparations to those that will be used for study samples; these may also become important experimental controls. Complementary methods can also help to gain confidence in staining. For instance, staining cell lines by flow cytometry with validated antibodies can inform the expected frequency and characteristics of positive staining, and Western blotting can be used to identify antibody specificity outside of the potentially confounding features involved in immunohistochemistry. Thereafter, staining of control tissue, such as tonsil tissue to identify lymphocytes, tumor, or adjacent normal tissue, can be used to validate antibody sensitivity and specificity in complex tissues containing connective tissue, titrate antibodies, and evaluate staining protocols. Sensitivity can be confirmed through optimization experiments, including serial dilutions of antibodies on cell samples or tissues/cell pellets known to express various levels of the marker of interest.

Non-specific staining (staining in the negative sample) can reflect several challenges. Non-specific staining may occur when an antibody binds with a protein containing a similar epitope as the intended target, or when the antibody itself is poorly selective. Poor experimental technique, overdrying, or oxidized tissue sections may all contribute to autofluorescence. Often overlooked is the possibility that the target of interest could be variably expressed within intact tissues and during differential states. Non-specific staining may manifest also as staining at the inappropriate cellular location (membrane vs. intracellular vs. nuclear) or an overrepresentation of the proportion of the cells of interest. While validation steps may not all be possible based on the marker of interest and the availability of resources, positive and negative staining controls, antibody isotype controls, and a titration series should be conducted at a minimum to ensure accurate staining.

If non-specific or background staining remains after antibody specificity and sensitivity are validated, it may be autofluorescence. All tissues have some level of autofluorescence, and some are more autofluorescent than others, especially tumor tissues with high levels of collagen/elastin or vascularization [5, 6]. There are several approaches to reducing the autofluorescence intrinsic to tissues, including Eriochrome black T [7], Sudan black B [8] (which we refer to in this protocol), sodium borohydride [9], Tris-glycine, ammonium chloride, ammonia ethanol, and UV transillumination [10]. Each approach has disadvantages and poses some risk to interfere with immunofluorescence staining itself; selections should be made for compatibility with antibodies and tissues of interest. In highly autofluorescent tissues, even with quenching methods, some background will remain. In these cases, ensuring a high signal-to-noise ratio, by using bright fluorophores or secondary antibodies to amplify the signal, may help to identify true staining.

Monoplex validation of antibodies is an important first step toward building multiplex panels. Staining signal can change as a function of the fluorophores associated with each marker and concurrent staining for other features. There are several options for fluorophores and adjacent platforms that can be used for immunofluorescence staining and subsequent quantification. Optimizing staining methodology and selecting ideal fluorophores become especially important when a number of markers will be queried on the same sample [11]. Here we provide a protocol for using classical secondary fluorescent antibodies that bind directly to the primary antibody of interest. This protocol can easily be adapted to use several primary antibodies, of different species, with unique secondary fluorophores that correspond to each species of origin. Recursive, multi-round staining or cyclic amplification-based immunofluorescence can be built from this platform, to enable more thorough interrogation of FFPE tissues. Empiric testing to determine the proper conditions for staining is useful in this context, but generally staining the most prominent markers first, or with dimmer secondary antibodies, is advisable, as they remain robustly visible despite multiple rounds of antigen retrieval and co-staining or staining with other fluorophores. Here, we aim to provide a beginner-friendly and highly accessible method for immunofluorescence of FFPE tissues, hoping to enable more laboratories to take on this highly informative technique and empower them to begin IF analysis in their own FFPE tissues of interest.

2 Materials

2.1 Tissue and Equipment

1. FFPE tissue sections (up to 20 μm; *see* **Note 1**).
2. Positively charged glass microscope slides (our recommendation: Fisherbrand™ Superfrost™ Plus Microscope Slides).
3. Six slide staining jars.
4. An antigen retriever or microwave (*see* **Note 2**).
5. Processing staining jar for washes (our recommendation: DWK Life Sciences Wheaton™ Glass Staining Dish and Slide Rack, Fisher Scientific).
6. Incubation Chamber: Simport™ Scientific StainTray™ Slide Staining System), cost-conscious alternative: make your own incubation chamber (Fig. 1). Materials required: 10 cm culture dish, bench paper or any other absorbent paper material, parafilm, or plastic wrap.
   (a) Step 1: Cut absorbent paper to fit within a 10 cm culture dish, and place at the base of the clean culture dish.
   (b) Step 2: Dampen cloth with H_2O.
   (c) Step 3: Cut parafilm or plastic wrap to fit within the 10 cm culture dish and place atop the damp paper.
   (d) Step 4: Place the slides to be incubated atop the parafilm/plastic wrap.
   (e) Step 5: Place the lid atop the container and use tin foil to protect the slide and staining from direct light.
7. Delicate task wipes.
8. Hydrophobic Barrier PAP Pen, cost-conscious alternative: use clear nail polish.
9. DAPI (4′,6-diamidino-2-phenylindole, dihydrochloride).
10. ProLong™ Gold Antifade Moutant, cost-conscious alternative: make your own mounting media (Mowiol). Mowiol 4-88 is an excellent and cost-conscious alternative mounting media. Importantly, Mowiol mounting medium is not practical for long-term storage of fluorescent samples and does not maintain fluorescent signal as efficiently as hard-fixing mounting solutions such as ProLong mounting mediums; however, DABCO (1,4-diazabicyclo[2.2.2]octane) can be added to mounting solutions to help prevent fluorescent fading.

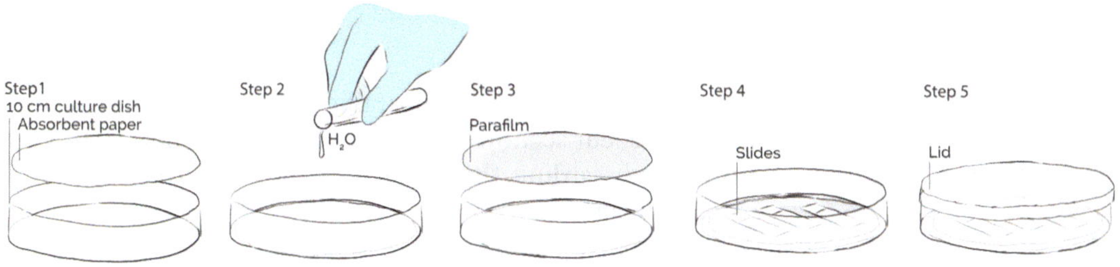

Fig. 1 Create a humidity chamber for incubation of slides with common lab equipment

11. Glass coverslips appropriate size for tissue being stained. For routine immunofluorescence: any glass coverslips with a #1H thickness that will cover your tissue generously is appropriate. For any samples with which you wish to conduct more detailed analysis with high-performance microscopes, glass coverslips with a #1.5H thickness that will cover your tissue generously are appropriate.

2.2 Buffers and Solutions

1. Histological grade xylenes (and appropriate waste disposal).
2. 100% Ethanol.
3. 95% Ethanol.
4. 70% Ethanol.
5. Fixation solution, 10% neutral buffered formalin solution (10% NBF).
6. dH_2O.
7. 1 M TRIS-HCl: dissolve 121.14 g Tris Base in 800 mL of dH_2O, and adjust pH to 7.5 and top to final value of 1 L with dH_2O (store at RT for up to 12 months).
8. Wash buffer TBST (Tris-buffered saline, 0.5% TWEEN®20 detergent): 25 mM TRIS-HCl pH 7.5 (dilute from 1 M TRIS-HCL, pH 7.5), 150 mM NaCl, 0.05% Tween®20.
9. Antigen retrieval buffers either sodium citrate buffers or EDTA buffer (empirically test these options for compatibility with antigen of interest; others are commercially available; *see* **Note 3**).
10. Antibody blocking solution: 5% bovine serum albumin (BSA), 0.1% Triton™ X-100 in PBS (store at 4 °C for up to 1 week).
11. Antibody dilution buffer: 1% BSA, 0.3% Triton™ X-100 in PBS (store at 4 °C for 1 week).
12. Sudan Black staining solution: create a concentration of 1.5% Sudan Black B diluted in 70% ethanol, mix, and then filter using a 0.2 μm filter before use.

3 Methods

3.1 Tissue Preparation

1. Cutting tissue sections from paraffin-embedded blocks: using a microtome, cut sections to a maximum of 20 μm thickness (*see* **Note 1**), discarding the first few tissue slices that have been exposed to air.
2. Mount the tissue on a positively charged histology slide using a water bath.
3. Dry the slides thoroughly by orienting the slides vertically in a staining rack or dish.
4. Bake tissues onto slides in a static incubator or oven at 60 °C for a maximum of 1 h.

3.2 Deparaffinization and Rehydration (Fig. 2) (from this step forward, ensure that slides are not left without liquid atop the tissue (see Note 2))

1. Deparaffinize tissue sections by placing slides into the three sequential slide staining containers filled with xylene solution for 10 min each.
2. Rehydrate tissues sections by placing slides first in 100% and then 95% ethanol for 10 min each.
3. Rinse the tissue sections by dipping them in the 70% ethanol containers.

3.3 Tissue Fixation

1. Place slides in processing staining jar, and cover in 10% NBF for 20 min (ensure all tissue sections are fully submerged).
2. Following fixation, completely remove all 10% NBF to appropriate waste containers, and refill the staining jar with dH_2O to rinse.
3. Repeat another thorough rinse with dH_2O.

3.4 Antigen Retrieval

1. Antigen retrieval can be completed using a commercial apparatus (follow the manufacturer's instructions). For a cost-conscious alternative, a microwave may also be used (for microwave specifications, *see* **Note 4**).
2. Place the processing slide rack in a pipet tip box that fits snugly.
3. Fill the tip box completely with appropriate antigen retrieval buffer (*see* Subheading 2.2, **item 4**, Fig. 3).
4. Place in microwave on highest power setting for 2 min, or until boiling point is reached.
5. Change the microwave power setting to 20% and microwave for 15 min.
6. After antigen retrieval (using either the microwave or antigen retriever), rinse slides twice in wash buffer TBST in a processing jar.

Fig. 2 Deparaffinization and rehydration solutions setup

Fig. 3 Microwave-based antigen retrieval setup

3.5 Blocking

1. Using a delicate cloth wipe, carefully remove excess liquid surrounding tissue on the surface to ensure that the area where you plan to use PAP pen, or nail polish, is completely dry while maintaining moisture on the tissue itself (Fig. 4).
2. Using a PAP pen or nail polish, surround the tissue creating a hydrophobic barrier to contain the liquid (Fig. 4; *see* **Note 5** if tissues are too close to the edge to apply barrier).
3. Carefully pipette enough antibody blocking solution to cover the tissue completely without overcoming the hydrophobic barrier (Fig. 4).
4. Allow for antibody block solution to incubate for 30 min at room temperature.

3.6 Antibody Incubations (all the following steps should be protected from direct light exposure when possible)

1. Dilute primary antibody(ies) in antibody dilution buffer at optimized concentration (*see* **Note 6**).
2. Use a delicate cloth wipe to carefully remove excess liquid surrounding tissue on the surface, and ensure your hydrophobic barrier is intact. If the barrier is lost or broken, dry the area completely while maintaining moisture on the tissue itself.

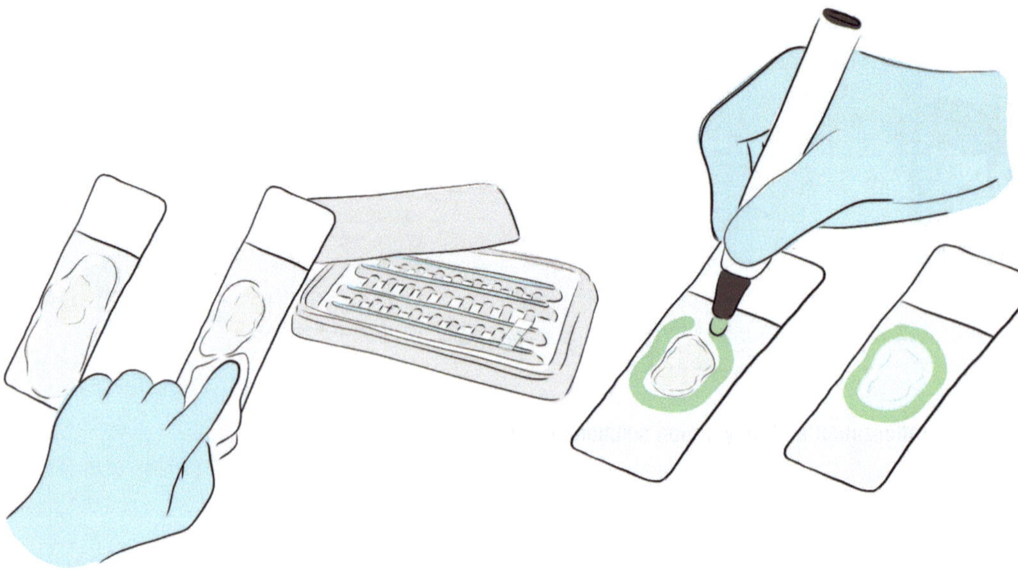

Fig. 4 Preparing and incubating microscope slides

3. Place slide in the incubation chamber of your choice.
4. Using a PAP pen or nail polish, surround the tissue to create a hydrophobic barrier.
5. Without direct contact to the tissue, carefully pipette enough primary antibody solution (usually a minimum of 50 μl) to cover the tissue completely without overcoming the hydrophobic barrier in a dropwise fashion.
6. Allow primary antibody to incubate for 1 h at room temperature (*see* **Note 6**).
7. Following primary antibody incubation, place the slides back in processing jar, and perform three rinses in TBST, for 2 min each.
8. Dilute secondary antibody(ies) in antibody dilution buffer at optimized concentrations (*see* **Note 7**).
9. Again, using a delicate cloth wipe, carefully remove excess liquid surrounding tissue on the surface.
10. Place slide in the incubation chamber.
11. If some of the PAP pen or nail polish has been washed away, re-apply to restore the hydrophobic barrier.
12. Without direct contact to the tissue, carefully pipette enough secondary antibody solution to cover the tissue completely without overcoming the hydrophobic barrier in a dropwise fashion.
13. Cover the incubation chamber, and allow secondary antibody to incubate for 1 h at room temperature.

14. Following primary antibody incubation, place the slides back in processing jar, and perform three rinses in wash buffer TBST, leaving each in the buffer for 2 min.

3.7 Quenching

Quenching is recommended for tissues where autofluorescence is expected (e.g., highly vascularized tissue).

1. Place slides in prepared Sudan Black solution for 1 h.
2. Rinse thoroughly with wash buffer TBST over Sudan Black waste container thrice.
3. Follow this with three additional rinses in wash buffer TBST for 2 min each.

3.8 Counterstaining

DAPI (nuclear) staining is recommended for cellular visualization.

1. Dilute DAPI in recommended or optimized concentration in antibody dilution buffer.
4. Using delicate cloth wipe to carefully remove excess liquid surrounding tissue on the surface.
5. Place slide in the incubation chamber.
6. Without direct contact to the tissue, carefully pipette enough DAPI solution to cover the tissue completely without overcoming the hydrophobic barrier, and place cover on the incubation chamber in a dropwise fashion.
7. Incubate DAPI for 15 min at room temperature.
8. Following DAPI staining place the slides back in processing jar for three rinses in wash buffer TBST for 2 min each.

3.9 Mounting

1. Use a delicate cloth wipe to carefully remove excess liquid surrounding tissue on the surface.
2. Ensure that no particles or containments are present anywhere on the tissue surface or surrounding area. If containments are present, repeat rinse steps until they are removed.
3. Apply 10–50 μL of mounting media (*see* **Note 8**).
4. Using forceps, place one long edge of the glass coverslip on the slide, just past the edge of the tissue. Carefully lower the slide to mount so that it is laid over the tissue and displaces any bubbles.
5. Allow mounting media to dry for 1 h at room temperature in a covered container before imaging.
6. Slides should be stored in a light protected container at 4 °C to preserve fluorescence and imaged as soon as possible after staining (*see* **Note 9**).

3.10 Imaging

1. After drying, slides can be readily imaged. To troubleshoot immunofluorescent staining too weak, *see* **Note 10**, or for high immunofluorescent background, *see* **Note 11**.

4 Notes

1. The thickness of FFPE tissues needs to be optimized for the specific tissue being analyzed by immunofluorescence. To obtain a single layer of cells, most tissues should be sectioned at 5 μm; however, this is not always possible with the tissue being sectioned, and thicker sections may be required to keep the tissue intact. Tissues that are sectioned too thick can cause problems during fixation and antigen retrieval, yielding suboptimal staining results. We recommend not exceeding 20 μm.
2. Once tissues have been placed in xylene to be deparaffinized, they are no longer preserved; drying of the tissue will result in antigen loss. Therefore, it is crucial to ensure tissues are not left without liquid on them at any point during the protocol. Tissues that have dried out cannot be reliably stained.
3. Sodium citrate buffer: 10 mM Sodium Citrate, 0.05% Tween®20, and adjust to pH 6 (store at 4 °C for up to 6 months) *or* EDTA buffer: 10 mM Tris Base, 1 mM EDTA solution, 0.05% Tween®20, and adjust to pH 9 (store at 4 °C for up to 6 months).
4. To conduct successful microwave-based antigen retrieval, there are options to purchase laboratory grade microwaves. However, we have had success in our laboratory using a 1000 W kitchen microwave. We first reach boiling by setting the microwave on 100% power for 2 min and then follow this with 15 min at 20% power to maintain temperature while not continuously boiling. Adjustments to power and time will likely be necessary when using microwaves with a different wattage. Specific tissues may require longer microwave or antigen retrieval processes. Care should be taken that no metal is included in the apparatuses used in the microwave and that the liquid does not boil over or dry out during retrieval.
5. In a situation where tissues are close to the edges of the microscope slide and hydrophobic barriers can't be placed on the slide, slight alterations are required for incubations. Firstly, ensure the chamber and surface where incubation will take place are level. This is crucial to ensure the incubating liquid remains on the tissue for the entire time of incubation and subsequently prevent the tissue from drying. When removing liquid from the microscope slide, rather than using a delicate cloth wipe around the tissue, place the microscope slide

horizontally on a lab bench covered with a delicate cloth wipe. Allow the liquid to be absorbed from the microscope slide to the delicate cloth wipe by capillary action. Carefully dry the edges that are away from the tissue with another delicate cloth wipe. Once the excess liquid is removed, place the microscope slide into the incubation chamber, and ensure that liquid is not running in any direction. Use the pool of liquid that is currently atop the tissue as a target to place incubating liquid. Add a minimum of 100 μL of incubation liquid to ensure tissue is appropriately covered. The liquid should stay around the tissue due to surface tension. Check on your tissue during incubations regularly to ensure liquid has not run off the tissue.

6. When staining with a new antibody or adapting one previously used for another protocol, it is important to first conduct validation for antibody specificity. This can be achieved through Western blotting, positive and negative tissue controls, or other validated protocols already developed in your laboratory. Serial dilutions ranging well below and above the recommended dilution for IF/IHC are suitable for initial optimization steps. Typically a higher concentration of antibodies is required from IF than for flow cytometry. Importantly, optimization is required for each type of tissue being stained (i.e., tonsil tissue used frequently for optimization may stain differently from tumor tissue – especially those that are cut at various thicknesses).
7. Secondary antibodies should be optimized by using reliable primary antibodies to ensure specific binding and antibody serial dilutions.
8. Mounting can be optimized through practice and with the use of several common lab tools. Sharp-ended tweezers are useful to handle and place glass coverslips over mounting media. The blunt end of the tweezers is also useful to gently push bubbles out of the area immediately covering the tissue, but this should be done with caution to avoid disrupting the tissue. Place one edge of the coverslip on the slide, past the outside edge of the tissue, and carefully lower down so that excess mounting media is edged out on the opposite side. The quantity of mounting media depends on the type of media and the viscosity. Aim to have enough liquid to adequately fill the entire area underneath the coverslip being used with a little liquid surrounding the coverslip itself. The coverslip should be able to shift slightly but shouldn't have enough liquid underneath to move if the slide is angled slightly.
9. To ensure mounting media stays moist and does not dry out, a seal can be applied surrounding the coverslip. Apply a thin coat of clear nail polish, ensuring the edges of the coverslip are

entirely covered to prevent the drying out of mounting media and subsequent loss of immunofluorescent signal. This is an optional step if slides will be imaged once within 1 week of staining and not archived. If slides will be imaged later than 1 week after staining, sealing the coverslips is essential.

10. Troubleshoot weak staining by adjusting the steps in the IF staining protocol. Weak staining can result from overfixation of tissues, low primary antibody or secondary antibody concentrations, short incubation times, exposure of fluorochromes to light, or insufficient hydration of the tissues (drying out) at any point.

11. High background interfering with staining can be troublesome and can result from autofluorescence inherent in the tissue being used, tissue being too thick, or non-specific binding. Autofluorescence of tissue should be first addressed through quenching techniques including Sudan Black staining. If quenching does not address the autofluorescence in the tissue on non-stained controls, attempt to cut sections that are thinner. If controls do not have high background, autofluorescence may be in fact due to non-specific binding. Non-specific binding can be reduced through implementing a longer blocking step, decreasing the primary or secondary antibody concentrations, and increasing the length of TBST wash steps or may require selection of a different antibody.

References

1. Taube JM, Akturk G, Angelo M, Engle EL, Gnjatic S, Greenbaum S, Greenwald NF, Hedvat CV, Hollmann TJ, Juco J, Parra ER, Rebelatto MC, Rimm DL, Rodriguez-Canales J, Schalper KA, Stack EC, Ferreira CS, Korski K, Lako A, Rodig SJ, Schenck E, Steele KE, Surace MJ, Tetzlaff MT, von Loga K, Wistuba II, Bifulco CB, Society for Immunotherapy of Cancer Pathology Task F (2020) The Society for Immunotherapy of cancer statement on best practices for multiplex immunohistochemistry (IHC) and immunofluorescence (IF) staining and validation. J Immunother Cancer 8(1):e000155. https://doi.org/10.1136/jitc-2019-000155
2. Uhlen M, Bandrowski A, Carr S, Edwards A, Ellenberg J, Lundberg E, Rimm DL, Rodriguez H, Hiltke T, Snyder M, Yamamoto T (2016) A proposal for validation of antibodies. Nat Methods 13(10):823–827. https://doi.org/10.1038/nmeth.3995
3. Lipman NS, Jackson LR, Trudel LJ, Weis-Garcia F (2005) Monoclonal versus polyclonal antibodies: distinguishing characteristics, applications, and information resources. ILAR J 46(3):258–268. https://doi.org/10.1093/ilar.46.3.258
4. MacNeil T, Vathiotis IA, Martinez-Morilla S, Yaghoobi V, Zugazagoitia J, Liu Y, Rimm DL (2020) Antibody validation for protein expression on tissue slides: a protocol for immunohistochemistry. Biotechniques 69(6):460–468. https://doi.org/10.2144/btn-2020-0095
5. Whittington NC, Wray S (2017) Suppression of red blood cell autofluorescence for immunocytochemistry on fixed embryonic mouse tissue. Curr Protoc Neurosci 81:2 28 21–22 28 12. https://doi.org/10.1002/cpns.35
6. Croce AC, Bottiroli G (2014) Autofluorescence spectroscopy and imaging: a tool for biomedical research and diagnosis. Eur J Histochem 58(4):2461. https://doi.org/10.4081/ejh.2014.2461
7. Kittelberger R, Davis PF, Stehbens WE (1989) An improved immunofluorescence technique for the histological examination of blood vessel

tissue. Acta Histochem 86(2):137–142. https://doi.org/10.1016/S0065-1281(89)80082-0

8. Baschong W, Suetterlin R, Laeng RH (2001) Control of autofluorescence of archival formaldehyde-fixed, paraffin-embedded tissue in confocal laser scanning microscopy (CLSM). J Histochem Cytochem 49(12):1565–1572. https://doi.org/10.1177/002215540104901210
9. Clancy B, Cauller LJ (1998) Reduction of background autofluorescence in brain sections following immersion in sodium borohydride. J Neurosci Methods 83(2):97–102. https://doi.org/10.1016/s0165-0270(98)00066-1
10. Neumann M, Gabel D (2002) Simple method for reduction of autofluorescence in fluorescence microscopy. J Histochem Cytochem 50(3):437–439. https://doi.org/10.1177/002215540205000315
11. Prost S, Kishen RE, Kluth DC, Bellamy CO (2016) Choice of illumination system & fluorophore for multiplex immunofluorescence on FFPE tissue sections. PLoS One 11(9):e0162419. https://doi.org/10.1371/journal.pone.0162419

Chapter 14

Western Blot Analysis of Lapatinib-Mediated Inhibition of the Epidermal Growth Factor Receptor 2 (HER2) Pathway in Breast Cancer Cells

Louisa L. Wiede and Sheila Drover

Abstract

Western blotting is an excellent technique to investigate aberrations and/or therapy-induced changes in signaling proteins in cancer. Using an in vitro system, we prepared whole cell lysates from HER2-overexpressing breast cancer cell lines, treated or not with the tyrosine kinase inhibitor, lapatinib, in the presence and absence of IFN-γ. Here we describe the protocol whereby proteins in the lysates were separated by SDS-PAGE, electrophoretically transferred to nitrocellulose membranes followed by an enzyme-linked immunoassay and chemiluminescence to reveal the relevant phosphorylated and dephosphorylated proteins. Herein, Western blot analysis confirmed lapatinib dephosphorylated HER2 and downstream signaling proteins and IFN-γ induced phosphorylation of STAT1.

Key words Sodium dodecyl sulfate polyacrylamide gel electrophoresis (SDS-PAGE), Western blotting, Enzyme-linked immunoassay, Breast cancer cells, Human epidermal growth factor receptor 2 (HER2), Lapatinib

1 Introduction

Many cancer therapies have immunomodulatory effects [1], which can be explored in vitro by treating cell lines with physiological concentrations of a test drug [2] and assaying for changes in cell surface immune molecules by flow cytometry, as described in (*see* Chapter 11. Provided that suitable and validated antibodies are available, analysis should include Western blotting [3–5], a technique that was first described more than three decades ago [6, 7]. Moreover, Western blotting has the advantage of comparing alterations in signaling pathways in cells treated or not with kinase inhibitors such as lapatinib [8, 9].

The first step for Western blotting is to prepare a cell lysate from treated and untreated cells. Depending on the intracellular location of the proteins of interest, a lysis buffer may be formulated for

Sherri L. Christian (ed.), *Cancer Cell Biology: Methods and Protocols*,
Methods in Molecular Biology, vol. 2508, https://doi.org/10.1007/978-1-0716-2376-3_14,

cytoplasmic or nuclear proteins or more typically for whole cell lysates using radiation immunoprecipitation assay (RIPA) buffer [10, 11]. Since proteins rapidly degrade on cell lysis, it is essential to add protease and phosphatase inhibitors to the lysis buffer [12], perform all manipulations on ice, and store the lysate at −20 °C, or for long-term storage at −80 °C. To ensure equal loading of protein from different lysates, total protein is measured using the bicinchoninic acid method [13] for which commercial kits are readily available. A sample buffer, initially described by Laemmli [14], is used to prepare the lysate for electrophoresis. Typically, the proteins are separated by size using sodium dodecyl sulfate polyacrylamide gel electrophoresis (SDS-PAGE) [15, 16]. Although the separated proteins in the gel may be transferred passively to nitrocellulose paper [17], most often this is done using the classic electrophoretic transfer [6, 7].

An enzyme-linked immunoassay is used to detect the relevant protein band (s) on the nitrocellulose paper [3]. This involves blocking nonspecific sites on the paper with a buffer containing a protein such as milk powder, usually referred to as Blotto [18], followed by incubation with a primary antibody that should be validated and optimally diluted [3–5]. The blot is then extensively washed in a neutral buffer, either phosphate-buffered saline (PBS) or Tris-buffered saline (TBS) containing a detergent such as Tween-20 to remove unbound proteins and reduce background. An appropriate secondary antibody, for example, goat anti-mouse (GAM) if the primary antibody is of mouse origin, labelled with horseradish peroxidase (HRP) is then added. After incubation and more extensive washing, GAM-HRP binding is detected using a suitable substrate (H_2O_2) and dye such as diamine-amino benzidine (DAB) or more sensitively by chemiluminescence [19, 20]. The size of the visualized band is compared to prestained protein markers that are electrophoresed and transferred simultaneously. A house keeping protein such as GAPDH or α-tubulin is also immunodetected on the same membrane. The protein density of the bands can be estimated using various tools such as Image J, reviewed in detail elsewhere [21, 22]. Membranes may be stripped and reprobed as described by others [3, 23], but in our hands, this has not been entirely successful.

As described in Chapter 11, we treated two breast cancer cell lines, BT-474 and SKBR3, that overexpress HER2 with lapatinib, a reversible tyrosine kinase inhibitor, employed in the treatment of HER2+ metastatic breast cancer [24–26]. Lapatinib attenuates HER2 signaling by targeting the intracellular domain of HER2, competing for the ATP-binding domain of protein kinases, and thereby, interrupting receptor phosphorylation and subsequent activation of downstream signaling pathways such as MAPK and PI3K/Akt [27–29]. To ensure lapatinib-inhibited HER2 signaling, we prepared cell lysates for Western blot analysis of phosphorylated

(p) HER2 and total (t) HER2 and downstream signaling proteins, pAkt, tAkt, pERK, and tERK and. Since the immunomodulator, IFN-γ, was also used in combination with lapatinib, we further analyzed for STAT1, a key component of the IFN-γ signaling pathway and critical for upregulation of many immune molecules [30].

2 Materials

2.1 Sample Preparation

1. $HER2^+$ breast cancer cell lines: SKBR3 (ATCC® HTB-30 and BT-474 (ATCC® HTB-20), cultured and treated as described in Chapter 11.
2. Distilled H_2O (dH_2O).
3. 1× Phosphate-buffered saline (PBS), diluted 1/10 from 10× Stock PBS: 160 g NaCl, 4 g KCl, 54.36 g Na_2HPO_4, dissolved in 1 l dH_2O, and the pH adjusted to 7.4.
4. Protease inhibitors: aprotinin solution (aprotinin saline solution, 3–7 TIU/mg protein, Sigma), leupeptin (10 mg/mL in dH_2O), pepstatin A (2 mg/mL in ethanol), phenylmethylsulfonyl fluoride (PMSF), (200 mM in isopropanol alcohol).
5. Halt phosphatase inhibitor cocktail.
6. Radioimmunoprecipitation assay (RIPA) buffer: 0.5% sodium deoxycholate, 0.1% sodium dodecyl sulfate (SDS), and 1% NP40 in PBS (pH 7.3).
7. Cell scraper.
8. 1.5 ml microfuge tubes.
9. Micro pipettors and tips – various sizes.
10. Bicinchoninic acid protein (BCA) assay kit.
11. Bovine serum albumin (BSA).
12. 5 ml polystyrene tubes.
13. Spectrophotometer.
14. 3× Reducing sample buffer (20 ml): Combine 7.8 ml 0.5 M Tris, pH 6.8 with 6 ml 30% glycerol and 3 ml 2-Mercaptoethanol (2-ME). Add 1.8 g SDS and 1.5 mg Bromophenol blue and make up to 20 ml with dH_20. Mix well and dispense 1 ml aliquots in Eppendorf tubes and store at −20 °C.

2.2 SDS PAGE

1. dH_2O.
2. 1.5 M Tris, pH 8.8.
3. 0.5 M Tris, pH 6.8.
4. 30% Acrylamide-Bis solution 29:1.
5. 10% SDS; 10% Ammonium persulfate (APS).

6. Tetramethylethylenediamine (TEMED).
7. Running Buffer: 25 mM Tris, pH 8.8, 0.1% SDS, 190 mM glycine.
8. Mini-PROTEIN® Cell Electrophoresis System Bio-Rad: tank, lid, electrode assembly and running module, casting frames and stand, glass spacers plates, silicone spacers, glass plates of various thicknesses and combs with different thicknesses and number of wells.
9. Glass plates for 1.5 mm gels.
10. Combs (10 or 15 wells).
11. 15 ml centrifuge tube.
12. Gel loading capillary tips.
13. Prestained protein ladder.
14. Bio-Rad Power Pack 3000.

2.3 Western Blotting

1. Mini trans Blot module.
2. Transfer sandwich cassettes.
3. Transfer Buffer: 25 mM Tris, pH 8.8, 190 mM glycine, 20% methanol.
4. Nitrocellulose paper (0.45 μm).
5. Filter paper (0.21 mm).
6. Foam pads.
7. Ice pack (10 cm × 12 cm).
8. Power supply.

2.4 Immuno-detection of Proteins

1. Tris-buffered saline with Tween 20 (TBS-T): 0.15 M NaCl, 0.05 M Tris, pH 7.4, 0.05% Tween 20.
2. Blotto-blocking buffer: 5% milk powder (w/v) in TBS-T.
3. Incubation and wash boxes (9.5 cm × 3.0 × 1.6 cm, 8.9 cm × 6.5 cm × 2.5 cm, or other sizes as appropriate).
4. Shaker.
5. Micropipettors and tips.
6. Serological pipettes (5 and 10 ml).
7. Primary antibodies, optimally diluted in TBS-T or blocking buffer: Phospho-HER2 (pTry 1248) – clone PN2A (ThermoFisher Scientific), mouse IgG1, diluted 1/1000; HER2 – clone e2-4001 (ThermoFisher Scientific), mouse IgG1, diluted 1/1000; Phospho-ERK – clone E4 (Santa Cruz), mouse IgG2a, diluted 1/500; ERK-1 – polyclonal K23 (Santa Cruz), rabbit IgG, 1/5000; Phospho-AKT (Ser473) XP, polyclonal D9E (Cell Signalling), rabbit IgG, diluted 1/1000; AKT (pan), polyclonal C67E7, (Cell Signalling) rabbit IgG, diluted

1/1000; Phospho-STAT1 (Tyr701) – clone 4a, (BD-Pharmingen), mouse IgG2a, diluted 1/1000; STAT1 – clone 1/Stat1, (BD-Pharmingen), mouse IgG1, diluted 1/1000; GAPDH – clone 6C5 (Abcam), mouse IgG1, diluted 1/20,000.

8. Secondary antibodies, optimally diluted in TBS-T: Horseradish peroxidase (HRP-conjugated affiniPure f(ab)$_2$ fragment goat anti-mouse (GAM) Fc specific; HRP-conjugated affiniPure f (ab)$_2$ fragment goat anti-rabbit (GAR) Fc specific.
9. Chemiluminescence HPR substrate – Clarity® Western ECL (Bio-Rad).

3 Method

BT-474 and SKBR3 are HER2-overexpressing breast cancer cells that were cultured and treated with lapatinib and IFN-γ as described in Chapter 11.

3.1 Preparation of Whole Cell Lysates

1. Remove treated 6-well plates from the 37 °C incubator (as described in Chapter 11) and aspirate medium from each well.
2. Wash adherent cells in each well with ice cold PBS.
3. Add 100 μl ice cold RIPA buffer containing protease inhibitor cocktail (aprotinin 1 μg/ml; leupeptin, 1 μg/ml; pepstatin A, 1 μg/ml; PMSF 10 μg/ml), and Halt phosphatase inhibitor cocktail, 10 μl/ml) (*see* **Note 1**).
4. Scrape the cells from the wells using a cell scraper.
5. Transfer the lysates to 1.5 ml Eppendorf tubes.
6. Centrifuge at 11,000 × *g*/10 min.
7. Transfer supernatant to a clean Eppendorf tube, appropriately labelled and store at −80 °C.

3.2 Protein Concentration Using BCA Assay

1. Prepare a set of protein standards using BSA: 500 μg, 250 μg, 125 μg, 62.5 μg, 31.25 μg 0 μg in distilled H_2O.
2. Dilute lysate 1/50 in distilled H_2O.
3. Perform BCA assay according to the manufacturer instructions.
4. Read using a spectrophotometer and a wavelength 582 nm.
5. Calculate the protein concentration of the lysate based on the standard curve.

3.3 Preparation of Sample for SDS-PAGE

1. Thaw lysate on ice if frozen and keep on ice.
2. Dilute the lysate in 3× sample buffer to 10 μg/20 μl (*see* **Note 2**).

3. Boil samples at 100 °C for 5–7 min prior to sample loading – freeze leftover samples at minus 80 °C after boiling.

3.4 SDS-PAGE

1. Clean plates with 70% alcohol.
2. Assemble the casting gel plates and apparatus according to manufacturer's instructions.
3. Prepare 8% resolving gel using the volumes of reagents listed in Table 1 (*see* **Note 3**) and ensuring APS and TEMED are added last (*see* **Note 4**).
4. Gently mix resolving gel in 15 ml centrifuge tube.
5. Immediately pour gel between the short and tall glass plates, leaving about 1.7–2.0 cm from top of the short plate.
6. Overlay with dH_2O or isopropanol and allow the gel to polymerize for about 20 min. Check leftover gel in tube to ensure the gel has polymerized before continuing to the next step.
7. Pour off dH_2O overlay, and remove excess with filter paper.
8. Mix stacking gel reagents using the volumes listed (Table 1) in centrifuge tube, ensuring TEMED and APS are added last (*see* **Note 4**). Mix and immediately pour it over the resolving gel. Quickly position comb in the stacking gel, ensuring no air bubbles are below the wells.
9. Let gel polymerize for about 20 min, again checking leftover gel in tube to ensure polymerization before continuing to the next step.
10. Remove the gel from gel casting apparatus and cassette and place in electrophoresis tank with the short plate facing inward (*see* **Note 5**).
11. Fill tank with 1× running buffer and slowly and carefully remove comb from stacking gel. Ensure all wells are filled with running buffer before sample loading.
12. If samples were frozen (*see* **Step 3**, Subheading 3.3), thaw quickly and place on ice.
13. Vortex and quickly centrifuge the samples in a microfuge.
14. Using a 20 μl pipettor and a gel loading capillary pipette tip, load the samples in the wells ensuring the tall plate is facing you. Samples are usually loaded left to right and according to a predetermined order; (*see* **Note 6**) with a prestained protein marker loaded at both ends (*see* **Note 7**).
15. Once the gel is loaded, secure tank cover and attach electrodes to the power supply – red to red and black to black.
16. Turn on the power supply and run for approximately 1 h at 20 mA/gel (40 mA for 2 gels) or until the dye has reached the

Table 1
Recipes for preparing two SDS-PAGE resolving gels with different percentages of acrylamide and stacking gels

	Percent resolving gels					
Reagents	**6%**	**8%**	**10%**	**12%**	**15%**	**Stacking gel**
dH20	993 μl × 4	862 μl × 4	3 μl	800 μl × 3	825 μl × 2	860 μl × 2
1.5 M Tris, pH 8.8	937 μl × 2	938 μl × 2	938 μl × 2	975 μl × 2	975 μl × 2	0 μl
0.5 M Tris, pH 6.8	0 μl	0 μl	0 μl	0 μl	0 μl	760 μl
30% Acrylamide	750 μl × 2	675 μl × 3	825 μl × 3	1000 μl × 3	937.5 μl × 4	500 μl
10% SDS	75 μl	75 μl	75 μl	75 μl	75 μl	30 μl
10% APS	75 μl	75 μl	75 μl	75 μl	75 μl	30 μl
TEMED	6 μl	6 μl	6 μl	6 μl	6 μl	5 μl

bottom of the gel. Alternatively, gel can also be run for up to 1 h at 100 V (*see* **Note 8**).

3.5 Western Blotting – Electrophoretic Protein Transfer

1. Cut 1 sheet of nitrocellulose (6 cm × 8 cm) and 4 sheets of absorbent filter paper (7 cm × 10 cm) per gel and soak in transfer buffer along with absorbent transfer pads.
2. Remove gels from gel running tank and place short plate facing down. Separate short and tall plates ensuring the gel remains attached to one of them. Cut off the stacking gel.
3. Immerse plate with attached gel into transfer buffer, and gently detach gel from glass plate.
4. Place gel holder cassette with black side on the bottom on workbench and assemble the sandwich arrangement as follows: transfer pad, 2 sheets of filter paper, gel, nitrocellulose, 2 sheets of filter paper, transfer pad (Fig. 1).
5. Remove any air bubbles after adding filter papers over nitrocellulose by moving gloved index fingers from center to side of transfer sandwich, thereby applying gentle pressure.
6. Place sandwich in transfer tank with black plastic grid facing black of apparatus (cathode (+)) and therefore, transferring negatively charged proteins from gel toward red of apparatus (anode (−)) onto the nitrocellulose (Fig. 1).
7. Fill tank with transfer buffer and add an ice pack and a small stir bar (*see* **Note 9**).
8. Transfer with stirring at 100 V for 1 h or at 70 V for 75 min.

3.6 Western Blotting – Immunodetection

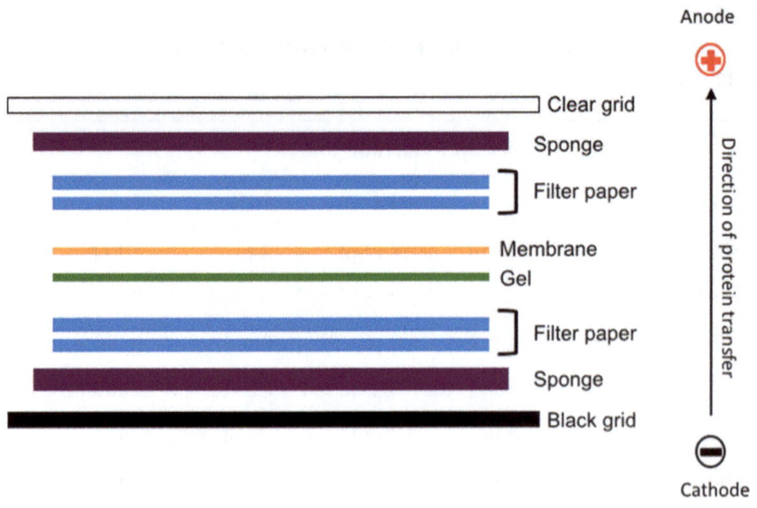

Fig. 1 Schematic of the assembly of a Western blot

1. Prepare Blotto buffer while the proteins are transferring.
2. Disassemble the sandwich apparatus, check to see that the prestained markers have completely transferred; then, using the markers as a guide, cut the membrane for immunoassays to detect the proteins of interest (Fig. 2).
3. Place the membrane in a wash box just big enough to hold membrane with 15 ml Blotto per whole membrane for 1 h at RT on a shaker.
4. Prepare primary antibodies (*see* **Note 10**). Dilute the primary antibodies in Blotto or TBS-T. Add sufficient antibody to fully cover the membrane.
5. Incubate membranes with primary antibodies overnight at 4 °C on shaker.
6. Pour off primary antibodies, rinse well with TBS-T, and discard the rinse.
7. Wash membrane 3 × 5 min in TBS-T with gentle agitation on a shaker at RT.
8. Prepare and add appropriate secondary antibody (*see* **Note 11**) diluted in Blotto using sufficient volume to completely cover the membrane.
9. Incubate 1 h on a shaker.
10. Pour off secondary antibody, and rinse the membrane with TBS-T.
11. Wash membrane 3 × 10 min in TBS-T with gentle agitation on a shaker at RT.

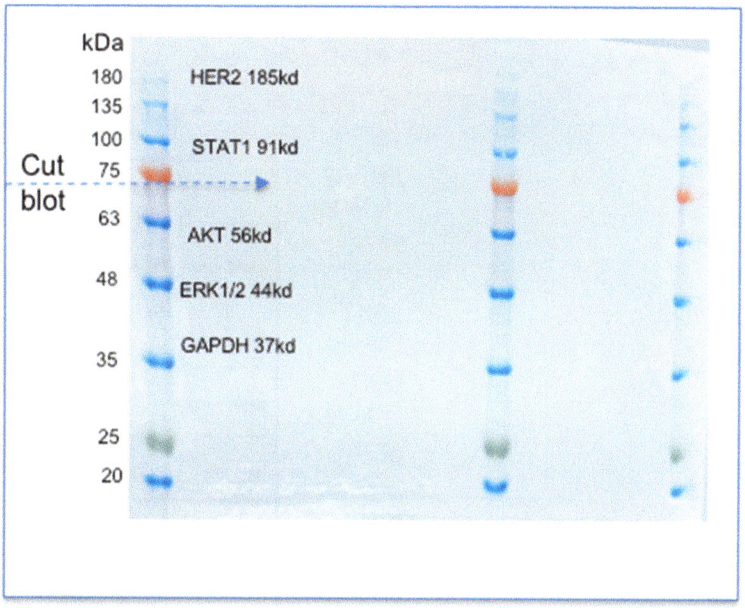

Fig. 2 A representative photograph of a membrane after electrophoretic transfer demonstrating transfer of prestained markers. The dashed line indicates where the gel is cut to analyze for indicated proteins

12. Prepare substrate: Add ½ reagent A + ½ reagent B (usually 2 ml A + 2 ml B per blot); for every additional blot, prepare an additional ml (i.e., 2 blots = 5 ml).
13. Add 4 ml substrate.
14. Place membrane protein side down in the shallow lid of an incubation box or on clear acetate.
15. For every additional blot, place blot in lid, pipette off 1 ml substrate, and add 1 ml of fresh substrate.
16. The signals can be detected using the ImageQuant Las 4000 station or other chemiluminescent system (e.g., Fig. 3), ensuring the detected bands are the expected size using the prestained protein ladder.
17. Protein band densities can be analyzed using the software provided by Image J (NIH) or another suitable program, and the protein of interest can be normalized to a housekeeping protein such as GAPDH or α-tubulin. A detailed explanation of how to quantify proteins by Western blotting can be found in a recent article by Pillai-Kastoori [22].

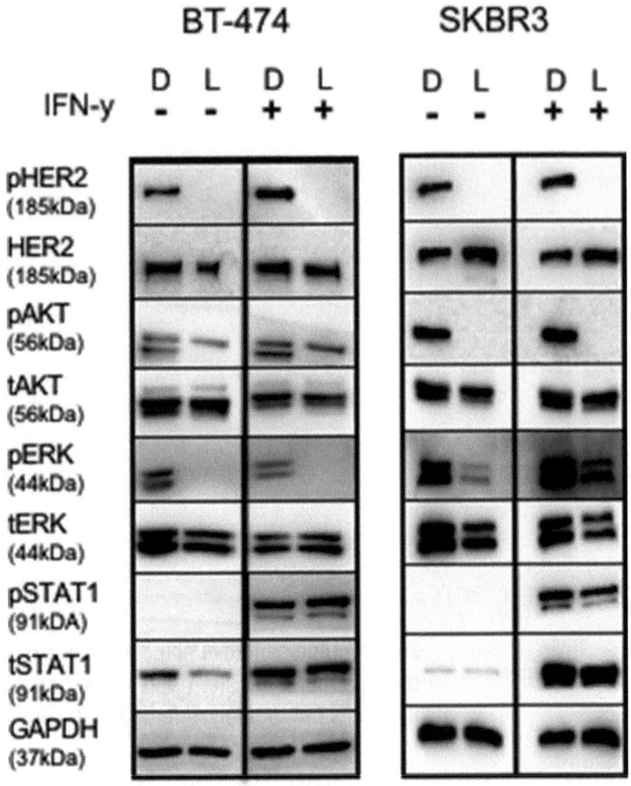

Fig. 3 Western blot analysis of whole cell lysates from BT-474 and SKBR3, treated with lapatinib (L) or DMSO (vehicle control), +/− IFN-γ (100 U/mL) for 24 h

4 Notes

1. The protease inhibitors and Halt phosphatase inhibitor should be added to RIPA buffer immediately prior to making the cell lysates.
2. Calculate the amount of lysate based on its protein concentration, the amount of protein and volume to be loaded per well. It is typically 10 μg in 20 μl (0.5 μg/μl) per well and a convenient volume to prepare is 90 μl, equivalent to 45 μg protein. Divide 45 μg by the protein concentration (μg/μl) of the lysate = μl of lysate. Add this volume to 30 μl 3× sample buffer and add dH_2O to make up 90 μl. If knowing the protein concentration is not essential, an alternative method is to simply dilute 2 parts lysate with 1 part 3× sample buffer and load 10 or 20 μl /well.
3. Choosing the optimal gel percentage depends on the size of the proteins you want to detect. Separation of the proteins is determined by the pore size in the gel, which in turn depends

on the percentage of acrylamide – the higher the percentage, the smaller the pore size. For example, a 20% gel is good for detecting proteins ranging from 4 to 20 kD, whereas the 8% gel used in this study can detect proteins ranging from 25 to 200 kD. Gels with lower than 6% acrylamide can be fragile and tear easily, so these need to be handled with extra care.

4. It is important to add APS and TEMED last as TEMED reacts quickly with APS to catalyze the polymerization of acrylamide and Bis-acrylamide. Thus, the gel should be poured immediately after these two reagents are mixed with the other reagents.
5. Gels can be prepared the day before use, covered securely with plastic wrap, and stored at 4 °C.
6. The amount of sample to add to each lane largely depends on the abundance of the protein you wish to detect. If it is present in amounts such as a that of housekeeping protein, 5–10 μg is sufficient; for low abundant proteins, load up to 50 μg or even 100 μg per lane. If you are not sure, it is best to determine this experimentally, by loading different amounts of proteins and assaying with a validated antibody that has been optimally diluted (*see* **Note 10** below).
7. Prestained protein markers should be loaded at each end of the sample wells as this make it easier to cut the membrane for immunodetecting different proteins. To create an asymmetric order to maintain orientation, the two lanes can contain different size or concentration of marker. This ensures that the orientation is obvious post transfer. It is also wise to omit adding sample to the end wells as sometimes they run slanted.
8. Running time will vary depending on the % of the resolving gel; aim for good separation of proteins using the prestained protein markers as a guide. It also is prudent to check the gel occasionally during the run to ensure there are no technical problems.
9. Electrophoretic transfer generates considerable heat; so, it is important to keep the transfer buffer cool. This may be done in a cold room, but a more efficient solution is to use a common ice pack (10 cm × 12 cm), which fits nicely into the transfer tank.
10. The choice and preparation of antibodies are critical considerations. Ideally, obtain a validated antibody, either polyclonal or monoclonal, from a vendor or colleague and perform a series of dilutions to find the optimal concentration. We typically start with the vendor-recommended dilution and make one or two dilutions on either side of it. Some antibodies such as phospho-antibodies do not work as well in Blotto due to interference from casein – in that case, try TBS-T or TBS-T with lower

percentages of milk powder, or alternatively, use 1–3% bovine serum albumin in TBS-T.

11. There are numerous choices for secondary antibodies – we generally find those prepared as Fab_2 fragments and specific for IgG of the primary antibody with minimal cross-reactions to give clean signals and low background. Again, it is it important to establish the optimal dilution and buffer conditions as described above in **Note 10**.

Acknowledgments

This work was funded by the Canadian Breast Cancer Foundation (BC-RG-13-1 grant #310985) and by a Dean's Fellowship, Faculty of Medicine, Memorial University of Newfoundland. We wish to acknowledge a former PhD student, Dr. Ahmed Mostafa, for his work to optimize Western blotting in our laboratory and Mr. Kaiwen Liu for continuous technical advice.

References

1. Petroni G, Buqué A, Zitvogel L, Kroemer G, Galluzzi L (2021) Immunomodulation by targeted anticancer agents. Cancer Cell 39(3): 310–345
2. Mostafa AA, Codner D, Hirasawa K, Komatsu Y, Young MN, Steimle V, Drover S (2014) Activation of ERα signaling differentially modulates IFN-γ induced HLA-class II expression in breast cancer cells. PLoS One 9(1):e87377
3. Gallagher SR, Winston SE, Fuller SA, Hurrell JG (2001) Immunoblotting and immunodetection. Curr Protoc Cell Biol Chapter 6:Unit 6.2. https://doi.org/10.1002/0471143030.cb0602s00
4. Simpson K, Browning M (2017) Antibodies that work again and again and again. Methods Mol Biol 1554:41–59
5. Pillai-Kastoori L, Heaton S, Shiflett SD, Roberts AC, Solache A, Schutz-Geschwender AR (2020) Antibody validation for Western blot: by the user, for the user. J Biol Chem 295(4):926–939
6. Towbin H, Staehelin T, Gordon J (1979) Electrophoretic transfer of proteins from polyacrylamide gels to nitrocellulose sheets: procedure and some applications. Proc Natl Acad Sci U S A 76(9):4350–4354
7. Burnette WN (1981) "Western blotting": electrophoretic transfer of proteins from sodium dodecyl sulfate--polyacrylamide gels to unmodified nitrocellulose and radiographic detection with antibody and radioiodinated protein A. Anal Biochem 112(2):195–203
8. Hegde PS, Rusnak D, Bertiaux M, Alligood K, Strum J, Gagnon R, Gilmer TM (2007) Delineation of molecular mechanisms of sensitivity to lapatinib in breast cancer cell lines using global gene expression profiles. Mol Cancer Ther 6(5):1629–1640
9. Carloni S, Fabbri F, Brigliadori G, Ulivi P, Silvestrini R, Amadori D, Zoli W (2010) Tyrosine kinase inhibitors gefitinib, lapatinib and sorafenib induce rapid functional alterations in breast cancer cells. Curr Cancer Drug Targets 10(4):422–431
10. Peach M, Marsh N, Miskiewicz EI, MacPhee DJ (2015) Solubilization of proteins: the importance of lysis buffer choice. Methods Mol Biol 1312:49–60
11. Harlow E, Lane D (1988) Antibodies: a laboratory manual. Cold Spring Harbor Press, Cold Spring Harbor
12. Kurien BT, Scofield RH (2015) Western blotting: an introduction. Methods Mol Biol 1312: 17–30
13. Smith PK, Krohn RI, Hermanson GT et al (1985) Measurement of protein using bicinchoninic acid. Anal Biochem 150(1):76–85
14. Laemmli UK (1970) Cleavage of structural proteins during the assembly of the head of

bacteriophage T4. Nature 227(5259): 680–685
15. Smith BJ (1984) SDS Polyacrylamide gel electrophoresis of proteins. Methods Mol Biol 1: 41–55
16. Al-Tubuly AA (2000) SDS-PAGE and western blotting. Methods Mol Med 40:391–340
17. Kurien BT, Scofield RH (2015) Multiple immunoblots by passive diffusion of proteins from a Single SDS-PAGE Gel. Methods Mol Biol 1312:77–86
18. Johnson DA, Gautsch JW, Sportsman JR, Elder JH (1984) Improved technique utilizing nonfat dry milk for analysis of proteins and nucleic acids transferred to nitrocellulose. Gene Anal Tech 1(1):3–8
19. Durrant I (1994) Enhanced chemiluminescent detection of horseradish peroxidase labeled probes. Methods Mol Biol 31:147–161
20. Alegria-Schaffer A (2014) Western blotting using chemiluminescent substrates. Methods Enzymol 541:251–259
21. Schneider CA, Rasband WS, Eliceiri KW (2012) NIH Image to ImageJ: 25 years of image analysis. Nat Methods 9(7):671–675
22. Pillai-Kastoori L, Schutz-Geschwender AR, Harford JA (2020) A systematic approach to quantitative Western blot analysis. Anal Biochem 593:113608
23. Litovchick L (2020) Stripping of the immunoblot for reprobing. Cold Spring Harb Protoc 2020(3):098491
24. Rimawi MF, Schiff R, Osborne CK (2015) Targeting HER2 for the treatment of breast cancer. Annu Rev Med 66:111–128
25. Duchnowska R, Loibl S, Jassem J (2018) Tyrosine kinase inhibitors for brain metastases in HER2-positive breast cancer. Cancer Treat Rev 67:71–77
26. Iqbal N, Iqbal N (2014) Human Epidermal Growth Factor Receptor 2 (HER2) in cancers: overexpression and therapeutic implications. Mol Biol Int 2014:852748. https://doi.org/10.1155/2014/852748
27. D'Amato V, Raimondo L, Formisano L, Giuliano M, De Placido S, Rosa R, Bianco R (2015) Mechanisms of lapatinib resistance in HER2-driven breast cancer. Cancer Treat Rev 41(10):877–883
28. Scaltriti M, Verma C, Guzman M, Jimenez J, Parra JL, Pedersen K, Smith DJ, Landolfi S et al (2009) Lapatinib, a HER2 tyrosine kinase inhibitor, induces stabilization and accumulation of HER2 and potentiates trastuzumab-dependent cell cytotoxicity. Oncogene 28(6): 803–814
29. Miricescu D, Totan A, Stanescu-Spinu II, Badoiu SC, Stefani C, Greabu M (2020) PI3K/AKT/mTOR signaling pathway in breast cancer: from molecular landscape to clinical aspects. Int J Mol Sci 22(1):173. https://doi.org/10.3390/ijms22010173
30. Krause CD, He W, Kotenko S, Pestka S (2006) Modulation of the activation of Stat1 by the interferon-gamma receptor complex. Cell Res 16(1):113–123

Chapter 15

Evaluating Cell Membrane Localization and Intracellular Transport of Proteins by Biotinylation

Eduardo Reyes-Alvarez, Timothy J. Walker, and Lois M. Mulligan

Abstract

Protein translocation to the cell membrane and transport through intracellular compartments are dynamic processes frequently altered in cancer cells. Abnormal protein localization can affect key cell functions, including transduction of extracellular signals and organization of the cytoskeleton, significantly affecting oncogenicity and therapeutic responses. In this chapter, we describe a surface protein biotinylation method that allows the study of membrane localization and endosomal transport of membrane-associated proteins. Surface biotinylation can be used to evaluate baseline protein levels at the membrane, and other processes such as internalization, recycling, and degradation of proteins in response to different treatments or as a consequence of oncogenic mutations. Further, the combination of this technique with other strategies, such as treatments with transport inhibitors, allows investigation of specific steps of protein trafficking through the cell.

Key words Cell membrane proteins, Biotinylation, Internalization, Recycling, Degradation

1 Introduction

A wide variety of proteins synthesized and processed in the endoplasmic reticulum (ER) and Golgi apparatus are translocated to the cell surface (Fig. 1) [1, 2]. Once at the plasma membrane, these proteins mediate cell environment-sensing, and responses required for cell function or survival [3]. Subsequently, these proteins must be physically removed from the cell membrane, either to "turn off" signaling triggered in response to environmental cues, or to relocalize the proteins to alternative cellular compartments [4, 5]. Membrane proteins may be internalized through different endocytic routes to endosomal compartments, where cargo is sorted and transported to the appropriate destination, such as to lysosomes for degradation or to recycling endosomes for return to the cell membrane (Fig. 1) [6, 7]. Intracellular trafficking of membrane proteins is important for the spatial and temporal regulations of

Sherri L. Christian (ed.), *Cancer Cell Biology: Methods and Protocols*,
Methods in Molecular Biology, vol. 2508, https://doi.org/10.1007/978-1-0716-2376-3_15,

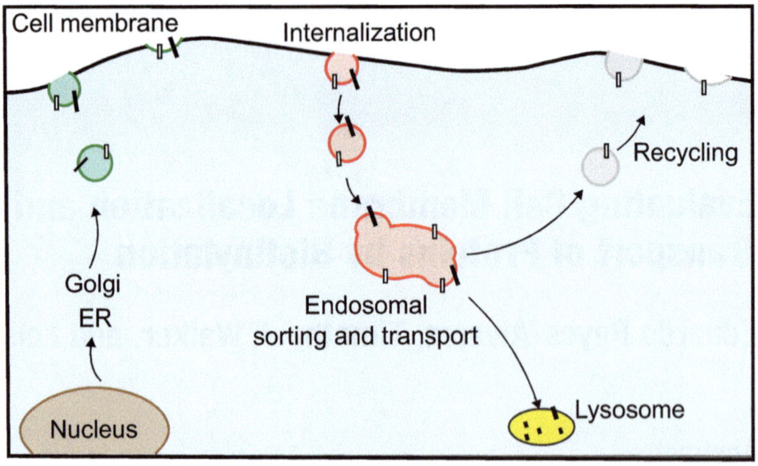

Fig. 1 Membrane receptors and other transmembrane proteins are synthesized and translocated to the cell membrane where they exert their functions. Environmental signals, such as ligand stimulation, trigger endocytic mechanisms that remove proteins from the cell membrane to intracellular compartments for different purposes. Internalized vesicles fuse with endosomes, where sorting processes occur, and target cargo for transport to different destinations, including the lysosome for degradation, or recycling back to the cell membrane. These changes in association of the proteins with the cell membrane and intracellular localization allow regulation of protein function at different levels, such as timing, type of interactions, and downstream effectors

protein function, allowing these proteins to associate with protein complexes localized to specific cellular compartments as signaling continues from these compartments after internalization [4, 5]. Consistent with its important roles in regulation of cell signals, intracellular trafficking is often altered in cancer cells, changing the nature or duration of interactions, or altering protein life cycle, leading to oncogenic phenotypes [8, 9].

Biotinylation of cell surface proteins, described here, is an effective method to investigate localization, distribution, and intracellular trafficking of cell membrane proteins in response to different stimuli [10–12]. Biotin derivatives can covalently bind to primary amine groups on cell membrane proteins but are impermeable to the cell membrane due to a sulfonate group and can be selectively removed by reducing agents (*see* Heading 2). Thus, the described technique allows exclusive labeling of proteins residing at the cell membrane, which can be immunoprecipitated from cell lysates using streptavidin-coated beads (*see* Heading 2) and analyzed by techniques such as SDS-PAGE and western blotting. An important advantage of biotinylation as a labeling method is the small size of the biotin label, which minimizes interference with protein–protein interactions, and allows addition of multiple biotin tags to a single protein, enhancing detection [10, 13]. Importantly,

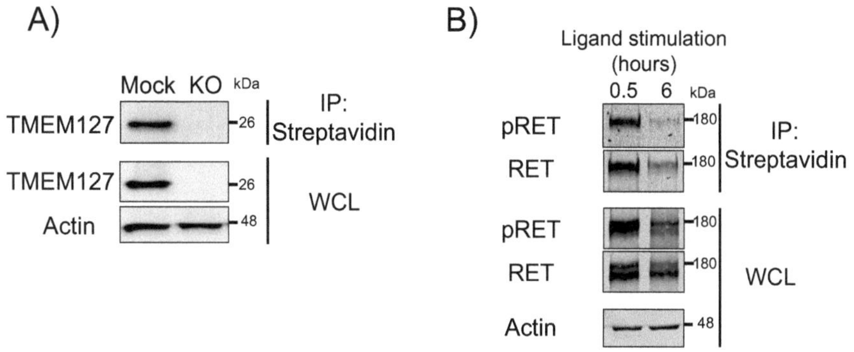

Fig. 2 (**a**) Surface biotinylation of cell membrane proteins to compare protein levels in different cell lines. Expression of cell surface protein TMEM127 in Mock knockout or TMEM127 knockout (KO) SH-SY5Y cells generated by CRISPR-Cas9. Membrane protein biotinylation confirms that TMEM127 was knocked out and no longer localized to the cell membrane. (**b**) Surface biotinylation of cell membrane proteins to evaluate the effect of prolonged ligand stimulation on total and membrane expression of the RET receptor tyrosine kinase. SH-SY5Y cells were treated for either 0.5 or 6 h with the RET ligand. After ligand incubation, membrane proteins were biotinylated to evaluate levels of total and phosphorylated RET localizing to the cell membrane. Prolonged ligand stimulation decreased levels of membrane (IP) and total (WCL) RET, consistent with RET degradation and attenuation of signaling over time after ligand stimulation [17]

since these assays use a single streptavidin immunoprecipitation to collect all biotinylated proteins in the cell, multiple surface proteins can be studied at once in a single experiment, by using specific antibodies for each protein of interest in the detection steps.

Several strategies can be combined with biotinylation of cell membrane proteins to study specific intracellular protein transport processes. Simple surface biotinylation can be used to detect levels of receptors or other proteins at the cell surface under various conditions or in different cell lines (Fig. 2). Biotinylation assays can also be used to identify location of different pools of a protein of interest, or distinguish a subset, for example, active receptors, in response to stimuli (Fig. 2). Figure 3 shows simple surface biotinylation to explore the effects of inhibitors of protein synthesis (cycloheximide) or transport (monensin, brefeldin A) on localization of a protein of interest to the cell membrane, allowing the study of protein maturation, degradation, and trafficking when these processes are impaired [14–16]. These studies can also allow us to compare wild-type and active mutant forms of the proteins of interest to identify differences in location or interactions associated with aberrant activity, for example, of an oncogenic cell surface receptor (Fig. 3).

Biotinylation of surface proteins is also a valuable tool to evaluate internalization and recycling of cell membrane proteins in response to stimuli (Fig. 4). After surface biotinylation, it is possible to alternate additional incubations at 4 °C or 37 °C with surface-stripping steps using reducing agents to remove any biotin-labeled proteins remaining on the cell surface. In this way, the proportion of the original surface protein internalized or subsequently returned to the cell membrane can be quantified.

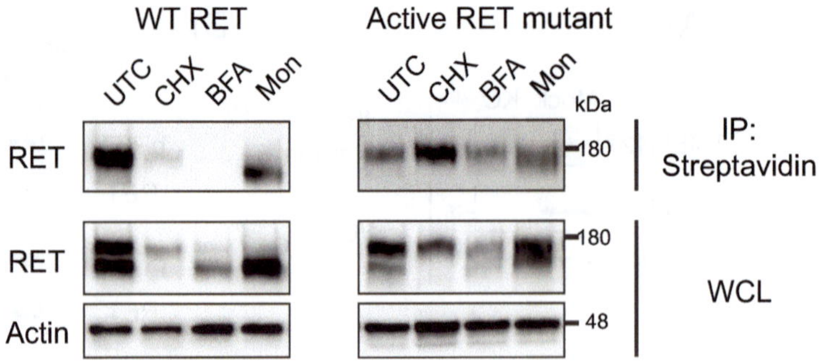

Fig. 3 Surface biotinylation of cell membrane proteins to evaluate effects of inhibitors of protein maturation and transport on cells expressing wild-type (WT) and active mutant RET receptors. Cells were treated 12 h with either cycloheximide (CHX) to inhibit protein synthesis, brefeldin A (BFA) to inhibit ER-to-Golgi transport, or monensin to inhibit Golgi-to-membrane export. Mutant RET receptors show increased localization to the cell membrane (IP) when protein synthesis is inhibited, and increased levels of mature (175 kDa) protein compared to WT-RET when ER and Golgi transport is blocked

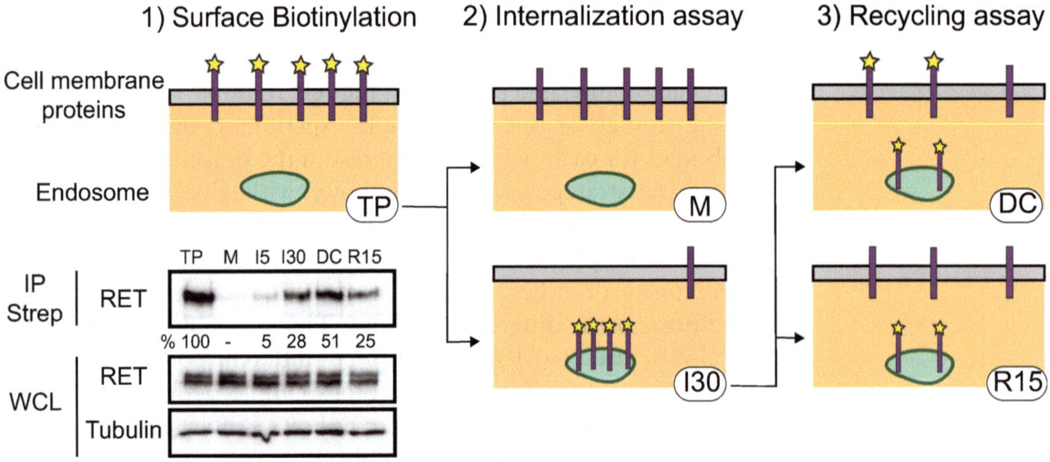

Fig. 4 Internalization and recycling assays to evaluate endosomal transport of the RET receptor. (1) All conditions undergo surface biotinylation. The total protein (TP) condition gives the baseline of initial amounts of RET protein at the cell membrane. (2) To evaluate internalization, cells are incubated for 5 or 30 min (I5 and I30) and then undergo biotin stripping to remove any residual membrane biotinylation and to evaluate the portion of the initial membrane RET that was internalized. Control cells undergo MESNA stripping (M) without the internalization step. (3) To evaluate recycling, cells treated as for internalization conditions (I30) undergo a further incubation (DC control) to allow protein recycling to the membrane, after which an additional biotin-stripping step is performed (R15). The portion of RET receptors recycled back to the cell membrane (relative to TP) can be calculated from the difference between DC and R15

Table 1 summarizes the steps and controls necessary for each of these types of biotinylation assays. Briefly, in a biotinylation assay, we need to include the following conditions and controls:

Table 1

Types and applications of biotinylation assays. Protocol steps required (starting in Subheading 3.2) and the necessary control conditions for each assay type are indicted. In the simple biotinylation assays, multiple TP plates can be prepared to evaluate protein levels under different treatments or with different cell lines. In the internalization assay, all plates are treated as above, but two additional conditions are used (*see* Subheading 3.4). These conditions include a control for the biotin-stripping step (M) and one or more conditions to evaluate internalization of the protein of interest at specific time points (I_). For the recycling assay, the TP, M, and I_ plates are treated as above, and two additional conditions are included (*see* Subheading 3.5). One of these conditions is a control for protein degradation (DC), and the other(s) evaluates recycling of the protein of interest after specific times (R_)

Steps needed	Type of assay	Application	Treatment conditions required
3.2, 3.3, 3.6	Simple biotinylation	Labeling of cell surface proteins. Detection/ quantification of membrane localization	TP 1–2 plates per condition
3.2, 3.3, 3.4, 3.6	Internalization	Internalization of surface-labeled proteins Accumulation of surface proteins	TP, M, I_ 1–2 plates per condition Single or multiple I_ time points, as needed
3.2, 3.3, 3.4, 3.5, 3.6	Recycling	Return of internalized surface proteins to the plasma membrane	TP, M, I_, DC, R_ 1–2 plates per condition Single or multiple I_ and R_ time points, as needed

(a) The total protein (TP) condition will indicate the total amount of protein labeled with biotin at the membrane. It is used to evaluate baseline protein levels at the cell membrane (simple biotinylation), and acts as a pretreatment control for protein internalization and recycling assays.

(b) The MESNA (M) condition is used as a control in internalization and recycling assays. MESNA is a reducing agent that strips all biotin labeling from surface proteins but does not affect labeled proteins that have been internalized and are no longer exposed on the cell surface. The M condition acts as a background control for residual biotin labeling under these experimental conditions. We expect to see no bands corresponding to our protein of interest in this control on western blots; otherwise, it is not possible to interpret the results (e.g., Fig. 4).

(c) For internalization assays, the cells for internalization (I_) condition(s) are surface-labeled with biotin, as for the TP condition, then incubated for a specific time (e.g., 15 min (I15)) or for a time course, using multiple plates of cells (e.g., I5 = 5 min, I15 = 15 min, I30 = 30 min etc.) at

37 °C to allow internalization of surface proteins from the membrane. After these incubations, the remaining portion of biotinylated proteins at the membrane is stripped off with MESNA treatment, as above, leaving only biotinylated proteins that had been internalized (e.g., Fig. 4).

(d) The Recycling (R_) conditions(s) are used to identify the proportion of proteins internalized into the cell and then returned to the cell surface by recycling pathways over specific times (e.g., 5, 15, or 30 min). Cells are labeled as for the TP condition, and incubated to allow internalization, as for the I_ condition(s). Residual biotin is then removed by MESNA stripping, and cells are reincubated at 37 °C for appropriate times to allow proteins to recycle (e.g., R5 = 5 min, R15 = 15 min, or R30 = 30 min). A second MESNA-stripping step following this incubation removes any biotinylated protein that has returned to the cell surface (e.g., Fig. 4).

(e) The Degradation Control (DC) is used in recycling assays as a control for the amounts of internalized proteins that may be lost nonspecifically during the recycling incubations. Cells are treated as for the R_ conditions, but the second MESNA stripping is omitted. The DC sample includes all biotin-labeled protein that remain in the cell. Using this control condition, the portion of the original membrane protein that underwent recycling can be determined by subtracting the intensity of the R_ band from the DC condition and normalizing to the TP condition. This represents the amount of protein that was labeled with biotin, internalized from the membrane, and then returned to the cell membrane, relative to the total amount of protein available (TP) (Fig. 4).

2 Materials

2.1 Reagents

1. Biotin [Sulfosuccinimidyl 2-(biotinamido)-ethyl-1-3-′-dithiopropionate, EZ-link Sulfo-NHS-SS Biotin].
2. MESNA [Sodium 2-mercaptoethanesulfonate].
3. Streptavidin beads [Pierce streptavidin agarose resins].

2.2 Stock Solutions

Prepare the following solutions using autoclaved, double-distilled water and store at room temperature (*see* **Note 1**):

1. 5 M NaCl.
2. 0.5 M HEPES, pH 7.6.
3. 0.25 M KCl.
4. 2 M $MgCl_2$.
5. 2 M $CaCl_2$.

6. 1.5 M Tris–HCl, pH 8.6.
7. Calcium–Magnesium-supplemented PBS (PBS*): 137 mM NaCl, 2.68 mM KCl, 10 mM Na_2HPO_4, 1.76 mM KH_2PO_4, 1 mM $MgCl_2$, 0.1 mM $CaCl_2$, pH to 7.4.

2.3 Biotinylating Solutions

Prepare 10× stock solutions of biotinylating buffer (BB), and MESNA buffer (MB) if needed, and store at room temperature. Prepare the working 1× solutions at least an hour before the experiment, by diluting the 10× stock solutions and adding the last components, as indicated below. All solutions are prepared with autoclaved, double-distilled water (*see* **Note 2**).

1. 10× BB: 1.54 M NaCl, 100 mM HEPES, 30 mM KCl, 10 mM $MgCl_2$, 1 mM $CaCl_2$, 100 mM glucose.
2. 10× MB: 1 M NaCl, 500 mM Tris–HCl pH 8.6, 10 mM $MgCl_2$, 1 mM $CaCl_2$.
3. 1× BB: Dilute the required volume of 10× BB, add 0.5 mg/mL biotin powder, and gently invert the tube multiple times until the biotin is dissolved completely.
4. 1× MB: Dilute the required volume of 10× MB, add 8.2 mg/mL MESNA powder, and gently invert the tube until the powder is dissolved completely.
5. PBS* + glycine: Prepare a small amount of PBS* with 100 mM glycine, sufficient for two washes for each plate of cells, using 5 mL per 10-cm plate. Only one wash step requires glycine to remove excess unbound biotin, all the other washing steps use 1× PBS*.

2.4 Cell Harvesting and Sample Preparation Solutions

A variety of standard lysing buffers and protein quantification protocols, such as those suggested below, can be used for these assays. Optimize the lysis buffer and quantification method for the cell line and proteins of interest using more or less stringent conditions as needed for the specific experiment.

1. Lysis buffer + protease and phosphatase inhibitors: 20 mM Tris–HCl pH 7.8, 150 mM NaCl, 1% IGEPAL CA-630, and 2 mM EDTA disodium salt in autoclaved, double-distilled water. Before lysing the samples, add 100 mM sodium orthovanadate, 100 mM PMSF, 10 mg/mL aprotinin, and 10 mg/mL leupeptin to prevent sample degradation.
2. Bicinchoninic acid (BCA) protein assay kit.
3. 3× Laemmli buffer + beta-mercaptoethanol: 187.5 mM Tris–HCl pH 6.8, 6% SDS, 30% glycerol, and 0.03% bromophenol blue in autoclaved, double-distilled water. Before use, add 15% beta-mercaptoethanol.

3 Methods

3.1 Experimental Design

Different types of biotinylation assays are appropriate for different research questions. Consult Table 1 for details of the experimental controls and conditions, and number of confluent plates of cells and treatment steps required for each type of assay.

3.2 Cell Preparation

1. Seed cells in 10-cm plates using the appropriate culture medium and grow until they reach 80–90% confluence. A separate plate(s) for each experimental condition and the required controls (Table 1) should be seeded at equal density (*see* **Note 3**).
2. Serum-starve the cells (optional). This may be required if evaluating a treatment response but may not be needed if evaluating steady-state levels of surface protein (e.g., Figs. 2a and 3). Cells will have different sensitivities to reduced serum; appropriate conditions should be established for each cell line (e.g., 3–24 h in low- (1%) or serum-free media, *see* **Note 4**).
3. The following day (day of experiment), prepare and chill biotinylating 1× working solutions, as described in Subheading 2.3.
4. Experiment-specific cell treatment (optional): Cells can be treated with growth factors, drugs, or other agents to promote or inhibit membrane localization and/or intracellular trafficking (e.g., Fig. 3; *see* **Note 4**). Treated cells may be returned to the incubator for appropriate times, and then proceed to Subheading 3.3.

3.3 Surface Biotinylation

1. Following the specific treatments, remove the plates from the incubator, and wash twice with 5 mL PBS* on ice.
2. Add 5 mL of cold 1× BB to each plate and incubate for 45 min, keeping the plates at 0–4 °C. It is critical to keep the plates flat due to the low volume used, and very cold to prevent any membrane protein internalization during this step (*see* **Note 5**).
3. Remove the 1× BB and wash twice with 5 mL of PBS* + glycine for 5 min on ice.
4. For a simple surface-protein biotinylation assay, wash once with 5 mL PBS* and harvest the cells with the appropriate amount of lysis buffer (e.g., 70–150 μL per 10-cm plate). Then proceed to Subheading 3.6 to process the samples. For internalization or recycling assays, harvest one plate(s) at this step to obtain the TP control condition and proceed to Subheading 3.4 with the remaining plates.

3.4 Internalization

1. Leave the M control plate(s) on ice. Remove the PBS* and add appropriate prewarmed culture media to all the other plates (I_, or I_ + DC + R_ if doing a recycling assay). Place these plates in the tissue culture incubator (37 °C) for the predetermined incubation times (e.g., 5, 10, or 30 min, *see* **Note 6**).
2. After incubation for the appropriate time, remove each plate from the incubator and place on ice. Remove the culture media and wash twice with 5 mL of cold PBS*. Leave these plates on ice until all internalization incubations are completed.
3. Remove the PBS* from all the plates including the M control plate(s). Add 5 mL of 1× MB and incubate 20 min on ice.
4. Remove the 1× MB from the plates, add 5 mL of fresh 1× MB, and incubate 20 min on ice.
5. Remove the 1× MB from all the plates and wash twice with 5 mL of PBS* (*see* **Note 7**).
6. If the experiment evaluates only internalization, remove the PBS*, harvest all plates, and proceed to Subheading 3.6 to process the samples. If the experiment is also evaluating protein recycling, remove the PBS* 1× only from the M and I_ plates, harvest those plates, and proceed to Subheading 3.5 with the remaining (DC and R_) plates.

3.5 Recycling

1. Remove the PBS* from the remaining plates (DC and R_), add prewarmed media, and return the plates to the tissue culture incubator (37 °C) for an appropriate time, based on the protein of interest. Shorter times (e.g., 5 min) are appropriate to assess fast recycling, while longer incubations may be needed for slow recycling of surface proteins (e.g., 15–30 min), (*see* **Note 8**).
2. Retrieve the DC and R_ plates from the incubator, place them on ice, and wash twice with cold PBS*.
3. The DC control plate(s) can now be harvested. Remove the PBS* from the R_ plates, add 5 mL of 1× MB, and incubate 20 min on ice.
4. Remove the 1× MB from the R_ plate(s), add 5 mL of fresh 1× MB, and incubate for an additional 20 min on ice.
5. Remove the 1× MB from the R_ plate(s), wash twice with PBS*, and harvest (*see* **Note 7**). Proceed to Subheading 3.6 with the cell lysates.

3.6 Sample Processing and Immunoprecipitation of Biotin-Labeled Proteins

1. Quantify the harvested whole cell lysate (WCL) collected in the previous sections by the BCA protein assay, following the provider recommendations for the kit.
2. For each treatment condition used, prepare two sample tubes, one for the WCL and one for the immunoprecipitation (IP) sample.

3. For the WCL tubes combine: 100 μg of WCL sample with loading buffer containing reducing agents (i.e., 3× Laemmli buffer + beta-mercaptoethanol). It is convenient to use a final volume of 100 μL with a final protein concentration of 1 μg/μL for each sample. Once prepared, store these samples at −20 °C (or −80 °C if assessing phospho-proteins).
4. For the IP samples: Prepare 500–1000 μg of WCL diluted to a final volume of 400 μL with lysis buffer. Add 30 μL of streptavidin beads and incubate overnight on a rotator at 4 °C (*see* **Note 9**).
5. The following day, centrifuge the IP samples for 1 min at 2000 × *g* at 4 °C to pellet the beads and perform the following washes keeping the tubes on ice as much as possible.
6. Discard the supernatant and add 300 μL of fresh lysis buffer per tube, gently invert tubes multiple times to resuspend the beads, and centrifuge for 1 minute at 2000 × *g* at 4 °C.
7. Repeat **Step 3.6.6** twice more, for a total of 3 washes.
8. After the last wash, remove the supernatant and add 40 μL of 3× Laemmli buffer used in **Step 3.6.3**. These IP samples can be stored together with the WCL samples overnight or 1–3 days, as mentioned in **Step 3**, but it is highly recommended to proceed to **Step 3.6.9** as soon as possible.
9. To analyze biotin labeling of the protein of interest, WCL and IP samples can now be boiled (5 min) to denature the proteins, separated by SDS-PAGE, and visualized by western blotting, using standard methods (*see* **Note 10**).

4 Notes

1. Before proceeding to seed the cells and start the experiment, it is highly recommended to prepare solutions 1–7 as indicated in Subheading 2.2. Volumes of 250 mL for solution 1 and 100 mL for solutions 2–6 will be sufficient for most experiments and may be stored at room temperature between uses. PBS* may need to be prepared more frequently, since it is used in all washing steps. 1 L is sufficient for several experiments and should be stored at 4 °C or precooled long enough to be very cold (0–4 °C) before starting the experiment.
2. It is recommended to prepare 50 mL or less 10× BB and 10× MB to prevent long periods of storage. These solutions can be stored at room temperature for a few weeks. Use the necessary amount to prepare the 1× solutions for each experiment. Working 1× biotinylating solutions should be prepared fresh from the 10× stocks to avoid biotin and MESNA degradation/inactivation. Incubate the final prepared 1× solutions at 0–4 °C

for at least an hour before the experiment to cool thoroughly and ensure that no membrane internalization happens during the biotinylation step. You will need 5 mL per 10-cm plate of 1× BB and 1× MB (if needed). Calculate the total number of plates and number of incubations required (1 incubation for 1× BB, 2 or more for 1× MB) to determine the appropriate volumes of solutions to prepare for each experiment.

3. Before attempting a surface biotinylation assay, it is strongly recommended to become familiar with culturing the cell line of interest. Experience with cell growth and adhesiveness will minimize loss of cells during the multiple washing steps required and allow timing of steps to be adjusted to account for cell stickiness and time needed for harvesting. When designing the experiment, consider the density of cell growth and amount of protein that can be retrieved from a confluent plate. Larger or multiple plates per condition may be needed for some cell types.
4. If the cells require long treatments with specific molecules (e.g., inhibitors, agonists), these can be done at the same time as the serum-starvation or overlapping the serum starvation period, as timing requires.
5. Plates can be placed on a flat surface inside a 4 °C refrigerator, if available (avoid repeated opening and closing of the refrigerator door during the incubation), or plates can be incubated on ice (check every 5–10 min that plates remain flat, and that BB is covering all cells evenly).
6. The media may contain ligand for a membrane receptor, other molecules to induce/block vesicular trafficking, or may be nonsupplemented media to evaluate constitutive processes.
7. If the cells adhere well and do not lift easily from the plates during washes or if this is the final step for some of the plates, a third wash is highly recommended, since traces of 1× MB can interfere with some protein quantification methods (such as an instant color change in the BCA protein assay). If this is not feasible, make sure to gently tilt the plate a few times during the two washes to completely wash the 1× MB off the plates.
8. If evaluating recycling of the protein of interest after ligand or drug stimulation, the media in this step should not contain the stimulatory molecule. If evaluating multiple time intervals for recycling, a pair of DC and R plates must be incubated for each time point.
9. The immunoprecipitation step with streptavidin beads works well with 500–1000 μg protein lysate per sample. Be sure to use the same amount of lysate for each of the samples within an experiment so that conditions can be compared. However, it is

highly recommended to set a standard of the minimum amount of lysate that can be used for the IP when harvesting one or more plates per condition. For example, if two 10-cm plates generally yield 700–900 μg of protein lysate, a standardized input of 600 μg per condition for the IP can reliably be prepared each time, instead of trying to adjust protein amounts for each experiment.

10. It is also important to optimize western blotting conditions for the cells of interest prior to the biotinylation experiments. An appropriate lysis buffer to collect proteins of interest should be used. Primary and secondary antibodies used for detection should be tested and immune detection protocols optimized by western blotting of whole cell lysates of the cell line of interest to confirm that the proteins can be detected.

References

1. Bard F, Malhotra V (2006) The formation of TGN-to-plasma-membrane transport carriers. Annu Rev Cell Dev Biol 22:439–455. https://doi.org/10.1146/annurev.cellbio.21.012704.133126
2. Parmar HB, Barry C, Kai F et al (2014) Golgi complex-plasma membrane trafficking directed by an autonomous, tribasic Golgi export signal. Mol Biol Cell 25(6):866–878. https://doi.org/10.1091/mbc.E13-07-0364
3. Alenghat FJ, Golan DE (2013) Membrane protein dynamics and functional implications in mammalian cells. Curr Top Membr 72:89–120. https://doi.org/10.1016/B978-0-12-417027-8.00003-9
4. Sigismund S, Confalonieri S, Ciliberto A et al (2012) Endocytosis and signaling: cell logistics shape the eukaryotic cell plan. Physiol Rev 92(1):273–366. https://doi.org/10.1152/physrev.00005.2011
5. von Zastrow M, Sorkin A (2021) Mechanisms for regulating and organizing receptor signaling by endocytosis. Annu Rev Biochem 90:709–737. https://doi.org/10.1146/annurev-biochem-081820-092427
6. Kumari S, Mg S, Mayor S (2010) Endocytosis unplugged: multiple ways to enter the cell. Cell Res 20(3):256–275. https://doi.org/10.1038/cr.2010.19
7. Thottacherry JJ, Sathe M, Prabhakara C et al (2019) Spoiled for choice: diverse endocytic pathways function at the cell surface. Annu Rev Cell Dev Biol 35:55–84. https://doi.org/10.1146/annurev-cellbio-100617-062710
8. Schmid SL (2017) Reciprocal regulation of signaling and endocytosis: implications for the evolving cancer cell. J Cell Biol 216(9):2623–2632. https://doi.org/10.1083/jcb.201705017
9. Sneeggen M, Guadagno NA, Progida C (2020) Intracellular transport in cancer metabolic reprogramming. Front Cell Dev Biol 8:597608. https://doi.org/10.3389/fcell.2020.597608
10. Crupi MJ, Richardson DS, Mulligan LM (2015) Cell surface biotinylation of receptor tyrosine kinases to investigate intracellular trafficking. Methods Mol Biol 1233:91–102. https://doi.org/10.1007/978-1-4939-1789-1_9
11. Hörmann K, Stukalov A, Müller AC et al (2016) A surface biotinylation strategy for reproducible plasma membrane protein purification and tracking of genetic and drug-induced alterations. J Proteome Res 15(2):647–658. https://doi.org/10.1021/acs.jproteome.5b01066
12. Kuhn BT, Zimmermann I, Egloff P et al (2020) Biotinylation of membrane proteins for binder selections. Methods Mol Biol 2127:151–165. https://doi.org/10.1007/978-1-0716-0373-4_11
13. Rennick JJ, Johnston APR, Parton RG (2021) Key principles and methods for studying the endocytosis of biological and nanoparticle therapeutics. Nat Nanotechnol 16(3):266–276. https://doi.org/10.1038/s41565-021-00858-8
14. Brahimi-Adouane S, Bachet JB, Tabone-Eglinger S et al (2013) Effects of endoplasmic

reticulum stressors on maturation and signaling of hemizygous and heterozygous wild-type and mutant forms of KIT. Mol Oncol 7(3): 323–333. https://doi.org/10.1016/j.molonc.2012.10.008
15. Kao SH, Wang WL, Chen CY et al (2015) Analysis of protein stability by the cycloheximide chase assay. Bio Protoc 5(1):e1374. https://doi.org/10.21769/BioProtoc.1374
16. Dayekh K, Johnson-Obaseki S, Corsten M et al (2014) Monensin inhibits epidermal growth factor receptor trafficking and activation: synergistic cytotoxicity in combination with EGFR inhibitors. Mol Cancer Ther 13(11): 2559–2571. https://doi.org/10.1158/1535-7163.MCT-13-1086
17. Crupi MJF, Maritan SM, Reyes-Alvarez E et al (2020) GGA3-mediated recycling of the RET receptor tyrosine kinase contributes to cell migration and invasion. Oncogene 39(6): 1361–1377. https://doi.org/10.1038/s41388-019-1068-z

Chapter 16

Multiplexed Quantitative Proteomic Profiling of Cancer Cells and Tissues Using Isobaric Labeling-Based Tags

Mukhayyo Sultonova, Devanand M. Pinto, and J. Patrick Murphy

Abstract

Comparing cancer proteomes across many samples offers a window into cancer cell biology and may reveal new treatment options for specific subsets of cancer. Here we describe a method using tandem mass tag (TMT) technology to multiplex up to 18 samples in a single analysis, paving the way for the analysis of large cohorts of tumors, cell lines, and perturbations thereof. The procedure we describe will result in samples ready for in-depth LC-MS/MS analysis in 3–4 days.

Key words Proteomic profiling, Isobaric labeling, LC-MS/MS, SP3

1 Introduction

As Winston and Gregory Timp recently stated, "Proteins can be the root cause of a disease, and they can be used to cure it" [1]. Quantitative proteomics, the systematic analysis of all expressed proteins, has become an essential tool in cancer research, yielding both novel protein targets and therapeutic proteins. Quantifying differences in protein abundance may reveal differences between healthy and cancerous tissues, response to drug treatment, and targetable biological phenotypes across cancers [2]. Cancer cell phenotype is often more directly related to the proteome than the genome due to the dynamics in protein posttranslational modifications, protein–protein, or protein–small molecule interactions [3–8].

Advances in mass spectrometry instrumentation have enabled a deeper mapping of the human proteome and, when coupled with sample multiplexing technology, large cohorts of cancer cells and tissues can be analyzed in order to reveal robust biomarkers panels and novel therapeutic targets [9, 10]. The most common approaches in quantitative proteomics utilize two-dimensional high-performance liquid chromatography (HPLC)-tandem mass spectrometry (2DLC-MS/MS) of complex peptide mixtures

Sherri L. Christian (ed.), *Cancer Cell Biology: Methods and Protocols*,
Methods in Molecular Biology, vol. 2508, https://doi.org/10.1007/978-1-0716-2376-3_16,

resulting from proteolytic digestion of the proteome [11, 12]. Two-dimensional HPLC analysis greatly increases the number of peptides that can be interrogated, allowing for comprehensive analysis of the cancer cell proteome. However, this increased depth comes at the expense of long analysis times. Sample multiplexing using isobaric tagging-based reagents, such as tandem mass tags (TMT), is thus highly beneficial and allows the deep proteomic analysis of multiple cancer proteomes that is needed in order to study complex cancer system. All TMT reagents within its set have the same nominal mass and chemical structure; however, each tag is unique in the localization of heavy isotopes in its structure [13]. The chemical structure of TMT reagents consists of three parts: a reporter ion, a mass normalization, and an amine-reactive group. The N-termini and Lysine-amino groups of the digested peptides from a sample gets labeled by the amine-reactive group of the TMT reagent. When examining the MS1 scan, the labeled peptides appear as a single peak with the same m/z value. The reporter ion peaks corresponding to each of the TMT reagents appear during the subsequent MS2 or MS3 scans, allowing quantification across samples. Current isobaric labeling approaches allow for the simultaneous analysis of up to 18 proteomes [13, 14].

In this chapter, we describe an economical and streamlined workflow for using tandem mass tag (TMT) reagents for isobaric labeling-based multiplexed quantitative proteomics of small amounts of cancer cell protein mixtures. This protocol can routinely achieve 7000–8000 proteins across up to 18 cancer cell or tissue samples in 3–4 days (as outlined in Fig. 1).

2 Materials

2.1 Lysate Preparation

1. RIPA Buffer: 50 mM Tris–HCl (pH 7.6), 100 mM NaCl, 0.1% SDS, 1% Triton ×100, 0.5% Sodium deoxycholate, complete Mini protease inhibitor cocktail tablet (*see* **Note 1** and **2**).
2. Sterile Phosphate-Buffered Saline (PBS): 1×, Without Calcium and Magnesium (e.g., Fisher Scientific, cat. no. MT21040CV).

2.2 Sample Preparation for Mass Spectrometry

1. Bovine Serum Albumin (BSA).
2. Bicinchoninic acid (BCA) Protein Assay Kit (e.g., Pierce™, cat. no. CAPI23227).
3. 500 mM Dithiothreitol (DTT).
4. 500 mM Iodoacetamide (IAA).
5. Sera-Mag SpeedBeads (50 mg/mL) (GE Healthcare, cat. no. 65152105050250).
6. 100% Ethyl alcohol (non-denatured).
7. 80% Ethyl alcohol (diluted from 2.2.6 with Milli-Q water).

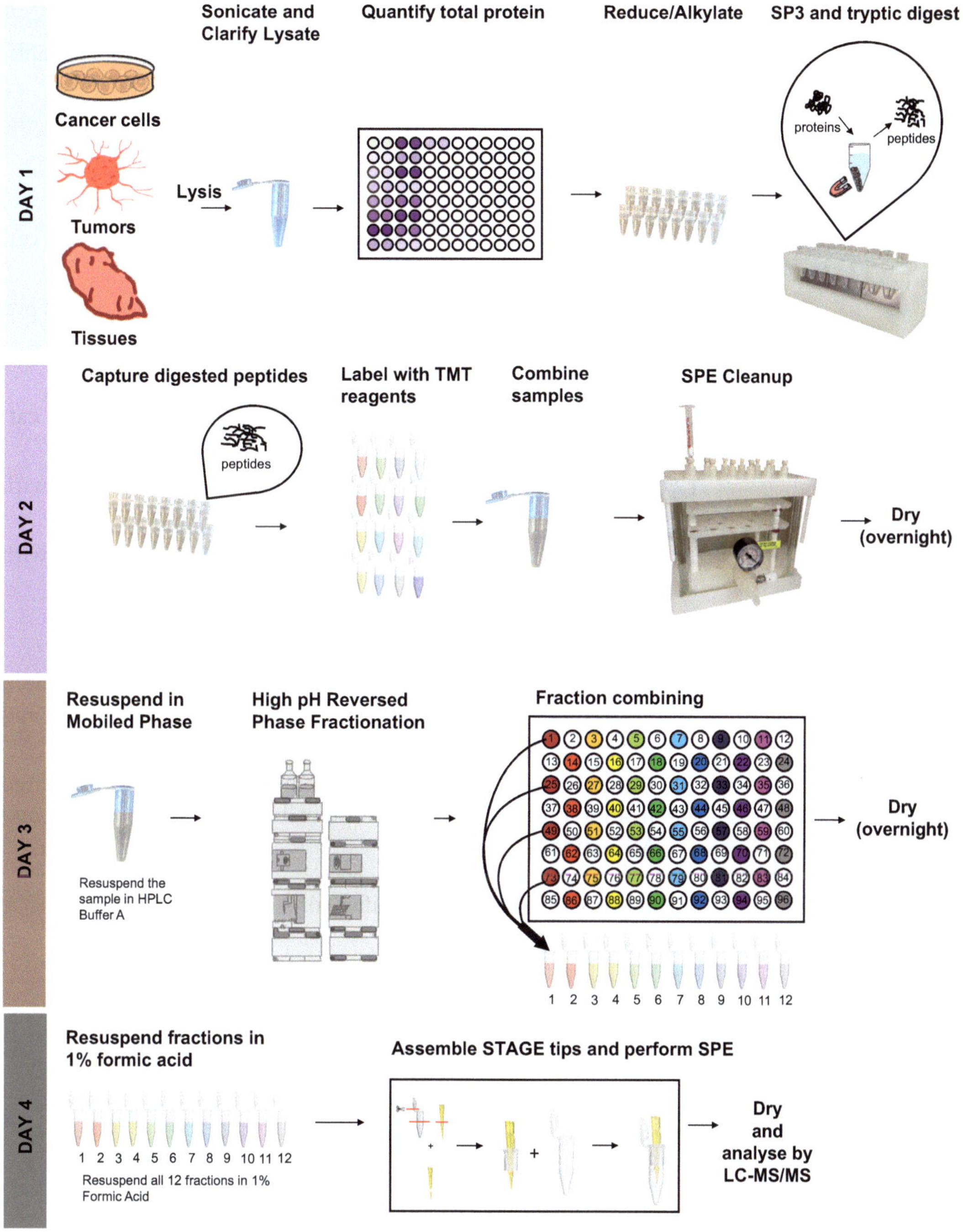

Fig. 1 Workflow for the isobaric tagging of cancer proteome samples. Samples are first lysed, typically in a detergent-containing buffer and then sheared by sonication and assayed for total protein content using a BCA assay. An aliquot containing 25 ug of protein is then selected, detergents are removed using an SP3 bead cleanup, and then digested overnight. The next day, peptides are labeled with TMT reagent and desalted using SPE. The following day, peptides are separated using high pH reversed phase chromatography and fractions are concatenated to 12 samples. Samples are then dried overnight and then desalted using STAGE tips and analyzed by LC-MS/MS

8. Trypsin (Promega, sequencing grade, 20 μg).
9. 200 mM *N*-(2-Hydroxyethyl)piperazine-N′-(3-propanesulfonic acid) (EPPS).
10. MS-grade acetonitrile (ACN).
11. MS-grade methanol.
12. MS-grade water.
13. Tandem Mass Tag™ reagent set (Thermo Scientific, cat. no. A44521) (*see* **Note 3**).
14. Anhydrous ACN ≥99.9% (AcroSeal® SC): for TMTpro reagent resuspension.
15. 5% (w/v) Hydroxylamine (prepared from powder).
16. MS-grade formic acid, concentrated (Fisher Scientific, cat. no. A11750).
17. 50% MS-grade ACN, 1% MS-grade formic acid (diluted with MS-grade water).
18. 70% MS-grade ACN, 1% MS-grade formic acid (diluted with MS-grade water).
19. 1% MS-grade formic acid (in MS-grade water).

2.3 High pH Reversed Phase HPLC

1. Buffer A: 5% ACN, 10 mM ammonium bicarbonate (prepared from a 1 M stock of ammonium bicarbonate (in water) and MS-grade ACN).
2. Buffer B: 90% ACN, 10 mM ammonium bicarbonate (prepared from a 1 M stock of ammonium bicarbonate (in water) and MS-grade ACN).

2.4 Equipment

1. Branson SLPe Digital Sonifier (or similar probe sonifier).
2. Centrifuge.
3. Vortex mixer.
4. Absorbance spectrophotometer.
5. Incubator (37 °C)
6. Magnetic rack for PCR tubes.
7. 16-gauge blunt needle.
8. Vacuum manifold.
9. Sep-Pak Vac 1 cc 100 mg tC18 Cartridges (Waters).
10. Vacuum concentrator (Speed-Vac).
11. Empore™ solid phases extraction (SPE) Disks C18, diam. 47 mm.
12. 250 uL glass conical insert with bottom spring (Canadian Life Science, cat. no. I025BS-629).

13. HPLC system (e.g., Agilent 1260 Infinity Quaternary LC System) equipped with a diode array detector (DAD) and automated fraction collector.
14. Agilent 300 Extend, 2.1 × 100 mm, 3.5-micron C18 column.
15. Orbitrap Fusion series mass spectrometer.
16. Accucore C18 resin column (2.6 μm, 150 Å, ThermoFisher Scientific).

3 Methods

3.1 Tissue/Cell Lysis

1. Rinse cells or tissues in PBS, prechilled to 4 °C.
2. For cells or tissues, lyse the sample with RIPA buffer prechilled to 4 °C (*see* **Note 4** and **5**).
3. Transfer the lysate suspension to 1.5 mL Eppendorf tubes.
4. Set the sonifier to a continuous mode at 20 amplitude. Sonicate the lysate three times for 20 s, resting for 1 min on ice between cycles.
5. Centrifuge the lysate at 20,000 × g for 15 min. Collect the supernatant and store on ice (*see* **Note 6**).

3.2 Total Protein Quantification (see Note 7)

1. Prepare a 10 mg/mL stock of BSA and dilute with 1× PBS to yield 2000 μg/mL concentration.
2. Serially dilute 2000 μg/mL BSA with 1× PBS to yield a total of 6 different BSA stocks: 2000 μg/mL, 1000 μg/mL, 500 μg/mL, 250 μg/mL, 125 μg/mL, and 62.5 μg/mL. 1× PBS can be used as a control (0 μg/mL).
3. In duplicate, load 20 μL of each BSA stock and lysate sample into 96-well plate wells in duplicates. Add 220 μL of BCA reagent to each well and mix thoroughly.
4. Incubate the plate in the incubator at 37 °C for 30 min.
5. Using an absorbance spectrophotometer, measure the absorbance at 595 nm and determine protein concentration of each lysate sample using the BSA standard curve.

3.3 Reduction and Alkylation of Proteins

1. Aliquot 25 μg of lysate samples into PCR tubes or strips (*see* **Note 8** and **9**).
2. Add sufficient 500 mM DTT to each sample so that final concentration of DTT is 5 mM. Mix thoroughly and incubate on the benchtop at room temperature for 30 min.
3. Add sufficient 500 mM iodoacetamide so that final concentration of iodoacetamide is12 mM (*see* **Note 10**). Mix thoroughly and incubate at room temperature in the dark for 30 min.

3.4 SP3 Cleanup of Protein Samples [15] and Tryptic Digestion (see Note 11)

1. Rinse the beads by first transferring the required amount of the bead stock into a PCR tube to process 25 μg of protein per sample as per manufacturer's instructions. A bead to protein ratio of 10:1 (wt/wt) is recommended (250 μg beads per sample = 4 mg per 16-plex experiment) (*see* **Note 12**). Place the tube on a magnetic rack and wait until the beads settle to the side of the tube. Remove and discard the supernatant. Remove the magnet, resuspend the beads in 200 μL of Milli-Q water, and mix well by pipetting up and down. Place the tubes back on the magnetic rack and allow beads to settle to the side of the tube. Remove and discard the supernatant. Resuspend the beads in the original volume of Milli-Q water for a stock of 50 mg/mL beads and then aliquot to PCR tubes (25 μg per tube).
2. Transfer 25 μg of the lysate sample into the PCR tube with beads and add 100% ethyl alcohol so that final concentration of ethanol in the sample is 50% (vol/vol) (*see* **Note 13**).
3. Shake tubes on a PCR tube-compatible shaker at 1000 rpm for 10 min.
4. Place the tubes on a magnetic rack and allow the beads settle to the side of the tube. Remove and discard the supernatant.
5. Remove the beads from the magnet, resuspend the beads in 200 μL of 80% ethyl alcohol, and mix well by gently pipetting up and down. On a magnetic rack, wait until the beads settle to the side of the tube. Remove and discard the supernatant. Repeat two more times for a total of three rinses (*see* **Note 14**).
6. Prepare trypsin stocks with concentrations 0.25 μg/μL and 25 ng/μL in 200 mM EPPS.
7. Add 10 μL of 0.25 ng/μL trypsin stock onto the beads and mix well. Incubate in the incubator overnight at 37 °C.
8. The next day, add 1 μL of 25 μg/μL trypsin stock onto the beads and mix well. Incubate for 2 h in the incubator at 37 °C. On a magnetic rack, wait until the beads settle to the side of the tube. Collect the supernatant into a new tube.
9. Remove the magnet, add 5 μL of 200 mM EPPS to the beads, and mix well. Place the tube on a magnetic rack, and wait until the beads settle to the side of the tube. Collect the supernatant. Repeat as necessary to achieve a volume of 17.5 μL.

3.5 TMT Labeling of Peptides

1. Add 7.5 μL of MS-grade ACN to the peptides so that the final concentration of acetonitrile is >30%.
2. Resuspend and aliquot the TMT reagents in anhydrous ACN (*see* **Note 15**).
3. Add peptides to each unique, prealiquoted TMT tag. Incubate on the benchtop at room temperature for 1 h.

4. Quench the reaction by adding 5 μL of 5% hydroxylamine to each sample. Incubate on the benchtop at room temperature for 10 min.
5. Combine all 16 samples in a 1.5 mL Eppendorf tube.

3.6 Cleanup of Sample Using SepPak Cartridge

1. Acidify the sample and dilute organic solvent by adding 500 μL of 1% formic acid. The pH of the mixture should be <3 (check with pH strips).
2. Place a 100 mg tC18 Sep-Pak cartridge on a vacuum manifold and a 15 mL waste tube under the cartridge inside the manifold. Load 500 μL MS-grade methanol into the cartridge. Let it through the cartridge slowly at the flow rate of around 1 drop per 5 s. Use suction if needed (*see* **Note 16** and **17**).
3. Load 1 mL of 50% ACN, 1% formic acid, to the cartridge and allow to flow through to waste using suction if needed. Try not to let the resin dry out.
4. Load 1 mL of 1% formic acid and flow through the cartridge keeping the resin wet. *Repeat once.*
5. Load the sample and let it through the cartridge to waste.
6. Load 1 mL of 1% formic acid and let it though the cartridge to waste. *Repeat once.*
7. Remove and discard the waste tube and place a 1.5 mL Eppendorf tube under the cartridge for collection.
8. Elute peptides by loading 1 mL 50% ACN, 1% formic acid onto the cartridge and allow it to flow through into the Eppendorf tube.
9. Further elute peptides by loading 500 μL of 70% ACN, 1% formic, and allow it to flow through into the Eppendorf tube.
10. Remove the Eppendorf from the vacuum manifold and dry the eluted peptides in a vacuum concentrator overnight.

3.7 Fractionation of Peptides

1. Resuspend the peptides in 90 μL of Buffer A.
2. Set the sonifier to a continuous mode at 20 amplitude. Sonicate twice for 12 s with resting for 1 min on ice in between the sonication cycles.
3. Centrifuge at 10,000 × G for 5 min, and transfer the supernatant into an HPLC vial.
4. Condition and set up the HPLC to monitor peptides at 214 nm using the diode array detector.
5. Inject 80 μL of the peptides onto an Agilent 300 Extend, 2.1 mm × 100 mm, 3.5 μm C18 column in starting conditions of 100% Buffer A (*see* **Note 18**).
6. Elute peptides using an elution gradient of 5–40% Buffer B at a flow rate of 0.25 mL/min.

7. Collect 96 peptide fractions over the peptide elution period.
8. Combine the 96 fractions into 12 fractions, based on the schematic in Fig. 1.
9. Dry the fractions in a vacuum concentrator until dry (typically overnight).

3.8 Purification of Peptides Using Stage Tips

1. Place C18 Empore™ Extraction Disk on a petri dish. With a blunt-ended needle, gently press the needle into the disk. When lifting the needle, ensure that the small piece of disk remains inside the needle. For a typical set of samples of 25 μg, use 5 "cookies" (*see* **Note 20**) per fraction. Place the needle into a p200 pipette tip and push the disk pieces into the tip with a straightened paper clip. Use the paper clip to gently pack the disks into the end of the pipette tip. Assemble the Stage tip based on the schematic in Fig. 1 (*see* **Note 21**) [16].
2. Resuspend the 12 dried peptide fractions in 100 μL of 1% formic acid.
3. Place 12 assembled stage tips into the centrifuge (one for each fraction).
4. Load 50 uL of MS-grade methanol into each stage tip and spin them for 3 min, at 650 × G. Aim to keep the resin from drying completely.
5. Load 100 uL of 50% ACN, 1% formic acid into each stage tip, and spin for 3 min at 650 × G.
6. Load 100 uL of 1% formic acid into each stage tip and spin them for 3 min at 650 × G. *Repeat once.*
7. Empty the 1.5 mL Eppendorf tubes of the Stage Tip apparatuses.
8. Load all 12 fractions into a separate Stage Tip and spin them for 3 min, keeping the resin wet.
9. Load 100 uL of 1% formic acid into each stage tip and spin them for 3 min at 650 × G. *Repeat once.*
10. Empty the 1.5 mL Eppendorf tubes of the stage tip apparatuses. Put a glass conical insert into each of the 1.5 mL Eppendorf tube of the stage tip apparatus.
11. Load 50 uL of 50% ACN, 1% formic acid into each stage tip and spin for 3 min at 650 × G, allowing peptides to elute directly into the conical insert.
12. Load 30 uL of 70% ACN, 1% formic acid into each stage tip and spin for 3 min at 650 × G, allowing peptides to elute directly into the conical insert. Allow resin to dry completely.
13. Remove the glass conical inserts from the stage tip apparatuses and put them into new 1.5 mL Eppendorf tubes.
14. Dry all 12 fractions in vacuum concentrator overnight.

3.9 Mass Spectrometry Analysis

The protocol for mass spectrometry analysis will differ, depending on the instrument used. Here we described the details when using Orbitrap Lumos mass spectrometer (ThermoFisher Scientific) coupled to a Proxeon EASY-nLC pump (ThermoFisher Scientific).

1. Resuspend each fraction in 5 μL of 5% acetonitrile/0.1% formic acid. Inject 2 μL of each sample for LC-MS/MS analysis.
2. Separate peptides on a 100 μm inner diameter microcapillary column packed in-house with a 35 cm of Accucore C18 resin (2.6 μm, 150 Å). Use a 90 min gradient (0–2 min: 3–8% acetonitrile with 0.1% formic acid, 2–5 min: 8–10%, 5–70 min: 10–25%, 70–80 min: 25–30%, 80–85 min: 30–100%, 85–90 min: 100%) at a flow rate of 575 nL/min.
3. Collect full MS1 scans using the Orbitrap at 120,000 resolution with maximum injection time 50 ms, and AGC 4e5.
4. Perform MS2 scans in the ion trap with CID fragmentation using an isolation window of 0.7 Da, 35% NCE, maximum injection time 35 ms, and AGC 2×10^4 (*see* **Note 22**).
5. (*Optional*) Trigger MS3 scans on the MS scans for increasing accuracy. Collect MS3 scans in the Orbitrap using a resolution of 50,000, NCE of 45%, maximum injection time of 86 ms, and AGC of 1.5×10^5.

3.10 Data Analysis

1. The details of the data analysis will be dependent on the search software used (*see* **Note 23**). We recommend MSFragger FragPipe [17] developed by the Nesvizhskii group. Some of these parameters such as digestion enzyme, variable and fixed peptide modifications, and peak mass tolerances may be similar, regardless of the algorithm used. A specific workflow for data filtering and conversion should be considered and will vary depending on the search software used. Specify trypsin as the proteolytic enzyme and allow, at most, two missed cleavages. Set fixed modification on lysine and N-termini of peptides to 304.2071 Da for 16-plex reagents or 229.1629 for 10-plex reagents and at 57.0215 Da on cysteines. Set a variable modification on methionine at 15.9949 Da. Set the precursor mass tolerance to ±10 ppm and fragment mass tolerance to 0.02 Da for high-resolution data (0.8 for low-resolution data).
2. For protein sequence database, an up-to-date protein FASTA sequence file from UniProt with common contaminants and decoys added should be used. Peptide and protein false discovery rate should be filtered to <1% using the desired software tool of choice (Protein Prophet is implemented in Fragpipe/MSFrgger).

3. The resulting quantitative proteome dataset can be analyzed using R Studio or Excel to normalize, manipulate, and visualize the data.

4 Notes

1. Alternative buffers may be used to lyse the cells for protein extraction. These include NP-40 lysis-based buffer, RIPA buffer, Urea/thiourea lysis buffer, and others [18]. Choosing the right one for the experiment will depend on the source of the cells. In our protocols, we have successfully used both RIPA and NP-40-based lysis buffers both on cells and tissues.
2. It is common to add a cocktail of protease inhibitors into the lysis buffer when harvesting and preparing cell lysates. When cells are lysed, proteins are vulnerable to various types of endogenous proteases. It is recommended to always keep the lysate on ice due to temperature sensitivity of protease inhibitors.
3. There are several versions of tandem mass tags that can be used, depending on how many different samples need to be analyzed in one experiment. These include 6-plex, 10-plex, 11-plex, and 16-plex. Recently, TMT multiplexing was expanded to TMTPro-18plex with addition of two new tags, 134C and 135N [14]. In our experiments, we have successfully used 10-plex and 16-plex reagents.
4. During cell culture, grow cells to a maximum of 90% confluency. We recommend rinsing cells at least 3 times with cold 1× PBS before lysis to remove FBS in the media. We also recommend using minimal lysis buffer to keep the protein in lysate concentrated. For example, use 500 μL of RIPA buffer for a 10 cm plate of cells.
5. For tissue lysis, a Bead Ruptor Elite Bead Mill Homogenizer (OMNI international) or equivalent device may be used to homogenize tissue lysates. Transfer about 20 mg of tissue into a tube containing beads and resuspend it in 500 μL of lysis buffer. Homogenize at 5 m/s for 20 s for three cycles with 30 s rest on ice between cycles. This step can also be done by hand and crushing the tissue with lysis buffer in an Eppendorf tube.
6. In some cases, the lysate might need to be centrifuged multiple times. Depending on harvesting methods and the type of lysate, insoluble parts might still be suspended and floating in the supernatant. This may be encountered when processing tissues with high fat content. In this case, transfer the supernatant and centrifuge again until the lysate is clear.

7. In addition to the BCA assay, there are several tests that can be used to determine protein concentration in your lysate. These include Biuret method, Lowry assay, Bradford assay, and UV absorption [19]. Regardless of the assay chosen, it should provide an accurate estimate of protein concentration in the lysate for accurate calculation of the amount of trypsin, SP3 beads, and TMT reagent to use.
8. This procedure uses PCR tubes as they are compatible with in-house designed magnetic racks for PCR tubes. Additionally, this allows work with small amounts of sample.
9. In our laboratory, we use 25 μg as a standard amount of lysate for SP3 and scale the rest of the experimental amounts accordingly. However, this can easily be scaled up or down, depending on how much sample is desired.
10. Iodoacetamide is light sensitive, and the stock solution should be prepared fresh.
11. Several suitable proteome sample preparation approaches, such as in-solution digestion and iST, FASP, and other affinity-based methods, can be used instead of SP3. Hughes et al. [15] compare the benefits and limitations of these methods. SP3 works well with small amounts of sample with minimal protein loss, and protein digestion efficiency is high.
12. In our experiments, we process up to 16 samples of 25 μg. To do this, we take out a total of 90 μL of 50 mg/ml beads from the stock and aliquot them into two tubes containing 45 μL each. After rinsing each with 200 μL of water, we combine both mixtures of beads back into 90 μL of water: 5 μL aliquots are the placed into 16 PCR tubes. Then, we add 25 μg of sample (made to 50% ethyl alcohol) into its corresponding PCR tube containing 5 μL of beads.
13. Besides ethanol, alternative solvents can also be used such as acetonitrile, acetone, isopropanol, trifluoroethanol, and xylene [15]. Regardless of solvent chosen, for optimal binding performance, it is recommended to have final organic solvent concentration ~50%. Ethanol is the most cost-effective and nontoxic out of the solvents listed above while still maintaining good performance.
14. It is recommended to remove the PCR tubes from the magnet between washes for the most thorough removal of contaminants and detergents from the sample.
15. Resuspend each of the TMT tags in 40–44 μL anhydrous ACN. For 10-plex reagents (0.8 mg/vial), aliquot 2.5 μL of each tag into separate tubes. For 16-plex (0.5 mg/vial), aliquot 5 μL of each tag into separate tubes. These amounts are sufficient to label 25 μg of peptides per sample to 97–99% completion. Store aliquots at −80 °C until ready for use.

16. Vacuum manifold is not essential. SepPak can be done under gravity; however, it may take longer. A vacuum manifold allows completion of a SepPak protocol in under an hour.
17. The flow rate should not be too fast especially when during the loading of your sample. There should be enough contact time between the sample and the sorbent. Adjust the vacuum as needed to get a slow and steady flow rate. (~1 drop per 5 s).
18. Other types of C18 columns can be used here. For us, Agilent 300 Extend, 2.1 × 100 mm, 3.5-micron C18 column provides sufficient separation and low overlap in peptide identifications between adjacent fractions.
19. The protocol for HPLC should be adjusted depending on which TMT reagents are used and whether your TMT reagents contain isotopes. Depending on the use of 10-plex or 16-plex reagents, fraction collection times should be adjusted to optimize even fractionation. We set it 15–55 min on our instrument for a16-plex experiment.
20. The technique is similar to using a "cookie cutter," making 1 flat circle (cookie) and continuously packing more circles into the needle. When the desired depth is achieved (~5 cookies), push the cookies into a p200 pipette tip.
21. For one Stage Tip apparatus, you will need one 0.6 mL Eppendorf tube, one 1.5 mL Eppendorf tube, two p200 pipette tips, and C18 Empore™ Extraction Disk. Cut off the cap and the bottom of the 0.6 mL Eppendorf tube as shown in Fig. 1 (red line). Also cut off the bottom of one of the p200 pipette tips as shown in Fig. 1. Pack the extraction disks into the other p200 pipette tip. Then place the pipette tip with extraction disk into the other pipette. Then, insert both into the 0.6 mL tube. Then, the whole thing goes into the 1.5 mL tube. Remember, you load all the buffers and solutions into the tip that has the disks. The 1.5 mL tube is there to be a catching tube for the waste. The other pipette tip and 0.6 mL tube will allow the pipette tip with disks to sit properly on the catching tube.
22. Depending on the desired accuracy of the mass spectrometry data, MS2 or MS3 scans that include synchronous precursor selection (SPS-MS3) can be triggered based on the MS1 scan as previously described [13].
23. Several alternative search tools and algorithms can be used such as Proteome Discoverer, MaxQuant, OpenMS, and others.

References

1. Timp W, Timp G (2020) Beyond mass spectrometry, the next step in proteomics. Sci Adv 6(2):eaax8978
2. Srinivas PR et al (2001) Proteomics in early detection of cancer. Clin Chem 47(10): 1901–1911
3. Aebersold R et al (2018) How many human proteoforms are there? Nat Chem Biol 14(3): 206–214
4. Smith LM et al (2013) Proteoform: a single term describing protein complexity. Nat Methods 10(3):186–187
5. Mertins P et al (2016) Proteogenomics connects somatic mutations to signalling in breast cancer. Nature 534(7605):55–62
6. Zhang B et al (2014) Proteogenomic characterization of human colon and rectal cancer. Nature 513:382–387
7. Zhang H et al (2016) Integrated proteogenomic characterization of human high-grade serous ovarian cancer. Cell 166(3): 755–765
8. Sinha A et al (2019) The proteogenomic landscape of curable prostate cancer. Cancer Cell 35(3):414–427
9. Nusinow DP et al (2020) Quantitative proteomics of the cancer cell line encyclopedia. Cell 180(2):387–402
10. Whiteaker JR et al (2014) CPTAC Assay Portal: a repository of targeted proteomic assays. Nat Methods 11(7):703–704
11. Patterson SD, Aebersold R (2003) Proteomics: the first decade and beyond. Nat Genet 33(3): 311–323
12. Macklin A, Khan S, Kislinger T (2020) Recent advances in mass spectrometry based clinical proteomics: applications to cancer research. Clin Proteomics 17:1–25
13. Li J et al (2020) TMTpro reagents: a set of isobaric labeling mass tags enables simultaneous proteome-wide measurements across 16 samples. Nat Methods 17(4):399–404
14. Li J et al (2021) TMTpro-18plex: the expanded and complete set of TMTpro reagents for sample multiplexing. J Proteome Res 20(5):2964–2972
15. Hughes CS et al (2019) Single-pot, solid-phase-enhanced sample preparation for proteomics experiments. Nat Protoc 14(1): 68–85
16. Rappsilber J et al (2003) Stop and go extraction tips for matrix-assisted laser desorption/ ionization, nanoelectrospray, and LC/MS sample pretreatment in proteomics. Anal Chem 75(3):663–670
17. Kong AT et al (2017) MSFragger: ultrafast and comprehensive peptide identification in mass spectrometry–based proteomics. Nat Methods 14(5):513–520
18. Peach M, Marsh N, MacPhee DJ (2012) Protein solubilization: attend to the choice of lysis buffer. Protein electrophoresis. Humana Press, Totowa, pp 37–47
19. Olson BJSC, Markwell J (2007) Assays for determination of protein concentration. Curr Protoc Protein Sci 48(1):3–4

Chapter 17

The Use of Seahorse XF Assays to Interrogate Real-Time Energy Metabolism in Cancer Cell Lines

Jenna K. Caines, David A. Barnes, and Mark D. Berry

Abstract

The propensity of cancer cells to preferentially undergo anaerobic metabolism despite oxygen being abundant is referred to as the Warburg effect. Measuring cellular metabolism is therefore central to understanding the cellular physiology of cancer cells. The Seahorse XFe Analyzer series allows real-time measurement of cellular metabolism. In the basic assay, two parameters, the oxygen consumption rate (OCR) and the extracellular acidification rate (ECAR), are used to determine real-time changes in the energy needs of live cells: OCR provides a measure of aerobic mitochondrial respiration; ECAR gives a measure of anaerobic glycolysis. Through the use of various respiration inhibitors, the Seahorse assay allows baseline respiration rate and total aerobic and anaerobic ATP production to be determined under a variety of experimental conditions. Here we describe the protocol for completing the Seahorse Real-Time ATP Rate Assay for adherent and suspension cancer cell lines. Depending on individual experimental results, more refined subsequent assays can then be performed to specifically determine, for example, the ability to utilize different substrates by the cell lines in the presence and absence of pharmacological and/or genetic interventions.

Key words ATP, Cancer cell metabolism, Energy metabolism, Glycolysis, Mitochondria, Seahorse

1 Introduction

The process by which normal cells develop into a tumorigenic state has long been established and described in terms of core hallmarks of cancer development. The six original hallmarks, namely, resisting cell death, sustaining proliferative signaling, evading growth suppressors, inducing angiogenesis, enabling replicative immortality, and activating metastasis, comprise a considerably complex series of changes that are ultimately enabled by alterations in cellular energy metabolism [1, 2]. The Warburg effect describes the propensity by which cancer cells preferentially convert glucose to lactate,

Jenna K. Caines and David A. Barnes contributed equally.

Sherri L. Christian (ed.), *Cancer Cell Biology: Methods and Protocols*,
Methods in Molecular Biology, vol. 2508, https://doi.org/10.1007/978-1-0716-2376-3_17,

undergoing anaerobic metabolism, even in the presence of abundant oxygen, sometimes described as aerobic glycolysis [3, 4]. Such metabolic reprogramming is therefore a critical component in cancer cell development and maintenance [2]. Studying the Warburg effect, and more generally cancer cell metabolism and the importance it holds in supporting the anabolic processes associated with cancer cell growth and proliferation, has been enhanced by recent methods for precisely measuring cellular metabolism.

The Seahorse XF instrumentation (Agilent Technologies, Santa Clara, California, USA) is a highly robust and sensitive piece of equipment that provides a series of simple, label-free, real-time assays that allow direct measurement of the underlying metabolic functions driving cellular activation, proliferation, signaling, and toxicity. The basic parameters measured in a Seahorse assay are the oxygen consumption rate (OCR) and the extracellular acidification rate (ECAR) (Fig. 1) [5]. OCR data measures mitochondrial oxidative phosphorylation, as oxygen is expended as the final electron acceptor during aerobic metabolism. The ECAR essentially measures the production of pyruvate and its subsequent conversion to lactate (lactic acid) and hence the rate of glycolytic metabolism. This pyruvate to lactate conversion occurs under anaerobic conditions to regenerate NAD^+ and allow continued glycolytic ATP production. Through the use of oligomycin (an ATP synthetase (complex V) inhibitor) and rotenone/antimycin A (respiratory chain complex I and complex III inhibitors) during assays, the Seahorse XF software interpolates separate measures of ATP production via glycolysis and the mitochondrial respiratory chain [6].

In addition to the ATP Rate Assay (above), a wide range of further assays are available for a more detailed study of the energy metabolism profile of cells under defined experimental conditions of interest. For example, a more detailed examination of oxidative phosphorylation is available with the Cell Mito Stress Test, which uses FCCP (carbonyl cyanide-*p*-trifluoromethoxyphenylhydrazone), an uncoupler of oxidative phosphorylation, in combination with the aforementioned oligomycin and rotenone/antimycin A. Including FCCP allows the user to assess the maximal respiration of the cell line of interest. When combined with the observed basal respiration, this allows the user to determine the ability of a cell to meet an energetic challenge such as activation, differentiation, or proliferation [7]. Furthermore, the Cell Mito Stress Test can be readily modified by including inhibitors prior to proceeding with the conventional assay to prevent various pathways from fueling oxidative phosphorylation and observing how the basal and maximal respiration responds [8]. For example, UK5099, an inhibitor of the mitochondrial pyruvate carrier, can be included to prevent glycolysis from feeding into oxidative phosphorylation. Another option could be to include etomoxir, an inhibitor of carnitine palmitoyltransferase-1, to prevent fatty acid transfer into the

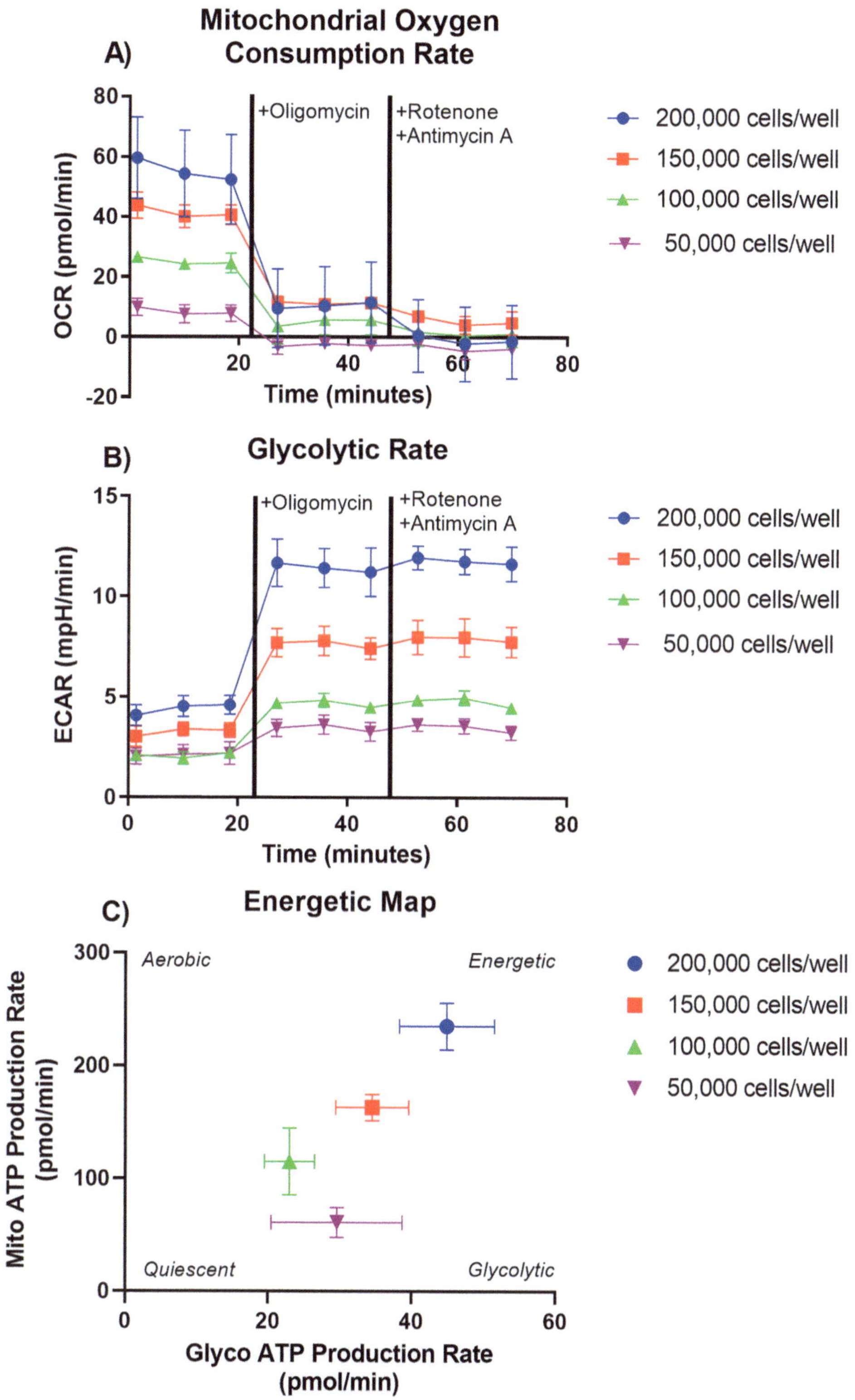

Fig. 1 Cell density optimization for THP-1 monocytes, showing an example of the readout obtained during a Seahorse XFe24 Analyzer Real-Time ATP Rate Assay. The measurements taken during the time course of the assay illustrate (**a**) OCR rates representative of mitochondrial oxidative phosphorylation metabolism and (**b**) ECAR rates representative of glycolytic metabolism. (**c**) Quiescent-active plot representative of the energy profile of the cells

mitochondria. Etomoxir thus prevents fatty acid oxidation from fueling oxidative phosphorylation. Finally, the user could utilize BPTES (*N*,*N*′-[thiobis(2,1-ethanediyl-1,3,4-thiadiazole-5,2-diyl)]bisbenzeneacetamide) to inhibit glutaminase 1, preventing glutamine from being incorporated into the citric acid cycle, another way that the cell fuels oxidative phosphorylation [8]. The use of these three metabolic inhibitors to assess the influence on basal and maximal respiration is the basis of the manufacturer's Substrate Oxidation Stress Test and allows for detailed elucidation of the interplay between metabolic pathways in the cells and conditions of choice [8].

Together, the many assays in the Seahorse repertoire, including the Cell Mito Stress Test and Substrate Oxidation Stress Test, provide an excellent tool for identifying the specific aspects of cellular energy metabolism that are altered in cancer cells under experimental conditions of interest. In each case, the assays are built upon the Real-Time ATP Rate Assay, the assay for which we will provide a step-by-step protocol in the following sections, as a gateway to projects and assessing the metabolic phenotype of cancer cells under user-defined conditions.

2 Materials

1. Cancer cells of choice.
2. Non-CO_2 incubator.
3. Centrifuge with a microplate rotor.
4. Seahorse XFe cartridge (Agilent, 102340-100).
5. Seahorse XF calibrant solution (Agilent, 100840-000).
6. Seahorse XFe24 V7 PS Cell Culture Microplates (Agilent, 100777-004).
7. Poly-L-lysine hydrobromide (Sigma Aldrich, P1399).
8. Seahorse assay medium: Mix 97 mL of Seahorse XF RPMI-based medium: 1 mM HEPES, pH 7.4 (Agilent, 103576-100) or Seahorse XF DMEM-based medium: 5 mM HEPES, pH 7.4 (Agilent, 103575-100) (*see* **Note 1**) with 1 mL of Seahorse XF 1.0 M glucose solution (Agilent, 103577-100), 1 mL of Seahorse XF 100 mM pyruvate solution (Agilent, 103578-100), and 1 mL of Seahorse XF 200 mM glutamine solution (Agilent, 103579-100).
9. Seahorse XF Real-Time ATP Rate Assay Kit (containing oligomycin and rotenone/antimycin A) (Agilent, 103592-100).

3 Methods

Carry out all procedures at room temperature in a sterile environment, unless otherwise specified.

3.1 Real-Time ATP Rate Assay with Suspension Cells

1. Culture suspension cells of choice as described in Chapter 2.
2. Hydrate one Seahorse XFe cartridge with 1 mL of XFe calibrant per well, 24 h prior to the start of the assay. To do this, remove the sensor cartridge and place it upside down next to the utility plate, then remove the hydro booster (Fig. 2), and gently add XF calibrant along the side of the wells. Return the sensor cartridge and hydro booster and store the cartridge in a 37 °C, non-CO_2 incubator (*see* **Note 2**).
3. Coat each well in the XFe microplate with 50 μL of 0.1 g/mL poly-L-lysine hydrobromide. Incubate the plate at room temperature for 1 h and then carefully aspirate poly-L-lysine solution from wells, taking particular care not to scratch the surface of the microplate. Wrap the plate in cling-film to seal and store at 4 °C for up to 1 week.
4. If applicable, pretreat the cells under study with pharmacological agents for the required time period. We find that plating the suspension cells in a regular 24-well cell culture plate at the desired density (*see* **Note 3**), in 500 μL of regular cell culture medium, and treating with 5 μL of the pharmacological agent works well. Incubate cells as normal at 37 °C in a CO_2 incubator.
5. Prepare the assay medium by mixing 48.5 mL of Seahorse XFe RPMI (or DMEM) medium, 0.5 mL of Seahorse XFe 1.0 M glucose solution, 0.5 mL of Seahorse XFe 100 mM pyruvate solution, and 0.5 mL of Seahorse XFe 200 mM glutamine solution in a graduated cylinder (*see* **Note 4**). Pour into a cell culture medium bottle and warm to 37 °C.
6. Exchange the growth medium for Seahorse assay medium and plate the cells in the Seahorse microplate as follows. Collect cells from each well into an appropriately labeled 1.5 mL microcentrifuge tube. Centrifuge the cells at 300 × *g* for 5 min, then remove the supernatants (*see* **Note 5**), and resuspend each pellet in 500 μL of warmed Seahorse assay medium. Add 100 μL of each cell suspension to appropriate wells in the Seahorse microplate previously coated with poly-L-lysine, leaving 4 wells empty for blank controls (Fig. 2). Centrifuge the microplate at 200 × *g* with zero braking for 1 min to embed the cells on the poly-L-lysine coating. Place the plate in a 37 °C, non-CO_2 incubator for 25–30 min to ensure cells have fully attached (*see* **Note 6**). Gently add an additional 400 μL of warm assay medium along the side of each well so as not to

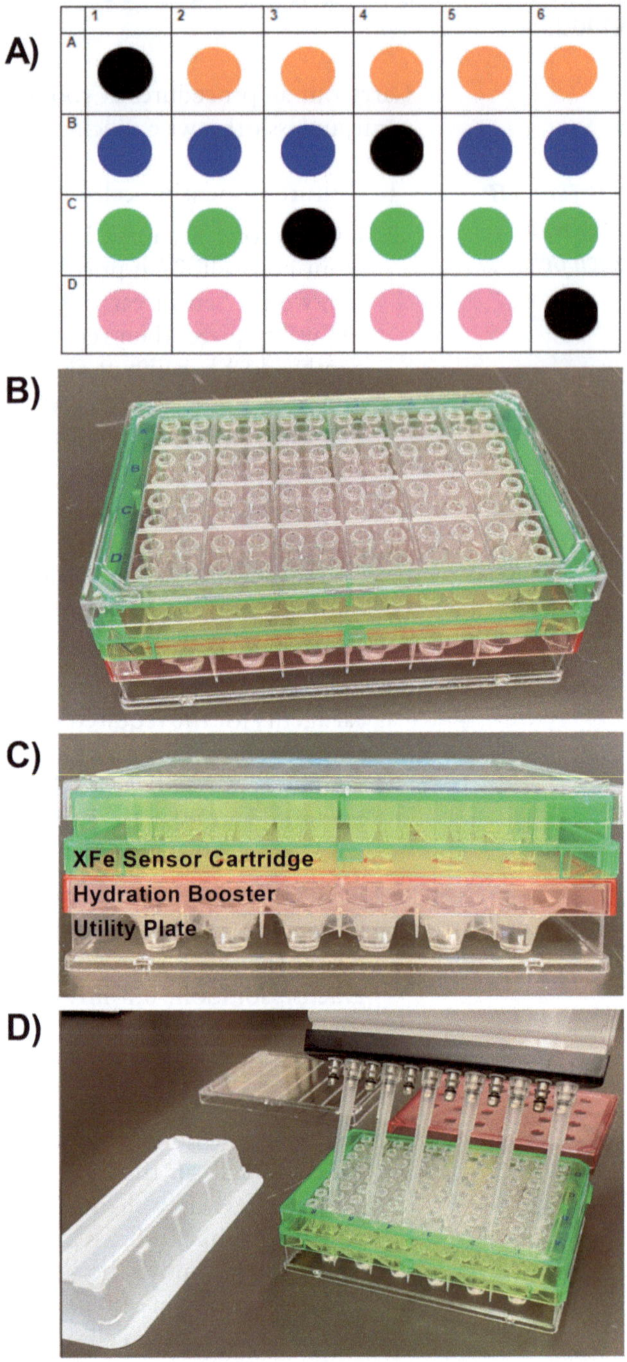

Fig. 2 Sample setup for the Real-Time ATP Rate Assay, showing (**a**) an example of a plate map illustrating the background wells in black and four different density options: 50,000 cells/well in orange, 100,000 cells/well in blue, 150,000 cells /well in green, and 200,000 cells/well in pink. (**b**) An image of the Seahorse XFe cartridge along with (**c**) the labeled components of the cartridge. (**d**) How to line up the multichannel pipette with the injector ports on the cartridge?

disturb the adhered cells. Add 500 μL of assay media to the 4 blank wells. Return to the incubator for 15–25 min.

7. Make up all compounds for the specific Seahorse assay during the previous incubation times as per the manufacturer's provided specifications. For the ATP Rate Assay, open the foil pack from the kit box and remove the vials containing the assay compounds (oligomycin and rotenone/antimycin A), ensuring the powder is at the bottom. Resuspend the powder in 420 μL and 540 μL of assay medium for the oligomycin and rotenone/antimycin A vials, respectively. Then vortex both tubes for 1 min to dissolve all of the drugs. To 2.7 mL of assay medium in labeled conical tubes, add 300 μL of the appropriate drug.
8. Load the assay components as follows: 55 μL of oligomycin into injector port A and 62 μL of rotenone/antimycin A into injector port B of the sensor cartridge. Note that this step does not have to be completed in a sterile environment. To properly load the drugs, position your pipette over the ports at a 45° angle and gently dispense in a single fluid motion (*see* **Note 7**).
9. Program the assay following the onscreen instructions using the Wave software package provided with the Seahorse XF instrumentation. To start the assay, first place the utility plate with the cartridge on top in the Seahorse tray in the correct orientation (well A1 positioned at the front left), making sure to remove the lid and hydro booster. Allow the calibration to complete and then replace the utility plate with the cell microplate.
10. From here, the assay will run automatically and take approximately 2 h to complete. Data can be exported as .pzfx, .xlsm, and .xlsx files for off-line analyses according to personal preference.

3.2 Real-Time ATP Rate Assay with Adherent Cells

1. As described in **steps 1–3** of Subheading 3.1, culture cells as per standard protocols (see also Chapter 3 for specifics on culturing adherent cell lines), prepare a poly-L-lysine-coated 24-well XFe microplate up to 1 week before your assay, and hydrate the Seahorse XFe cartridge 24 h prior to use.
2. At least 24 h before your assay, prepare a single cell suspension of cultured cells at the desired density (*see* Chapter 3) (*see* **Note 3**) and seed 100 μL of the suspension (*see* **Note 8**) into the XFe microplate, leaving 4 wells empty for blank controls (Fig. 2). Work with whichever cell culture medium is appropriate for your cells of choice. Let the plate incubate for 1 h at 37 °C with CO_2, allowing the cells to settle and attach to the surface of the XFe microplate wells.
3. Slowly add an additional 150 μL of cell culture growth medium to the filled wells. Additionally, add 250 μL of growth medium

to each of the blank wells. Incubate the plate overnight at 37 °C with CO_2 to allow the cells to firmly adhere and acclimate to the XFe microplate.

4. Pretreatments must be conducted in the XFe microplate when working with adherent cells. If time-points exceed 24 h, the process described above should be started earlier to allow the required time points to be attained.
5. On the morning of the Real-Time ATP Rate Assay, prepare and warm assay medium as described in **step 5** of Subheading 3.1 (*see* **Note 4**).
6. Remove the XFe microplate from the 37 °C CO_2 incubator, and check the cells to ensure that a consistent monolayer is present in each well.
7. Carefully remove all but 100 μL of the cell culture medium from each well. The remaining medium will be replaced through a series of washes and dilutions with the assay medium (*see* **Note 9**). From our experience, we have found that the following protocol works well: Add 1000 μL of Seahorse assay medium, bringing the total volume up to 1100 μL. Remove 900 μL from each well, bringing the total volume down to 200 μL. Add 300 μL of assay medium, bringing the total volume up to 500 μL. Remove 300 μL from each well, bringing the total volume back down to 200 μL. Finally, add 300 μL of assay medium, bringing the final volume to 500 μL.
8. Incubate the plate for 60 min at 37 °C in a non-CO_2 incubator. During this step, prepare the oligomycin and rotenone/antimycin A solutions as described in **step 6** of Subheading 3.1.
9. Proceed with the Real-Time ATP Rate Assay including loading the compounds and preparing the Seahorse. The procedure from this point is identical to **steps 7–10** of Subheading 3.1.

4 Notes

1. It is critical to use the Seahorse medium as precisely defined. Low buffering capacity medium that is precisely defined is required to allow for accurate determination of ECAR and subsequent respiratory parameters.
2. The Seahorse sensor cartridge contains 24 pairs of fluorescent biosensors that measure oxygen and pH levels in the medium. These sensors are highly sensitive and designed to measure small changes in analyte concentrations [5]. To prevent changes in the pH of the XF calibrant and maintain the specific buffering capacity before the assay, store the cartridge and perform appropriate incubations in a non-CO_2 incubator.

3. Before starting experimental manipulations, determine the optimal cell density for your cell line of choice. To do so, search the Agilent cell analysis publication database for publications with your specific cell line and/or follow the protocol as listed using a variety of cell densities. From our experience, for the best results, the density chosen should provide a consistent monolayer on the surface of each well in the microplate at the time of Seahorse assay. Sample results that can be expected from a cell density determination assay are provided in Fig. 1. From this data, we selected 150,000 cells/well as appropriate for subsequent studies.
4. We have found that preparing 50 mL of assay medium provides ample volume for our suspension cell protocol; for adherent cells that require multiple washing steps, we found that doubling this recipe is necessary.
5. When removing the supernatants, rather than aspirating the medium we routinely collect them in microcentrifuge tubes and store them for subsequent analyses, for example, measured secretion of various signaling molecules. By saving the supernatants, it allows multiple analyses to be obtained from a single experiment, in addition to determining the metabolic profile.
6. To visually check for cell adhesion, view the microplate under an inverted microscope. From our experience, the best results are obtained when the cells form a complete monolayer on the culture surface and there are no cells left in suspension. If necessary, the centrifugation step can be repeated.
7. To be more efficient, use a multichannel pipette with a tip on every second ejector. By using the pipette this way, a tip will line up to a single injector port (A, B, C, or D) of each well in a 24-well microplate (Fig. 2).
8. When seeding the cells into the plate, position your pipette tip at a 45° angle on the inner lip of the microchamber well. Quickly expel the volume and ensure that the total volume of suspension is dispensed into the lower portion of the chamber.
9. Slowly pipette the medium in each wash, maintaining an intact monolayer is essential for consistent results.

References

1. Hanahan D, Weinberg RA (2000) The hallmarks of cancer. Cell 100:57–70
2. Hanahan D, Weinberg RA (2011) Hallmarks of cancer: the next generation. Cell 144(5): 646–674

3. Warburg O (1956) On the origin of cancer cells. Science 123:309–314
4. Warburg O (1956) On respiratory impairment in cancer cells. Science 124:269–270
5. Wu M, Neilson A, Swift AL et al (2006) Multiparameter metabolic analysis reveals a close link between attenuated mitochondrial bioenergetic function enhanced glycolytic dependency in human tumor cells. Am J Phys 292(1): C125–C136
6. Sure V, Sakamuri S, Sperling J et al (2018) A novel high-throughput assay for respiration in isolated brain microvessels reveals impaired mitochondrial function in the aged mice. Geroscience 40(4):365–375
7. Wen YA, Xiong X, Zaytseva YY et al (2018) Downregulation of SREBP inhibits tumor growth and initiation by altering cellular metabolism in colon cancer. Cell Death Dis 9(3):265
8. Ajaz S, McPhail MJ, Singh KK et al (2021) Mitochondrial manipulation by SARS-CoV-2 in peripheral blood mononuclear cells of patients with COVID-19. Am J Phys 320: C57–C65

Chapter 18

Using CRISPR-Cas9 to Dissect Cancer Mutations in Cell Lines

Shady Sayed, Duran Sürün, Jovan Mircetic, Olga Alexandra Sidorova, and Frank Buchholz

Abstract

The CRISPR-Cas9 technology has revolutionized the scope and pace of biomedical research, enabling the targeting of specific genomic sequences for a wide spectrum of applications. Here we describe assays to functionally interrogate mutations identified in cancer cells utilizing both CRISPR-Cas9 nuclease and base editors. We provide guidelines to interrogate known cancer driver mutations or functionally screen for novel vulnerability mutations with these systems in characterized human cancer cell lines. The proposed platform should be transferable to primary cancer cells, opening up a path for precision oncology on a functional level.

Key words CRISPR-Cas9, Mutations, Cancer cell lines

1 Introduction

All cancers have in common that they arise due to accumulating changes in the DNA of malignant cells [1]. These changes can be classified according to their consequences on cancer development. "Driver" mutations confer a selective advantage and will lead to clonal expansion, while "Passenger" mutations do not alter the fitness of the cells and are viewed as by-products of the mutagenesis process [2]. Recent advances in genome sequencing technologies provide the means to identify all mutations present in a given tumor. However, our ability to interpret the functional consequences of these mutations lags far behind [3]. Several computational methods have been described to classify relevant cancer mutations based on their recurrence on a population scale [4–10]. However, while powerful, these approaches do not provide definitive proof of driver or passenger mutations in individual cancers. Information about the functional consequences these mutations have for tumorigenicity would be important for the

Sherri L. Christian (ed.), *Cancer Cell Biology: Methods and Protocols*,
Methods in Molecular Biology, vol. 2508, https://doi.org/10.1007/978-1-0716-2376-3_18,

development of optimized targeted therapies and to guide best possible medical decisions [11].

Having a system that can functionally test the relevance of mutations for cancer cell growth and proliferation in individual tumor samples would be a much appreciated advance in the field of precision oncology. Programmable nucleases, such as the CRISPR-Cas9 system, enable the targeting of specific genomic sequences including cancer mutations [12–15]. The CRISPR-Cas9 system has been adapted for use in mammalian cells from a naturally occurring genome editing system in bacteria. The system consists of two parts: the Cas9 enzyme that introduces DNA double-strand breaks (DSBs) and a single guide RNA (sgRNA) that directs the Cas9 protein to a specific site in the genome [13]. In principle, CRISPR nucleases can be designed to target almost any mutation that occurred during the transformation process. Hence, these genetic tools are optimally suited to functionally dissect the role of mutations in cancer cell lines. Additionally, Base Editors (BEs) are a recent addition to the CRISPR arsenal and are particularly useful to target point mutations without the introduction of double-strand breaks [16, 17]. Base editors chemically change a DNA base to another one without breaking the DNA backbone. This is achieved by fusing a nickase version of Cas9 to a deaminase enzyme. This enzyme deaminates a specific base, which is then converted to another base by the host cell, ultimately leading to a precise change in the DNA sequence [16].

Nevertheless, the minimal differences between the cancer and the wild-type sequences pose considerable challenges. In this chapter, we provide protocols to utilize CRISPR-Cas systems for targeting cancer mutations with the aim to identify and scrutinize cancer driver mutations in cell lines. We guide the reader through the process of screening public cancer genome databases, analyzing the genome of cancer cell lines to identify the most suitable mutations that can be addressed. We show how to identify cancer mutations and design efficient and specific sgRNAs against these mutations. We further provide step-by-step protocols from cloning of the sgRNAs to lentiviral particle production, infection, genomic DNA (gDNA) isolation, and determination of the editing ratio. Finally, we describe how to generate isogenic human-induced-pluripotent stem cells (hiPSCs) carrying cancer mutations.

2 Materials

2.1 Buffers and Reagents

Lysogeny broth (LB-Lennox version): Bacto tryptone, Yeast extract, NaCl.

For 1000 mL LB, weigh 10 g Bacto tryptone, 5 g Yeast extract, and 5 g NaCl. Dissolve all components in filter-sterilized, distilled water and adjust the pH to 7.5 using NaOH. Fill the volume up to 1000 mL, and autoclave the medium for 20 min at 121 °C prior to use.

Phosphate-Buffered Saline (PBS): 43 mM Na_2HPO_4, 14 mM KH_2PO_4, 27 mM KCl, 1.47 M NaCl, pH 7.4.

For 1000 mL 10X PBS, weigh 7.6 g Na_2HPO_4 ($Na_2HPO_4H_2O$; MW = 178.0 g/mol), 1.9 g KH_2PO_4 (MW = 136.1 g/mol), 2.0 g KCl (MW = 74.6 g/mol), 85.9 g NaCl (MW = 58.4 g/mol). Dissolve all components in distilled water. Adjust the pH of this solution to 7.4 using NaOH. If PBS is used for cell culture, autoclave for 20 min at 121 °C prior to use.

Tris-Borate-EDTA buffer (TBE): 89 mM Tris-HCl, 89 mM H_3BO_3, 2 mM EDTA.

For 1000 mL 1X TBE, weigh 10.8 g Tris-HCl (MW = 121.1 g/mol), 5.5 g H_3BO_3 (MW = 61.8 g/mol), and 4 mL EDTA (0.5 M stock solution, pH 8.0). Dilute all components in distilled water.

2.2 Cell Culture

1. HEK293T cells.
2. Dulbecco's Modified Eagle's Medium DMEM, Glutamax (Blank medium).
3. Full Medium: Dulbecco's Modified Eagle's Medium DMEM, Glutamax supplemented with 10% (v/v) fetal bovine serum (FBS) and 1% Penicillin-streptomycin (P/S).
4. DMEM supplemented with 20% FBS.
5. Consumables: 10 cm dish, T-175 flask, 6-well plate, falcon tubes (15 mL, 50 mL).
6. psPAX2 (Addgene plasmid #12260).
7. p.MD2.G (Addgene plasmid #12259).
8. pL-CRISPR.EFS.GFP.gRNA (Addgene plasmid #57818).
9. Polyethylenimine (PEI) solution, filtered PEI stock (1 mg/mL).
10. 0.45 μm sterile filter.
11. Plate centrifuge.
12. Protamine sulfate solution, filtered stock (5 mg/mL).
13. Flow cytometer (MACS quant).
14. Fluorescent microscope (EVOS).
15. OptiMEM.
16. Complete DEF-CS growth media.

2.3 Kits, Reagents, and Equipment

1. Competent *E. coli* cells (e.g., XL1-blue).
2. Agarose.
3. Thermo Fisher GeneJET Plasmid Miniprep.
4. QIAGEN-tip 20 Maxiprep.
5. QIAamp® DNA Blood Mini kit.
6. Nuclease-free water.
7. dNTP mix.
8. RNA MessengerMax reagent.
9. Phusion High-fidelity DNA polymerase.
10. Taq polymerase.
11. Herculase II Fusion DNA Polymerase.
12. 100-bp ladder.
13. RedSafe.
14. Gel imager.
15. Thermocycler.
16. Nanodrop spectrophotometer.
17. HiScribe T7 ARCA mRNA Kit/Monarch RNA Cleanup Kit (NEB).

3 Methods

3.1 sgRNA Design to Inactivate a Specific Mutation of Interest

All experiments start with the identification of cancer mutations of interest in a cancer cell line. Several online tools list cancer mutations in established cell lines [18–23]. Here, we use the database COSMIC, the Catalogue Of Somatic Mutations in Cancer, as the largest and most comprehensive resource of somatic mutations in human cancer. COSMIC catalogs mutation profiles of over 1000 cell lines curated from the literature, made publicly available, and free to download. To start an experiment aimed at inactivating cancer mutations:

1. Go to COSMIC database (https://cancer.sanger.ac.uk) and choose Cell Lines Project.
2. Use the search engine to search for a mutation (e.g., *BRAF* V600E) (*see* **Note 1**).
3. Click the Syntax link of the gene mutation in question to view a general overview of the mutation. Scroll down to view the cell lines, which harbor the mutation and pick a cell line for experimental validation (*see* **Note 2**).
4. For sgRNA design targeting a mutation (e.g., *BRAF* V600E): Import the genomic locus of interest flanking the mutation (ca. 50–100 bp) into a sequence-viewing software of choice.

We frequently use SnapGene (https://www.snapgene.com). Manually change the sequence to include the mutation. Annotate the mutation and locate PAM sequences that are downstream (i.e., NGG sequences toward the 3′ end of the strand containing the target mutation). The target mutation can be on either strand, as long as there is a suitable PAM on the same strand.

5. Identify the 20-bp sequence upstream from the PAM (The spacer sequence of the sgRNA should include the mutation and should not include the PAM) (*see* **Note 3**).

   For example: WT *BRAF* allele: 5′-TAGCTACAG**T**-GAAATCTCGATGG-3′.

   Mutant *BRAF* V600E: 5′-TAGCTACAG**A**GAAATCTCGA TGG-3′.

6. If the protospacer does not start with guanine, add a 5′ G to create a 21-nt spacer in the sgRNA. Adding a guanine boosts sgRNA transcription from the pol III U6 promoter, yielding higher editing efficiencies [24].

7. Design the spacer as a custom top primer oligo with the sequence 5′-**CACC**[spacer sequence]-3′. Design the bottom primer oligo with the sequence 5′-**AAAC**[Reverse complement- spacer sequence]-3′ (e.g., *BRAF*V600E top and bottom oligos: 5′-**CACC**GTAGCTACAGAGAAATCTCGA-3′ and 5′-**AAAC**TCGAGATTTCTCTGTAGCTAC-3′). The 5′-CACC-3′ and 5′-AAAC-3′ preceding the oligos are the complementary overhangs for cloning using BbsI/BsmbI sites, which are commonly used in vectors for cloning spacer sequences. However, one should note that other enzymes will perhaps require other overhangs. It is also recommended that a nontargeting spacer be cloned as a control (e.g., 5-′-GAATTGTCTTGTCCCTATCGA-3′).

8. Order the oligos of choice for custom sgRNA cloning using BbsI/BsmbI sites:

   Phosphorylate the 5′ of each primer by combining the following mixture in Table 1.

9. Run the mixture in a thermocycler with the following profile: 37 °C for 30 min; 95 °C for 5 min; ramp down to 25° at 0.1 °C/min.

10. Use the annealed oligos to perform golden gate sgRNA cloning following the protocol in Table 2.

11. Run the mixture in a thermocycler with the following protocol in Table 2.

12. Transform ~2 μL of the ligated product into competent *E. coli* cells of choice (e.g., XL1-blue). Plate 10–50 μL cells on an agar plate containing 100 μg/mL ampicillin in a manner that will yield single colonies, and grow overnight at 37 °C.

Table 1
Annealing oligos

Amount (μL)	Component
To 10	Nuclease-free water
0.5	T4 polynucleotide kinase
1	T4 DNA ligase buffer (10×)
100 pmol	sgRNA top oligo
100 pmol	sgRNA bottom oligo
Σ 10	

Table 2
Restriction/ligation Golden gate protocol and cycler program

Amount (μL)	*Component*
To 10	Nuclease-free water
50 ng plasmid DNA	sgRNA expression vector (e.g., pL-CRISPR.EFS.GFP.gRNA)
1	Tango buffer (Thermoscientific)
1	ATP (10 mM)
2	Annealed oligos (*diluted 1:250*)
0.5	DTT (10 mM)
0.5	BsmBI (Esp3I)
0.25	T4 ligase
Σ 10	
Cycler step	*Condition*
6 cycles	37 °C/5 min + 23 °C/5 min
1 cycle	37 °C for 15 min
1 cycle	80 °C for 5 min

13. Pick 1–2 colonies and grow in liquid LB medium with 100 μg/mL ampicillin overnight. Save the plates (seal using parafilm) at 4 °C and mark the colonies picked.

14. When cultures reach saturation, purify plasmid DNA using a miniprep kit (e.g., Thermo Fisher GeneJET Plasmid Miniprep). We recommend saving the original culture tubes with unlysed bacteria at 4 °C to create starter cultures for the correct clones.

15. Quantify the DNA concentration (e.g., Nanodrop), and then run an aliquot on an agarose gel (1.5%) alongside a DNA ladder to check the quality and size of the plasmid.
16. Submit preps to a sequencing company for Sanger sequencing using a sgRNA sequencing primer (e.g., hU6-Fwd, 5-′-GAGGGCCTATTTCCCATGATTCC-3′) to confirm the presence of the new spacer sequence.
17. Once validated, we recommend purifying the DNA using plasmid maxiprep kits (e.g., QIAGEN-tip 20 Maxiprep Kit) to provide endotoxin-free transfection-level DNA for all cell culture experiments.

3.1.1 Traffic-Light Reporter Assay

To examine the efficiency and specificity of a designed sgRNA, we recommend using the traffic light reporter assay [13], where the mutant or their corresponding wild-type sequences are cloned between a RFP coding sequence and an out-of-frame-GFP sequence. Cas9-induced indels activate GFP expression in transiently transfected mammalian cells. Sequences bearing the different cancer mutations can be cloned into pRG-HBV-S [25] via EcoRI and BamHI sites and subsequently co-transfected into easy-to-transfect cells (e.g., HeLa) with a Cas9 expression plasmid that also expresses the cancer mutation–specific sgRNA. Successful cleavage can be observed for constructs bearing cancer mutations through measuring the percentage of GFP-expressing cells. In case a sgRNA cuts the target sequence, it creates indels in the reporter plasmids, which will bring the GFP sequence into the correct reading frame. Ideally, no to few GFP positive cells should be observed when the cancer-mutation-specific sgRNA is combined with the wild-type sequence cloned into the pRG-HBV-S vector, while a significantly higher proportion of GFP-positive cells should be observed when the cancer-mutation-specific sgRNA is combined with the mutant sequence (Fig. 1).

3.2 Lentivirus Production

Once a sgRNA is validated for the efficient and specific cleavage of the target cancer mutation, the tested sgRNA can be cloned into a lentiviral vector expressing Cas9 and subsequently used to infect the target cells to investigate the functional relevance of the mutation. Lentiviral constructs bearing a fluorescent marker (e.g., GFP cassette) allow for simple evaluation of the infection rate and can be used as a straightforward transduction efficiency readout. Moreover, because lentiviruses integrate into the genome of host cells, measuring the percentage of fluorescent cells over time can be used to deduce if the targeted mutation affects cell growth/proliferation (i.e., if the GFP+/GFP− ratio decreases over time in culture). For the production of lentiviral particles, HEK293T cells are used as a packaging cell line and are routinely cultured in full medium.

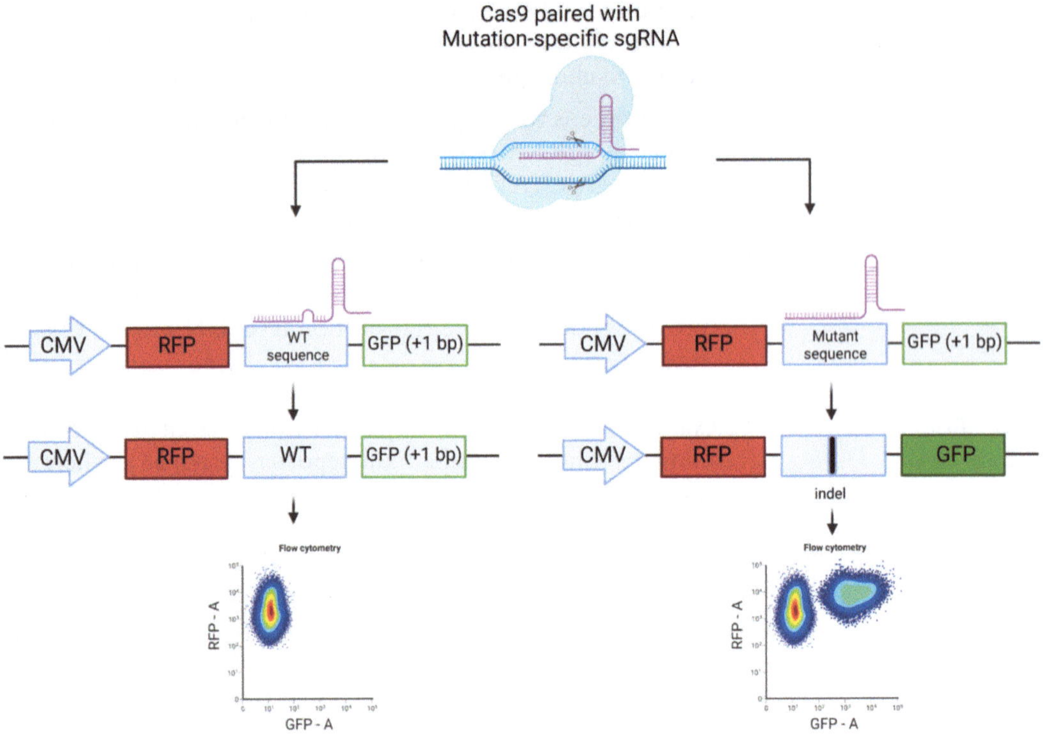

Fig. 1 Overview of the traffic light reporter assay. The reporter plasmid and sgRNA-Cas9 expressing plasmid are cotransfected into tissue culture cells. If the mutation-specific sgRNA does not lead to cleavage of the wild-type sequence on the reporter (left panel), no GFP-positive cells are observed. In contrast, if the reporter with the mutant spacer sequence (right panel), which perfectly matches the mutation-specific sgRNA, triggers Cas9-mediated cleavage, GFP-positive cells can be identified, because a portion of the induced indels brings GFP into frame, allowing its expression. These events can be visualized by a flow cytometric analysis, where the successful cleavage and repair events are represented by RFP^+/GFP^+ double-positive cells. This assay can be used to estimate the efficiency of the on-target and the rate of off-target cleavage of designed sgRNAs

Outlined below is the transfection protocol for one 10 cm dish for one transfer vector:

One day prior to transfection, seed HEK293T cells according to format:

- 6-well plate: 1.2–1.5 million (in 3 mL Full medium)
- 10 cm dish: 6–7 million (in 5 mL Full medium)
- T-175 flask: 15–20 million (in 20 mL Full medium)

1. Before transfection, ensure HEK293T cells are at ~80% confluency (*see* **Note 4**). In order to prepare plasmid transfection solutions for one 10 cm dish, prepare the following mixture: 10 μg Transfer vector (e.g., pL-CRISPR.EFS.GFP.gRNA, Addgene plasmid #57818), 6 μg psPAX2 (Addgene plasmid

#12260), 2 μg p.MD2.G (Addgene plasmid #12259). Add DMEM blank medium (without FBS or antibiotics) up to 1 mL.

2. Dilute 45 μL polyethylenimine (PEI) stock (1 mg/mL) in 955 μL blank medium.
3. Add 1 mL PEI solution dropwise to 1 mL plasmid solution and incubate the transfection mix for 30 min at room temperature.
4. In the meantime, aspirate the old medium from the dish, carefully add 4 mL DMEM supplemented with 20% FBS, then add the transfection mix, swirl gently, and then place cells back in tissue culture incubator.
5. After ~20 h, aspirate transfection medium (Biosafety level S2 procedure, *see* **Note 5**) and add 5 mL fresh prewarmed full medium. If the transfer vector expresses a fluorescent protein such as GFP, transfection efficiency can be deduced by monitoring the percentage of GFP-positive HEK293T cells.
6. After 48 h, collect the supernatant (~5 mL) into a falcon tube, centrifuge briefly for 5 min at 300 g, then pass through 0.45 μm filter, aliquot, and freeze at −80 °C (*see* **Note 6**).

3.3 Virus Particles Concentration (>100×)

If high virus titers are desired, scaling up the number of transfections (up to 7× 10-cm dishes) followed by ultracentrifugation/Ultrafiltration is recommended and could be done as follows:

1. On harvest day, 72 h post transfection, collect 35 mL of the virus-containing supernatant from seven dishes in a falcon tube and spin for 5 min at 300 g to pellet cellular debris.
2. Pass the supernatant through 0.45 μm filters into 50 mL falcons.
3. For ultraconcentration using Amicon® Ultra Centrifugal filters (Merckmillipore), apply 15 mL to the provided falcon and centrifuge following the manufacturer's instruction. Centrifuge until you reach the desired retention volume or concentration (i.e., 25×, 100×, etc.).
4. For ultraconcentration using ultracentrifugation, apply the supernatant from step2 into a sterile Open-Top Thickwall Polycarbonate Tube, compatible with a rotor for ultracentrifugation (e.g., JS-24.38, Beckman Coulter).
5. Place the labeled polycarbonate tubes containing the viral supernatant into the aluminum swinging-buckets. Weigh the buckets with high precision (to a hundredth of a gram). Balance buckets.
6. Centrifuge for 2 h at 100,000 g, with slow acceleration/deceleration at 4 °C.

7. After centrifugation, carefully decant the supernatant and keep the tubes upside down to drain media on a fiber-free tissue for 10 min.
8. Gently add 300 μL PBS to the inner tube wall, do not resuspend, and place on the shaker in a 4 °C room for at least 4 h.
9. Aliquot 50 μL into labeled cryovials. Store at −80 °C.

3.4 Transductions and Flow Cytometry Analysis

1. Seed target cells in a 96 well plate one day in advance, so that they are ~80% confluent at the time of infection.
2. Thaw the virus on ice. Based on the titer analysis on the target cells, add the amount of virus required to infect ~50% of the cells.
3. Add protamine sulfate to 5 μg/mL final concentration.
4. Fill up every well to 200 μL with full medium, and then spinoculation for 1 h at 1000 g at 37 °C in a plate centrifuge.
5. After 24 h (Biosafety level S2 procedure, *see* **Note 5**), aspirate virus-containing medium, wash twice with PBS, and then add prewarmed full medium 200 μL/well.
6. Transduced cells can be analyzed by flow cytometry for fluorescent marker expression (e.g., GFP) at 3 days post infection. To track the growth dynamics of the infected cells, follow the steps below; otherwise, proceed to Subheading 3.5.
7. Aspirate the medium completely using a multichannel pipette. Gently rinse cells with 200 μL sterile PBS. Discard PBS.
8. For adherent cells, dissociate cells by adding 30 μL/well trypsin to cover the adherent layer of cells. Incubate for 5 min at 37 °C.
9. Add 170 μL full DMEM medium directly to cells, homogenize cell solution by pipetting up and down vigorously 10 times. Transfer 30 μL (∼1:8) to a new 96 well plate, add 170 μL prewarmed full medium, and place back in the incubator for later acquisition during the time course.
10. Take 50 μL for FACS acquisition. Spin the remaining 120 μL down for DNA isolation and/or freezing.
11. Repeat **steps 7–10** every 2–3 days to monitor the growth rate of GFP-positive cells. Keep the cells in culture for 15 days (or ~5 population doublings, at least). It is expected to see a stable GFP percentage in the controls. A reduction of GFP percentage over time in the target cells carrying the mutation is a sign of a targeted cancer vulnerability.

3.5 Primer Design for Amplification of the Targeted Genomic Locus

Using a primer design tool (e.g., Primer3, [26]), generate a pair of primers to amplify a 500–1000 base pairs (bp) region containing the protospacer, for Sanger sequencing. These primers should ideally be 18–24 base pairs (bp) long and have 40–60% GC content and melting temperatures of 60–65 °C. The primer sequences can further be analyzed using the publicly available tool from IDT: https://www.idtdna.com/pages/tools/oligoanalyzer. To ensure that at least one primer is appropriately positioned to yield high-quality sequencing data for the protospacer region of the amplicon, the target sequencing region should be at least 100 bp away from the primer used for sequencing.

3.6 Harvesting Genomic DNA and Genotyping Using Sanger Sequencing

Use available kits to isolate genomic DNA (e.g., QIAamp® DNA Blood Mini kit), followed by performing targeted PCR. Sanger sequencing is usually sufficient to confirm that a sgRNA in combination with Cas9 is cutting/editing at the locus of interest. Indels are presented as overlapping peaks on the chromatogram following successful cleavage, 3-bp upstream of the PAM sequence. However, accurate quantification of editing efficiencies with Sanger sequencing is only possible when efficiencies are between 10% and 90% and when using high-quality sequencing traces. Targeted PCR amplification can be performed on isolated genomic DNA using Phusion high-fidelity DNA polymerase using the following protocol:

1. For PCR of genomic DNA (*see* **Note 7**), combine nuclease-free water up to 50 μL, 10 μL HF Buffer (supplied in the Phusion polymerase kit), 1 μL dNTP mix (10 mM each), 1.25 μL 20 μM (Forward Primer), 1.25 μL 20 μM (Reverse Primer), 250 ng gDNA, and 0.5 μL Phusion DNA polymerase.
2. Mix, quickly spin (~5 s at 2000 g) at room temperature, and run in a thermocycler under the following cycling conditions:
   - 1 cycle: 30 s 98 °C (initial denaturation).
   - 30 cycles: 10 s 98 °C (denaturation).
   - 20 s 65 °C (annealing, specific to primer pairs, please adjust as necessary).
   - 30 s 72 °C (extension).
   - 1 cycle: 5 min 72 °C (final extension).
   - 1 cycle: ∞ 8 °C (hold).
3. Run a 10-μL aliquot of each PCR reaction on a 2–3% agarose gel using a 100-bp ladder for size comparison to confirm primer specificities and correctly sized bands.
4. Once a correctly sized PCR product has been confirmed, the samples should be prepared for sequencing. Perform PCR purification using a kit (e.g., Bioline) following the manufacturer's instructions and measure the DNA concentration after PCR cleanup of each sample using a Nanodrop spectrophotometer. Prepare the appropriate amount of DNA along with the sequencing primer (we recommend using either the forward or reverse primer from the PCR reaction that is positioned at least 100 bp away from the target base of interest)

Table 3
Timeline of CRISPR-Cas9-based targeting of cancer mutations in cell lines

Procedure	Timing
1. Define the mutation or cell line of interest using the COSMIC database	1 day
2. Design the spacer for targeting the mutation	
3. Estimate the efficiency/specificity of the sgRNA, e.g., with the traffic-light reporter assay	7 days
4. Clone the spacer into a lentiviral vector expressing Cas9 and a selection marker (e.g., GFP)	2 days
5. Produce lentiviral particles for sgRNA-Cas9 delivery 5a. Optional: determine the lentiviral titer to be able to control the infection rate	5 days 5 days
6. Transduce the cell line of interest with the sgRNA-Cas9 lentivirus	5 days
7. Analyze the infection rate by flow cytometry	
8. Collect cells for gDNA isolation. Amplify and sequence the targeted locus	2 days
9. Culture the remaining cells and track the growth dynamics by flow cytometry	14(+) days

in the same tube according to company protocol (*see* **Note 8**). An overview of the aforementioned experiments aimed at inactivating cancer mutations with time considerations is summarized in Table 3.

3.7 CRISPR-Cas9 Dropout Screen to Identify Novel Cancer Vulnerability Mutations

The CRISPR-Cas9 system can be leveraged to functionally profile mutations in a given cancer cell line to identify novel cancer vulnerabilities. In order to screen for driver mutations in a given cell line, first extract the complete set of mutations in a particular cell line:

1. Go to the COSMIC database (https://cancer.sanger.ac.uk).
2. Type the cell line name in the search panel (some cell lines have ambiguous terminology, try spaced/hyphenated variants of the cell line name: e.g., U2OS is written U-2-OS).
3. Choose the cell line under sample name in the search results.
4. An overview of the given cell line is given including tissue origin, age, and gender of the patient from whom it was derived, mutation spectrum, and variants.
5. Go to variants tab showing the mutations observed in the cell line, export all mutations by downloading the CSV or TSV tables (*see* **Note 9**).
6. Potential Cas9 target sites can be searched using the Azimuth 2.0 algorithm [27, 28]. Extract 30 bp of sequence upstream and downstream of the reported mutations to identify potential sgRNAs in sense and antisense orientation using a sgRNA Designer algorithm (e.g., [27, 28]).

7. Discard all sgRNAs that have a perfect match in the wild-type genome from further analysis (selected sgRNA should only cut the cancer mutation).
8. Perform an off-target search for the obtained 20-bp long sgRNAs using short sequence aligner GEM ver 1.376 [29], with parameters set to report all alignments with up to four mismatches (*see* **Note 10**).
9. Include sgRNAs targeting essential genes as positive controls, as well as nontargeting sgRNAs negative controls [30].
10. Clone the sgRNAs individually into an all-in-one lentiviral Cas9-expression vector that is also expressing GFP (e.g., pL-CRISPR.EFS.GFP.gRNA, Addgene plasmid #57818). Purify the plasmid DNA using miniprep kit (e.g., Thermo Fisher GeneJET Plasmid Miniprep).
11. Quantify the miniprepped DNA and pool equimolar concentrations from the individually prepped plasmid DNA. Retransform the pooled plasmids (make sure that the number of colonies exceeds 100X the coverage of the pooled sgRNA clones to ensure equal representation) for Maxi prep and virus production.
12. Infect the target cell line carrying the mutations with the lentiviral sgRNA library at multiplicity of infection (MOI) of 0.3. This low infection rate is instrumental to ensure that most cells are infected with a single Cas9-sgRNA complex targeting the mutant sequence, making it possible to investigate the inactivation of individual mutations. Infect another cell line that does not carry the mutations as the negative control at the same MOI.
13. Three days after infection, split the cells and isolate genomic DNA from 50% of the cells. These samples are used as the baseline sample to quantify sgRNA abundance at the beginning of the experiment.
14. After 15 days of continuous culture and splitting (or ~5 population doublings, at least), isolate genomic DNA again from both cell lines, PCR-amplify the sgRNA cassettes, and quantify sgRNAs counts by means of deep sequencing.
15. sgRNAs depleted over the experiment course in target cells, but not in the control cell line, are targeting candidate mutations important for growth/proliferation of the cancer cell line.
16. It is essential that potential hits are validated through individual infections with the original sgRNA and a second sgRNA, if possible, and monitoring the infected cells over time. It is recommended to include the appropriate positive controls such as sgRNAs targeting core essential genes (e.g., sgRNA against PolR1C-1 5′-GAAGAATCTCATCCTGAACAA-3′).

3.8 Generation of Isogenic hiPSCs with Cancer Mutations

To study molecular and cellular disease mechanisms, as well as drug screening of different cancer-associated mutations of the same gene, it is desirable to perform experiments in the same genetic context. A major disadvantage of using cancer cell lines for modeling human diseases is their diverse genetic background and epigenetic memory [31, 32]. This makes it difficult to compare data obtained from different cell lines. Elimination of genetic background effects is therefore beneficial for loss-of-function and gain-of-function studies or drug screening. Here we describe the application of adenine and cytosine base editor (BE) systems to provide a fast and easy way for generating different cancer-relevant point mutations in an isogenic genetic background.

Bearing in mind, that BEs, especially cytosine base editors, show off-target effects on DNA as well as on RNA [33–36], it is recommended to reduce the expression time of the BEs in the cells [37]. Therefore, to minimize potential side effects, the BEs and sgRNAs are delivered as (m)RNAs into the cells. This transient, DNA-free approach reduces off-target editing as well as the risk of unwanted DNA integration. In comparison to the nuclease approach described above, these mutations have to be typically 13–17 nt upstream of a suitable PAM. The target base can be on either strand, as long as there is a suitable PAM on the same strand (PAM is always toward the 3′ end of the strand containing the target nucleotide). An example of this application is the introduction of cancer-associated *TP53* mutations into human iPSCs using both adenine and cytosine BEs with up to 90% efficiency [17]. Below is a stepwise protocol to introduce mutations with NGG PAMs, but the approach should also be transferable to other Cas9 variants (e.g., NG-Cas9).

1. Go to the COSMIC database (https://cancer.sanger.ac.uk) and choose your gene of interest. In the category "Mutation distribution," all types of substitution mutations are displayed.
2. Download the substitution mutations C → T, A → G, T → C, G → A.
3. Select the mutations of interest and annotate them on the gene sequence manually.
4. Check for NGG PAM sequences 13–17 nt downstream of the mutation.
5. Mark all cancer-associated mutations, which have a suitable NGG PAM.
6. Order synthetic sgRNAs (e.g., *Synthego* https://www.synthego.com) (*see* **Note 11**).

3.9 In Vitro mRNA Transcription

Prepare in vitro transcribed (IVT) mRNA from a PCR amplicon of adenine base editor (ABE) or cytosine base editor (CBE), which

Table 4
Primer design for mRNA production

Primer	Sequence
Common forward sequence	5′-GCTAATACGACTCACTATAGGGAGA-3′ Plus annealing sequence or alone in case your template starts with T7 promoter (blue sequence T7 promoter)
Common reverse sequence	5′-TTTTTTTTTTTTTTTTGGTTTATTC-3′ Plus annealing sequence (orange sequence PolyA tail)

can be generated using Herculase II Fusion DNA Polymerase (Agilent). Primers for mRNA production can be designed following Table 4. The PCR to amplify the plasmid region is done as follows:

1. Dilute 100 μM primers to 10 μM.
2. Dilute BE plasmid to 100 ng/μL. Prepare PCR reaction following Table 5.
3. Prepare at least 2 × 50 μL/mRNA to get high concentration of purified PCR.
4. Load 1 μL of the PCR with 6× Orange DNA loading dye (Thermo Fisher Scientific) on an agarose gel to verify the bands are of correct size.
5. Clean the rest of the PCR with PCR-purification Kit (ISOLATE II PCR and Gel Kit, Bioline Meridian Bioscience) over 1 column and elute in 50 μL nuclease-free water.
6. Generate IVT mRNAs according to the manufacturer's guidelines using the HiScribe T7 ARCA mRNA Kit (NEB) followed by purification using Monarch RNA Cleanup Kit (NEB).
7. Assemble the reaction for standard mRNA synthesis in a PCR tube at RT in the order stated in Table 6.
8. Mix thoroughly and pulse-spin in a microfuge, incubate at 37 °C for 30 min.
9. Add 2 μL of DNase I to 20 μL reaction in the PCR tube, mix, and incubate at 37 °C for 15 min.
10. Save 1 μL of leftover for gel electrophoresis (refer to **step 4**).
11. Set up tailing reaction in a new PCR tube following Table 7.
12. Mix thoroughly and pulse-spin in a microfuge, incubate at 37 °C for 30 min.

Table 5
PCR to amplify plasmid region

Amount (μL)		Component	
1		Plasmid (100 ng/μL)	
10		Herculase buffer (10×)	
0.5		dNTPs for Herculase (25 mM each dNTP)	
1.25		Forward primer (10 μM)	
1.25		Reverse primer (10 μM)	
1		Herculase	
35.5		Nuclease-free water	
Σ 50			
95 °C	2 min		
95 °C XX °C 72 °C	20 s 20 s 30 s/1 kb	25×	For unknown primers, gradient PCR might be necessary.
72 °C	5 min		
8 °C	∞		

13. Save 1 μL for gel electrophoresis (refer to **step 4**).
14. For RNA cleanup, add reagents to PCR tube following Table 8 and mix by pipetting.
15. Insert column in collection tube, load sample (max. 900 μL), spin at 16,000 g for 1 min, and discard flow-through.
16. Add 500 μL RNA Cleanup Wash Buffer (same collection tube), spin at 16,000 g for 1 min, and discard flow-through.
17. Add 500 μL RNA Cleanup Wash Buffer (same collection tube), spin at 16,000 g for 1 min, and discard flow-through.
18. Spin at 16,000 g for 5 min, transfer column in 1.5 mL eppi tube, and elute with 40 μL nuclease-free Water.
19. Elute a second time with 20 μL.
20. Combine both elutions.
21. Measure concentrations with nanodrop "RNA-40."
22. Freeze in aliquots of 5 μL at −80 °C.
23. Run 0.8% Agarose gel with 1 μL RNA sample with RNA ladder (Riboruler SM1823 high range) to verify the bands are of correct size.
24. Run at 90 V for ≥30 min (*see* **Notes 12** and **13**).

Table 6
mRNA synthesis reaction mixture

Amount (μL)	Component
To 20	Nuclease-free water
10	2× ARCA/NTP-Mix
3–8	1 μg PCR (step I)
2	T7 RNA Polymerase Mix
Σ 20	

Table 7
Tailing reaction

Amount (μL)	Component
20	Nuclease-free water
20	IVT reaction (step 2)
5	10× Poly(A)Polymerase Reaction Buffer
5	Poly(A)Polymerase
Σ 50	

Table 8
Monarch RNA Cleanup

Amount (μL)	Component
50	Tailing reaction (step3)
100	RNA Cleanup Binding Buffer
150	EtOH
Σ 300	

3.10 TP53 Base Editing in Human iPS Cells

Maintain iPS cells in Cellartis DEF-CS 500 culture system (Takara Bio) according to manufacturer's instructions. It is advisable to transfect iPS cells at 30–40% confluency in a 6-well plate. To transfect the cells, these steps can be followed:

1. Mix 5 μg ABE mRNA or 3.5 μg CBE-GFP mRNA with corresponding sgRNA in 1:5 molar ratio (BE-sgRNA) in total of 125 μL OptiMEM.
2. Dissolve 3.75 μL RNA MessengerMax reagent in 121 μL Opti-MEM, mix with complex in step1. Incubate the mix at room

temperature for 15 min, and then supplement it with 2.25 mL complete DEF-CS growth media.

3. Keep this mix on cells for 5 h, then aspirate it from cells and add fresh medium.
4. To analyze CBEs and ABEs editing efficiency, isolate total genomic DNA 96 h post transfection. PCR can be performed using Taq or Phusion polymerase and primers amplifying the targeted genomic locus (*see* **Note 14**).

3.11 Quantifying Base-Editing Efficiencies

We recommend measuring the base-editing efficiency of the harvested cells using an online software, for example, EditR (https://moriaritylab.shinyapps.io/editr_v10/) [38]. EditR is a free simple tool, which uses the .ab1 Sanger sequencing file of the potentially edited region (~500 bp) and the spacer sequence (~20 bp) to predict where a base edit has occurred. Once the .ab1 file and DNA protospacer sequence are correctly uploaded, EditR generates a plot displaying editing efficiencies at each base within the protospacer. "An editing efficiency of ~10%" means that ~10% of all harvested cells have a successful edit at the target base. This will be indicated by overlapping chromatogram peaks at the targeted nucleotide.

The following steps have been adapted from the online protocol found at: https://moriaritylab.shinyapps.io/editr_v10/:

1. Upload your .ab1 file of the sequenced region.
2. Enter the sgRNA spacer sequence. (If the sgRNA is antisense to the sequence in the .ab1 file, check the "Guide sequence is reverse complement" box).
3. Click the "Predicted Editing" tab.
4. Examine the protospacer chromatogram and underlying tile plot to determine whether base editing occurred. All colored tiles represent base calls that are deemed significant—that is, if there are multiple colored tiles under a single base call, base editing likely occurred.
5. If you wish to download a report of the operations performed on your data, click the "Download Report" tab on the top of the page.

4 Notes

1. Typing in *BRAF* will yield all the mutations reported in the *BRAF* gene, while typing in V600E will show all cell lines that carry a V600E variant. A given mutation could be searched for as amino acid change (e.g., V600E) or DNA coding sequence mutation (e.g., c.1799T>A). Importantly, clicking on the cell

line name opens a new window providing a comprehensive summary of known facts about this line, including a downloadable list of all described mutations in this cell line.

2. While the data quality of the COSMIC database is very high, we recommend validating that the chosen cell line really carries the mutation via PCR, followed by DNA sequencing.

3. A sgRNA consists of a user-defined ~20 nt spacer sequence that designates the genomic target to be modified and a ~80 nt tracrRNA scaffold sequence that is necessary for Cas9 binding. The protospacer refers to the target DNA region and is identical to the spacer sequence of the sgRNA, and followed by a protospacer adjacent motif (PAM). For sgRNA design, we refer to modifying the spacer sequence only. It is known that nucleotide differences in the seed region close to the PAM sequence improve selective cleavage [39–41]. Hence, for targeting point mutations, choosing protospacers that contain the mutation close to the PAM sequence is preferred over choosing protospacers where the mutation is far from the PAM (Fig. 2). Targeted point mutations in the sgRNA far from the PAM might lead to cleavage of the wild-type allele (see example in Fig. 3). Other variables affecting the occurrence of off-target editing include the number of mismatches (e.g., insertion or deletion mutations are more divergent from wild-type allele, hence likely to exhibit higher specificity). Additionally, extremely high or low GC content and stretches of the same nucleotide identity should be avoided [42, 43]. Furthermore, recent protein engineering of Cas9 has revealed mutant enzymes with higher specificities [27, 28, 44, 45]. These enzymes might be advantageous for improved on-/off-target ratios, and in addition come with alternate PAM preferences.

4. The quality and amount of DNA used for virus production is an important parameter, since higher transfection efficiency and dosage usually yield better-quality virus and subsequently higher genome-targeting efficiencies.

5. Following the production of virus particles (24 h after transfection), Biosafety Level 2 (BSL-2) must be followed. Additional safety precautions are taken to avoid the potential risk hazard to personnel and the environment where only laboratory personnel who have had a specific training in handling BSL-2 grade pathogenic agents are allowed to handle BSL-2 work. The precautions are maintained throughout the virus production protocol.

6. We recommend to determine how many viral particles can be obtained from one well/dish/flask by doing a titration experiment on the cell line of interest. This can be done by adding the

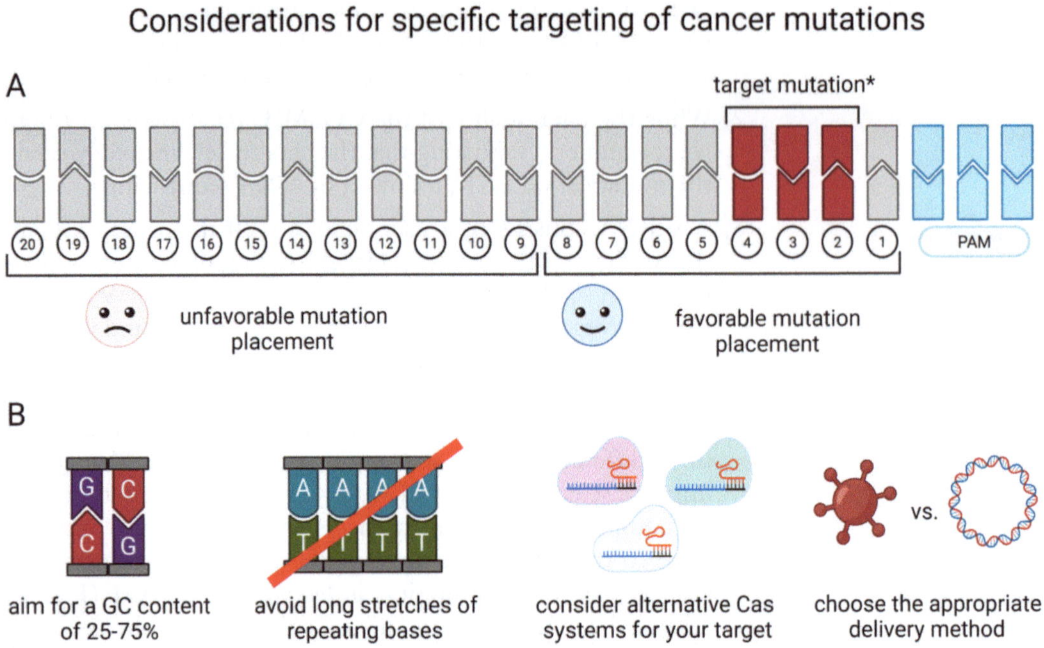

Fig. 2 Considerations for CRISPR-Cas9-mediated targeting of cancer mutations. (**a**) The first eight positions proximal to the PAM sequence are called the seed region. The seed region has a very low tolerance for mismatches and is crucial for on-target sgRNA binding. The mutation of interest should ideally be placed within this region. The mutation of interest should preferably not be located distal from the PAM (bp positions 9–20) because of the increased mismatch tolerance in this region * – the binding specificity is increased further, when the mutation of interest varies from the wild type sequence by >1 nucleotide, that is, sgRNAs against indel mutations will likely have lower off-target rates than sgRNAs against point mutations. (**b**) When spacer sequences are designed manually, care must be taken to avoid disproportionate guanine and cytosine (GC) content and long stretches of repeating bases. In addition, information about the available Cas systems and their respective PAM preference need to be acquired before starting the design process. Finally, long- or short-term expression of sgRNA-Cas9 might be desired, depending on the application and the final goal. There are a variety of delivery forms and methods available, e.g., sgRNA-Cas9 on a DNA vector for transfection or transduction, cotransfection or electroporation of Cas9 mRNA and sgRNA, transfection or electroporation of ribonucleoprotein (RNP) complexes

produced virus in serial dilutions and transductions can be performed in 96 well plates, followed by flow cytometry. This allows scaling the virus production up and down and controlling the ratio of cells to viral particles during transduction, thereby regulating the infection rate in order to do the experiment to inactivate the cancer mutation. If the cells are difficult to infect, try to improve the transduction rate in the presence of protamine sulfate (final concentration 5 μg/mL) and with spinoculation for 1 h at 1000 g at 37 °C in a plate centrifuge.

7. It is crucial to include a negative control sample every time when amplification from gDNA is performed. In this sample,

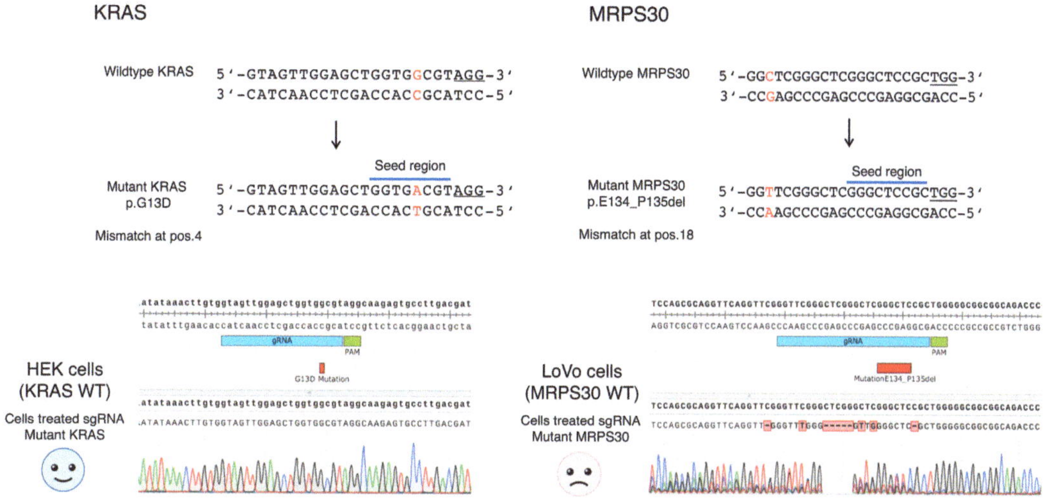

Fig. 3 Mutation position influences CRISPR-Cas9 target specificity. Representative examples of targeting mutations in different positions. In the case of KRAS p.G13D (left panel), the missense G > A mutation can be placed within the seed region of the spacer, because a suitable PAM sequence (AGG, underlined) is available 4 bp downstream. This spacer sequence is very specific, as proven by the clean wild-type (WT) sequencing read obtained after treating KRAS WT HEK cells with the respective sgRNA and Cas9. Note that the sequence chromatogram still shows the wild-type sequence, indicating that no cleavage has taken place. In contrast, the spacer designed for targeting the E134-P135 deletion in the MRPS30 gene (right panel) displayed off-target cutting in the MRPS30 WT LoVo cell line, as can be seen on the noisy sequencing chromatogram. Here, the WT and the mutant genomic sequences vary by one nucleotide outside of the seed region (position 18). This mismatch is not sufficient to prevent cleavage of the wild-type sequence

2 μL water is added instead of the gDNA to control for reagent contamination.

8. T7 Endonuclease 1 (T7E1) Assay to validate CRISPR-Cas9 Editing: as an alternative validation to Sanger sequencing, a T7E1 assay can be used to detect CRISPR-mediated gene editing. Genomic DNA isolated from transfected/infected cells can be used as substrate for a targeted PCR at the locus of Cas9 targeting, as described in [46]. Briefly, after purification and quantification of 500-bp PCR product, hybridize 500 ng of purified PCR product using a thermocycler, then digest with T7 endonuclease I at 37 °C for 20 min, and load on a 1% agarose gel containing 0.01% RedSafe. Run the gel at 90 V for 30 min and visualize the DNA bands with a gel imager.

9. Several free sgRNA design softwares have been described to generate a knockout and/or to predict the frameshift efficiency of a sgRNA, such as CHOPCHOP ([30, 31], https://chopchop.cbu.uib.no), CCTop ([32], https://cctop.cos.uni-heidelberg.de), GPP sgRNA Designer ([28], https://portals.broadinstitute.org/gpp/public/analysis-tools/sgrna-design), or InDelphi ([33], http://indelphi.giffordlab.mit.edu).

However, since we adopt a mutation-selective approach targeting point mutations, sgRNAs should be manually modified to match the mutant sequence in question and manually assessed (e.g., IDT CRISPR-Cas9 guide RNA design checker; https://eu.idtdna.com/site/order/designtool/index/CRISPR_SEQUENCE) to estimate on-/off-target potential. To design sgRNA libraries targeting a complete set of mutations in a particular cell line, one can make use of bioinformatic tools as described by Doench et al. [27].

10. Frequently, sgRNAs differ in only one nucleotide from the wild-type allele sequence, making the wild-type allele the most likely off-target site. False-positive hits could be enriched especially in a pooled screen setting, arising from unspecifically cleaving the WT allele. Therefore, sgRNAs libraries that primarily target cancer mutations other than point mutations provide better candidates for large-scale screens. For point mutations, criteria such as PAM proximity, excluding protospacers with up to two mismatches to the reference genome, especially those residing outside the seed region, in addition to avoiding high/low %GC content and stretches of same sequence identity [47], should be considered. To control for false-positive sgRNAs cleaving the wild-type allele, the sgRNA library should also be delivered to a control cell line, which does not carry the mutation. Only sgRNAs that score in the cell line of interest, but do not score in the control cell line, should be considered further.
11. Free software programs such as the BE-designer [48] or BE-Hive [49] can be used for automated sgRNA design and scoring. As most automated programs lack flexibility in deaminase and Cas variants, we recommend designing your spacers manually, including the mutation and checking predicted editing efficiencies in BE-Hive [49]. If bystander edits are unavoidable, choose the protospacer with bystander nucleotides farthest from the center of the window or one that will incorporate silent mutations, as long as it is not a splice site. If multiple BE-sgRNA options are available, we recommend experimentally validating all of them to identify the optimal combination that maximizes target base editing and minimizes bystander editing. In all base editing experiments, the spacer sequence will always contain the target base, because direct nucleobase modification occurs on ssDNA of the "nontarget strand" that is not duplexed with sgRNA.
12. Low editing efficiency/low cell viability after transfection: Sometimes, a low editing efficiency is due to low transfection efficiency. It is generally advisable to optimize the ratio of transfection reagent to mRNA. However, transfection of Cas9/BE mRNA together with synthetic sgRNA (Synthego) has shown to be well-tolerated with minimal toxicity, in various

cell types. Alternatively, Cas9-sgRNA ribonucleoprotein (RNP) can be used in cells, which are difficult to transfect, such as primary cells.

13. Base editor considerations and bystander editing: Bystander edits occur when As or Cs beyond the desired target nucleotide are also present in the editing window and become mutated concurrently with the target base. Bystander editing can be avoided by sliding the protospacer a few bases and thereby rendering potential bystander bases out of the editing window. Bystander edits may also be mitigated by using deaminase enzymes with alternative BE architectures, such as rigid linkers [35], or those that are modified to be less processive or sequence-specific. Moreover, ABEs, containing the mutations V106W, F148A, and V82G, show almost no off-target RNA editing [50]. In addition, the ABE8e variant significantly increases editing efficiencies; however, it also has a wider editing window, increasing the chance of bystander mutations [51]. In some cases, bystander edits are acceptable, such as in gene-knockout experiments or if the bystander creates a silent mutation, as long as it is not a splice site. When attempting to knock out genes via premature stop codon introduction or splice-site disruption, disruptions targeted to the first and last few exons can be inefficient because of translation reinitiating and/or alternative splicing. To achieve the highest level of protein knockout, it is advisable to mutate splice sites in the middle of the gene [52].
14. At higher cell densities, transfection efficiencies might drop, but the toxicity will likely be lower. For the more toxic CBEs, we recommend transfecting the cells at 70% confluency and use less mRNA, while for the ABEs, we recommend transfecting at 30% confluency and use more mRNA.

Acknowledgments

We thank Martina Augsburg for the excellent technical assistance. Figures were created using www.BioRender.com.

References

1. Stratton MR, Campbell PJ, Futreal PA (2009) The cancer genome. Nature 458:719–724. https://doi.org/10.1038/nature07943
2. Martincorena I, Campbell PJ (2015) Somatic mutation in cancer and normal cells. Science 349:1483–1489. https://doi.org/10.1126/science.aab4082
3. Gnad F, Baucom A, Mukhyala K, Manning G, Zhang Z (2013) Assessment of computational methods for predicting the effects of missense mutations in human cancers. BMC Genomics 14(Suppl 3):S7. https://doi.org/10.1186/1471-2164-14-S3-S7
4. Weghorn D, Sunyaev S (2017) Bayesian inference of negative and positive selection in human cancers. Nat Genet 49:1785–1788. https://doi.org/10.1038/ng.3987

5. Gonzalez-Perez A, López-Bigas N (2012) Functional impact bias reveals cancer drivers. Nucleic Acids Res 40:e169–e169. https://doi.org/10.1093/nar/gks743
6. Lawrence MS, Stojanov P, Polak P et al (2013) Mutational heterogeneity in cancer and the search for new cancer-associated genes. Nature 499(7457):214–218. https://doi.org/10.1038/nature12213
7. Mularoni L, Sabarinathan R, Deu-Pons J et al (2016) OncodriveFML: a general framework to identify coding and non-coding regions with cancer driver mutations. Genome Biol 17(1):128. https://doi.org/10.1186/s13059-016-0994-0
8. Martincorena I, Raine KM, Gerstung M et al (2017) Universal patterns of selection in cancer and somatic tissues. Cell 171:1029–1041.e21. https://doi.org/10.1016/j.cell.2017.09.042
9. Buljan M, Blattmann P, Aebersold R, Boutros M (2018) Systematic characterization of pan-cancer mutation clusters. Mol Syst Biol 14:e7974-19. https://doi.org/10.15252/msb.20177974
10. Dietlein F, Weghorn D, Taylor-Weiner A et al (2020) Identification of cancer driver genes based on nucleotide context. Nat Genet 52: 208–218. https://doi.org/10.1038/s41588-019-0572-y
11. Malone ER, Oliva M, Sabatini PJB et al (2020) Molecular profiling for precision cancer therapies. Genome Med 12(1):1–19. https://doi.org/10.1186/s13073-019-0703-1
12. Koo T, Yoon A-R, Cho H-Y et al (2017) Selective disruption of an oncogenic mutant allele by CRISPR/Cas9 induces efficient tumor regression. Nucleic Acids Res 45:7897–7908. https://doi.org/10.1093/nar/gkx490
13. Gebler C, Lohoff T, Paszkowski-Rogacz M et al (2017) Inactivation of cancer mutations utilizing CRISPR/Cas9. J Natl Cancer Inst 109:1546. https://doi.org/10.1093/jnci/djw183
14. Yang M, Wei H, Wang Y et al (2017) Targeted disruption of V600E-mutant BRAF gene by CRISPR-Cpf1. Mol Ther Nucleic Acid 8: 450–458. https://doi.org/10.1016/j.omtn.2017.05.009
15. Kim W, Lee S, Kim HS et al (2018) Targeting mutant KRAS with CRISPR-Cas9 controls tumor growth. Genome Res 28:374–382. https://doi.org/10.1101/gr.223891.117
16. Rees HA, Liu DR (2018) Base editing: precision chemistry on the genome and transcriptome of living cells. Nat Rev Genet 19:770–788. https://doi.org/10.1038/s41576-018-0059-1
17. Sürün D, Schneider A, Mircetic J et al (2020) Efficient generation and correction of mutations in human iPS cells utilizing mRNAs of CRISPR base editors and prime editors. Genes (Basel) 11:511–519. https://doi.org/10.3390/genes11050511
18. Barretina J, Caponigro G, Stransky N et al (2012) The Cancer Cell Line Encyclopedia enables predictive modelling of anticancer drug sensitivity. Nature 483(7391):603–607. https://doi.org/10.1038/nature11003
19. Ghandi M, Huang FW, Jané-Valbuena J et al (2019) Next-generation characterization of the Cancer Cell Line Encyclopedia. Nature 569(7757):503–508. https://doi.org/10.1038/s41586-019-1186-3
20. Tym JE, Mitsopoulos C, Coker EA, Razaz P, Schierz AC, Antolin AA, Al-Lazikani B (2016) canSAR: an updated cancer research and drug discovery knowledgebase. Nucleic Acids Res 44(D1):D938–D943. https://doi.org/10.1093/nar/gkv1030
21. Mitsopoulos C, Di Micco P, Fernandez EV et al (2020) canSAR: update to the cancer translational research and drug discovery knowledgebase. Nucleic Acids Res 47(D1):D917–D922. https://doi.org/10.1093/nar/gkaa1059
22. Bamford S, Dawson E, Forbes S, Clements J, Pettett R, Dogan A et al (2004) The COSMIC (catalogue of somatic mutations in cancer) database and website. Br J Cancer 91(2): 355–358. https://doi.org/10.1038/sj.bjc.6601894
23. Tate JG, Bamford S, Jubb HC et al (2018) COSMIC: the catalogue of somatic mutations in cancer. Nucleic Acids Res 47:D941–D947. https://doi.org/10.1093/nar/gky1015
24. Ran FA, Hsu PD, Wright J et al (2013) Genome engineering using the CRISPR-Cas9 system. Nat Protoc 8:2281–2308. https://doi.org/10.1038/nprot.2013.143
25. Karimova M et al (2015) CRISPR/Cas9 nickase-mediated disruption of hepatitis B virus open reading frame S and X. Sci Rep. Nature Publishing Group 5:13734. https://doi.org/10.1038/srep13734
26. Untergasser A, Cutcutache I, Koressaar T et al (2012) Primer3--new capabilities and interfaces. Nucleic Acids Res 40:e115

27. Slaymaker IM, Gao L, Zetsche B, Scott DA, Yan WX, Zhang F (2016) Rationally engineered Cas9 nucleases with improved specificity. Science 351(6268):84–88. https://doi.org/10.1126/science.aad5227
28. Chen JS, Dagdas YS, Kleinstiver BP et al (2017) Enhanced proofreading governs CRISPR–Cas9 targeting accuracy. Nat Publ Group 550:407–410. https://doi.org/10.1038/nature24268
29. Marco-Sola S, Sammeth M, Guigó R, Ribeca P (2012) The GEM mapper: fast, accurate and versatile alignment by filtration. Nat Methods 9:1185–1188. https://doi.org/10.1038/nmeth.2221
30. Evers B, Jastrzebski K, Heijmans JPM et al (2016) CRISPR knockout screening outperforms shRNA and CRISPRi in identifying essential genes. Nat Biotechnol 34:631–633. https://doi.org/10.1038/nbt.3536
31. Ben Jehuda R, Shemer Y, Binah O (2018) Genome editing in induced pluripotent stem cells using CRISPR/Cas9. Stem Cell Rev Rep 14(3):323–336. https://doi.org/10.1007/s12015-018-9811-3
32. Rouhani F, Kumasaka N, de Brito MC, Bradley A, Vallier L, Gaffney D (2014) Genetic background drives transcriptional variation in human induced pluripotent stem cells. PLoS Genet 10(6):e1004432. https://doi.org/10.1371/journal.pgen.1004432
33. Jin S, Zong Y, Gao Q et al (2019) Cytosine, but not adenine, base editors induce genome-wide off-target mutations in rice. Science 36: eaaw7166–eaaw7167. https://doi.org/10.1126/science.aaw7166
34. McGrath E, Shin H, Zhang L et al (2019) Targeting specificity of APOBEC-based cytosine base editor in human iPSCs determined by whole genome sequencing. Nat Commun 10: 1–9. https://doi.org/10.1038/s41467-019-13342-8
35. Tan J, Zhang F, Karcher D, Bock R (2019) Engineering of high-precision base editors for site-specific single nucleotide replacement. Nat Commun 10:1–10. https://doi.org/10.1038/s41467-018-08034-8
36. Zuo E, Sun Y, Wei W et al (2019) Cytosine base editor generates substantial off-target single-nucleotide variants in mouse embryos. Science 148:eaav9973–eaav9977. https://doi.org/10.1126/science.aav9973
37. Rees HA, Komor AC, Yeh W-H et al (2017) Improving the DNA specificity and applicability of base editing through protein engineering and protein delivery. Nat Commun 8:1–10. https://doi.org/10.1038/ncomms15790
38. Kluesner MG, Nedveck DA, Lahr WS et al (2018) EditR: a method to quantify base editing from sanger sequencing. CRISPR J 1:239–250. https://doi.org/10.1089/crispr.2018.0014
39. Kuscu C, Arslan S, Singh R et al (2014) Genome-wide analysis reveals characteristics of off-target sites bound by the Cas9 endonuclease. Nat Biotechnol 32(7):677–683. https://doi.org/10.1038/nbt.2916
40. Zheng T, Hou Y, Zhang P et al (2017) Profiling single-guide RNA specificity reveals a mismatch sensitive core sequence. Sci Rep 7: 1–8. https://doi.org/10.1038/srep40638
41. Anderson EM, Haupt A, Schiel JA et al (2015) Systematic analysis of CRISPR–Cas9 mismatch tolerance reveals low levels of off-target activity. J Biotechnol 211:56–65. https://doi.org/10.1016/j.jbiotec.2015.06.427
42. Wang X, Huang X, Fang X et al (2016) CRISPR-Cas9 system as a versatile tool for genome engineering in human cells. Mol Ther Nucleic Acid 5:1–9. https://doi.org/10.1038/mtna.2016.95
43. Wang T, Wei JJ, Sabatini DM, Lander ES (2014) Genetic screens in human cells using the CRISPR-Cas9 system. Science 343:80–84. https://doi.org/10.1126/science.1246981
44. Kleinstiver BP, Pattanayak V, Prew MS et al (2016) High-fidelity CRISPR–Cas9 nucleases with no detectable genome-wide off-target effects. Nature 529(7587):490–495. https://doi.org/10.1038/nature16526
45. Acharya S, Mishra A, Paul D et al (2019) Francisella novicida Cas9 interrogates genomic DNA with very high specificity and can be used for mammalian genome editing. Proc Natl Acad Sci U S A 116:20959
46. Sayed S, Paszkowski-Rogacz M, Schmitt LT, Buchholz F (2019) CRISPR/Cas9 as a tool to dissect cancer mutations. Methods 164–165:36–48. https://doi.org/10.1016/j.ymeth.2019.05.007
47. Wilkins BJ, Rall NA, Ostwal Y et al (2014) A cascade of histone modifications induces chromatin condensation in mitosis. Science 343:

77–80. https://doi.org/10.1126/science.1244508

48. Hwang G-H, Park J, Lim K et al (2018) Web-based design and analysis tools for CRISPR base editing. BMC Bioinformatics 19:1–7. https://doi.org/10.1186/s12859-018-2585-4
49. Arbab M, Shen MW, Mok B et al (2020) Determinants of base editing outcomes from target library analysis and machine learning. Cell 182(2):463–480.e30. https://doi.org/10.1016/j.cell.2020.05.037
50. Anzalone AV, Koblan LW, Liu DR (2020) Genome editing with CRISPR–Cas nucleases, base editors, transposases and prime editors. Nat Biotechnol 38(7):824–844. https://doi.org/10.1038/s41587-020-0561-9
51. Richter MF, Zhao KT, Eton E et al (2020) Phage-assisted evolution of an adenine base editor with improved Cas domain compatibility and activity. Nat Biotechnol 38(7): 883–891. https://doi.org/10.1038/s41587-020-0453-z
52. Webber BR, Lonetree C-L, Kluesner MG et al (2019) Highly efficient multiplex human T cell engineering without double-strand breaks using Cas9 base editors. Nat Commun 10:1–10. https://doi.org/10.1038/s41467-019-13007-6

Chapter 19

Single-Cell RNA Sequencing Analysis Using Fluidigm C1 Platform for Characterization of Heterogeneous Transcriptomes

Jinhong Kim and Paola A. Marignani

Abstract

Single-cell RNA sequencing (scRNA-seq) is an innovative technology that can be used to characterize transcriptome heterogeneity at the single-cell resolution. The Fluidigm C1 is an automated system for preparing complementary deoxyribonucleic acid (cDNA) samples that have been synthesized and preamplified from single cells for next-generation sequencing (NGS) analysis. Herein, we detail a workflow and protocol for scRNA-seq analysis of 3′-end enriched cDNA libraries and introduce strategies for validation of differential gene expression.

Key words scRNA-seq, Fluidigm C1 systems, Differential gene expression, Validation, qRT-PCR, Protein expression analysis

1 Introduction

Single-cell RNA-seq, an application that permits advanced transcriptome analysis, is used to characterize heterogeneous transcriptomes by profiling differentially expressed genes (DEGs) at the single-cell resolution [1]. Automated scRNA-seq applications are becoming more popular and are contributing to a better understanding of complex transcriptomes in various research areas (https://www.fluidigm.com/publications/c1), including cancers [2, 3]. Particular attention to sample preparation is a prerequisite to isolating high-quality total RNA from individual cells. Optimizing frontend workflow will ultimately result in high-quality datasets that will permit the identification of quantitative differences among aligned reads per gene among single cells.

Herein, we describe a simple protocol for scRNA-seq using the Fluidigm C1 advanced microfluidic platform with integrated fluidic circuit (IFC) composed of high-throughput chambers that are able to capture 800 single cells. The IFC device provides automated

Sherri L. Christian (ed.), *Cancer Cell Biology: Methods and Protocols*,
Methods in Molecular Biology, vol. 2508, https://doi.org/10.1007/978-1-0716-2376-3_19,

procedures from total RNA isolation to preamplification of synthesized cDNAs in C1 systems. For flexibility, IFCs capture cells in two different ranges of cell sizes, (1) 5–10 μm and (2) 10–17 μm, so that users can choose proper IFCs according to optimal cell size range. For further optimization, users need to identify ideal cell buoyancy that enables cells suspended in C1 suspension reagent to be uniformly distributed vertically and horizontally in multiple layers. Loading a mixture of cells in C1 suspension reagent at optimum buoyancy ratio increases single cell-capturing rates into chambers as cells evenly flow through microfluidic channels within the IFC. Successful DEG detection resulted from applying a combination of (1) popular bioinformatic tools that mine publicly available bulk RNA-seq datasets from raw read processing to sequence alignment procedures and (2) recently developed user-friendly bioinformatic tools specific to scRNA-seq analysis from normalization of expression values to single cell-clustering steps.

In this chapter, we introduce a protocol for scRNA-seq analysis using Fluidigm C1 systems following seven simple procedures that span 3′-end enriched cDNA library construction to the validation of detected DEG at gene and protein levels (Fig. 1).

2 Materials

Materials introduced in this chapter can be used as the protocol was developed for scRNA-seq analysis using Fluidigm C1 automated single-cell analysis systems [4]. Users can apply their own reagents or preferred bioinformatic tools for detection and validation of DEGs.

2.1 Single-Cell Preparation

1. Cell media: Mix 40 mL (8%) fetal bovine serum (FBS) and 460 mL of Dulbecco's Modified Eagle Medium (DMEM; 4.5 g/L D-Glucose, L-Glutamine) or media specific to cells of interest. Store at 4 °C.
2. Dulbecco's Phosphate-Buffered Saline (DPBS).
3. 50 mL V-bottom centrifuge tubes.
4. 40 μm cell strainer.
5. Water bath.
6. Trypan blue.
7. Hemocytometer and/or automated cell-counting systems.
8. Inverted microscope.
9. EasySep magnet (STEMCELL, Inc.) with EasySepTM Dead Cell Removal (Annexin V) kits (STEMCELL, Inc.).
10. Sterile 5 mL (12 × 75 mm) polystyrene U-bottom tubes (no caps required).
11. Dead cell-removal media: DPBS, 2% FBS, 1 mM $CaCl_2$.

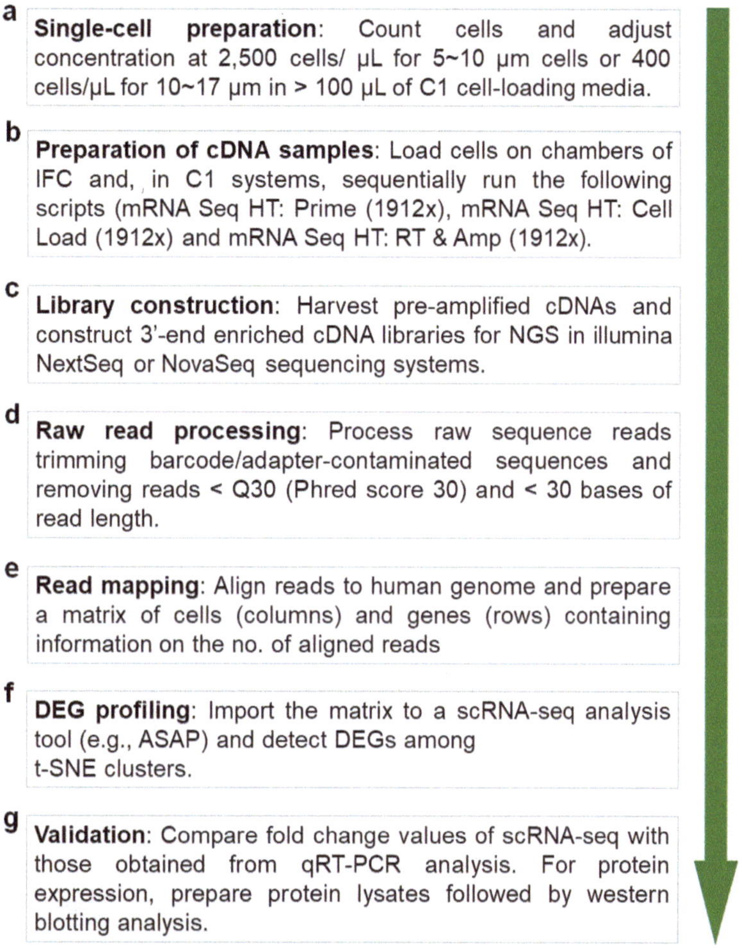

Fig. 1 Workflow of scRNA-seq analysis. (**a–g**) Seven primary procedural requirements for scRNA-seq analysis from single-cell preparation (**a**) to validation of profiled genes (**g**)

2.2 Buoyancy Test

1. C1 cell-loading media: Mix 400 µL 0.5 M EDTA, 5 mL (5%) FBS, and 94.6 mL DPBS. Store at 4 °C.
2. C1 suspension reagent (Fluidigm, Inc.).
3. 0.2 mL polymerase chain reaction (PCR) tubes.
4. Clear-flat bottom 384-well microplate.
5. Inverted microscope.

2.3 Construction of 3′-End Enriched cDNA Libraries (see Note 1)

1. C1 Manuals: C1 HT mRNA-seq technical note for 5–10 µm cells (PN 101-7557; Fluidigm, Inc.) or C1 HT mRNA-seq protocol for 10–17 µm cells (PN 101-4964; Fluidigm, Inc.).
2. IFCs: C1 Single-Cell mRNA Seq HT IFC for 5–10 µm cells (PN 101-0219; Fluidigm, Inc.) or 10–17 µm cells (PN 101-4981 or PN 101-4982).

3. ERCC RNA Spike-in control (invitrogen; Cat. No. 4456740).
4. Sterile nuclease-free PCR grade H_2O.
5. Low DNA/Protein-binding 0.2 mL PCR tube strips and caps.
6. Low DNA/Protein-binding 96-well PCR plates, and Microseal 'B' PCR plate sealing films (Cat. No. MSB1001; Bio-Rad Laboratories, Inc.).
7. Low DNA/Protein-binding 1.5 mL microcentrifuge tubes.
8. Low DNA/Protein-binding U-bottom 96-well microplates and magnet for 96-well plates.
9. Freshly made 75% ethanol: Mix 75 mL absolute ethanol and 25 mL sterile nuclease-free H_2O in a glass bottle. Measure the volume of ethanol and water separately in a 100 mL cylinder. Store at −20 °C.
10. High Sensitivity DNA Kit (Cat. No. 5067-4626; Agilent, Inc.) and Bioanalyzer 2100 systems (Agilent, Inc.), or gel-electrophoresis-based microfluidic DNA quantification systems.
11. Picogreen double-strand DNA assay and fluorescent DNA quantification systems.
12. 96-well microplates featured with black-colored wells and flat-opaque bottom.
13. Sterile filter tips (recommended) for various volumes from 0.1 μL to 1250 μL.

2.4 Bioinformatic Analyses

1. C1 mRNA Sequencing High-Throughput Demultiplexer Script (Fluidigm, Inc.)
2. sabre (RRID:SCR_011843).
3. BBDuk Trimmer (a part of Bestus Bioinformatics Tools; RRID:SCR_016968).
4. HISAT2 (RRID:SCR_015530).
5. FeatureCounts (a part of Subread; RRID:SCR_009803).
6. Spreadsheet software and text editor.
7. ASAP (https://asap.epfl.ch/).
8. Gene set enrichment analysis: Gene ontology (GO) through PANTHER classification system (http://pantherdb.org/), and KEGG pathway or MSigDB Oncogenic Signatures database through ASAP.
9. Morpheus by Broad Institute (RRID:SCR_017386).
10. GraphPad Prism (RRID:SCR_002798).

2.5 DEG Validation at Gene Level

1. Commercially available RNA isolation kits or TRIzol™ Reagent (invitrogen).
2. Chloroform.

3. Freshly made 70% ethanol. Mix 70 mL absolute ethanol and 30 mL sterile nuclease-free H_2O in a glass bottle. Measure the volume of ethanol and water separately in a 100 mL cylinder. Store at −20 °C.
4. DNase I or TURBO™ DNA-free kits (Invitrogen).
5. UV spectrophotometer.
6. One-step quantitative reverse transcription (qRT)-PCR kits or single-strand cDNA synthesis kits for two-step qRT-PCR.
7. SYBR- or probe-based qPCR kits corresponding to users' qRT-PCR systems.
8. Preferred primer-designing tool or existing qPCR primer databases.
9. Optical 96-well PCR plates and sealing films or optical 8-strip PCR tubes and caps.

2.6 DEG Validation at Protein Level

1. Cell lysis buffer: 50 mM Tris–HCI (pH 7.5), 150 mM NaCl, 5 mM EDTA, 0.1% NP40, 0.02% SDS, 0.05% DOC, 50 mM NaF, 1 mM sodium vanadate, 4 μg/mL protease inhibitors.
2. 5x Bradford protein assay reagent (Cat. No. 500-0006; Bio-Rad Laboratories, Inc.).
3. Ultracentrifuge.
4. 1.5 mL ultracentrifuge tubes.
5. Heating block.
6. Cuvettes.
7. Visual-light spectrophotometer.
8. 4× stacking gel buffer (pH 8.8): 1.4 M Tris–HCl, 0.4% SDS, sterile protease-free H_2O. Adjust pH with HCI.
9. SDS-PAGE gel (stacking area): 15% 37.5:1 acrylamide:bisacrylamide stabilized solution, 25% 4× stacking gel buffer (pH 8.8), 0.2% TEMED, 1% APS, 60% sterilized H_2O.
10. SDS-PAGE gel (resolving area): 25~50% 37.5:1 acrylamide to bisacrylamide stabilized solution, 25% 4× stacking gel buffer, 0.083% TEMED, 0.08% APS.
11. 5× sample loading buffer: 625 mM Tris–HCI (pH 6.8), 5% SDS, 25% glycerol, 50 mM DTT, 0.025% bromophenol blue, sterile protease-free H_2O. Make aliquots in 500 μL. Store at −20 °C.
12. 10× gel-running buffer: 1% SDS, 250 mM Tris, 1.92 M Glycine, sterile protease-free H_2O.
13. Protein size standard.
14. 10× Tris Buffered Saline (TBS; pH 7.5). 0.5 M Tris, 1.5 M NaCl, sterile protease-free H_2O. Adjust pH with HCl. Store at 4 °C.

15. 1× TBST: 10% TBS, 0.05% Tween-20, sterile protease-free H_2O.
16. Gel casting units and power supply.
17. Semidry transfer buffer: 48 mM Tris, 40 mM n-caproic acid, 1.3 mM SDS, 20% HPLC-grade methanol, HPLC-grade absolute methanol, sterile protease-free H_2O.
18. Hydrophobic PVDF transfer membrane.
19. Swing rocker systems.
20. Gel-transfer systems.
21. Blocking buffer: 3% milk powder, 1× TBST.
22. Primary and secondary antibodies for proteins of interest.
23. Bovine serum albumin (BSA) proteins.
24. Preferred membrane imaging systems.

3 Methods

3.1 Single-Cell Preparation

1. Cells should undergo repeated cycles of freeze-thaw at 37 °C in water bath for 3 min. Skip cell-thawing procedure when using cells in culture or freshly dissociated cells. Prepare 5 mL of 8% (v/v) FBS-containing DMEM in a 50 mL V-bottom centrifuge tube, and place a 40 μm cell strainer on the top of the tube. Add cells that have been completely thawed (or fresh cells) to the cell strainer to sieve. Collect residual cells from the original cell vial rinsing with 1 mL of the DMEM and filter through the cell strainer. Centrifugate cells at 4 °C and 200 × g for 5 min. Remove supernatant and wash cells in 5 mL of DPBS. Repeat centrifugation, and aspirate supernatant. Resuspend cells in 1 mL of DPBS.
2. Prepare a mixture of 10 μL of Trypan blue and 10 μL of washed cells in a 1.5 mL microcentrifuge tube. Gently mix by pipetting up and down >5 times. Load 10 μL of the cell mixture onto slide. Count cells using a hemocytometer on an inverted microscope or automated cell-counting systems. If the number of cells is <1 × 10^7 cells/mL, resuspend cells in 100 μL of dead-cell removal media following centrifugation and removal of supernatant. Collect live cells following manufacturer's protocol of EasySep Dead Cell Removal kits for negative magnetic separation approach.
3. After removal of dead cells, evenly divide live cells in ~2.5 mL of dead-cell removal media into two new 1.5 mL microcentrifuge tubes. Centrifugate at 4 °C and 200 × g for 5 min. Remove supernatant and then resuspend cells in the first 1.5 mL microcentrifuge tube with 500 μL of DPBS. Transfer resuspended cells to the second 1.5 mL microcentrifuge tube

and mix cells by gently pipetting up and down >5 times. Collect residual cells from the first 1.5 mL microcentrifuge tube with another 500 μL of DPBS and transfer to the second 1.5 mL microcentrifuge tube. Mix cells gently by pipetting up and down >5 times.

4. See the **step 2** of Subheading 3.1 to count cells and the step 3 of Subheading 3.1 to centrifuge. Following removal of supernatant, adjust cells at the concentration of 2500 cells/μL for 5–10 μm size cells or 400 cells/μL for 10–17 μm size cells in >100 μL of C1 cell-loading media (*see* **Note 2**).

3.2 Buoyancy Test

1. To identify optimal buoyancy ratio where cells are uniformly distributed in the C1 suspension reagent without horizontal and vertical bias (Fig. 2), prepare a mixture of cells and C1 suspension reagent at five different ratios (Table 1) in five 0.2 mL PCR tubes.
2. Transfer 20 μL of each titrated cell-reagent mixture to 384-well clear-flat bottom plate, and identify an optimum buoyancy using an inverted microscope (*see* **Note 3**).

3.3 Construction of 3′-End Enriched cDNA Library Pools for NGS

1. From Fluidigm C1 Support (https://www.fluidigm.com/productsupport/c1-support-hub), download C1 HT mRNA-seq technical note for 5–10 μm cells or C1 HT mRNA-seq protocol for 10–17 μm cells (*see* **Note 4**).
2. Follow the manufacturer's protocol or technical note with the minor modification: Prepare a diluent of ERCC RNA Spike-in control at 1:60,000. To make ERCC RNA Spike-in diluent, take four steps of consecutive dilutions as shown in the Table 2 (*see* **Note 5**).
3. Check for bubbles with an illuminated magnifier immediately after loading reagents into the IFC. When bubbles are found, reload reagents or remove bubbles by gently pipetting. Bubbles will disrupt the flow within IFC microfluidic channels (*see* **Note 6**).
4. Following a cell-loading step, examine and capture image of individual chambers using a camera-equipped inverted microscope (Fig. 3) (*see* **Note 7**).
5. Once after cDNA preamplification is done, unload IFC from the C1 systems and transfer 20 preamplified cDNA samples that contain 40 single-cell libraries per preamplified cDNA sample ($20 \times 40 = 800$ single-cell libraries in total) to a low DNA-/Protein-binding 96-well PCR plate following the order described in the Fluidigm C1 HT mRNA-seq protocol or technical notes. Quantify individual cDNA samples using a Picogreen-based assay in fluorescence-based DNA

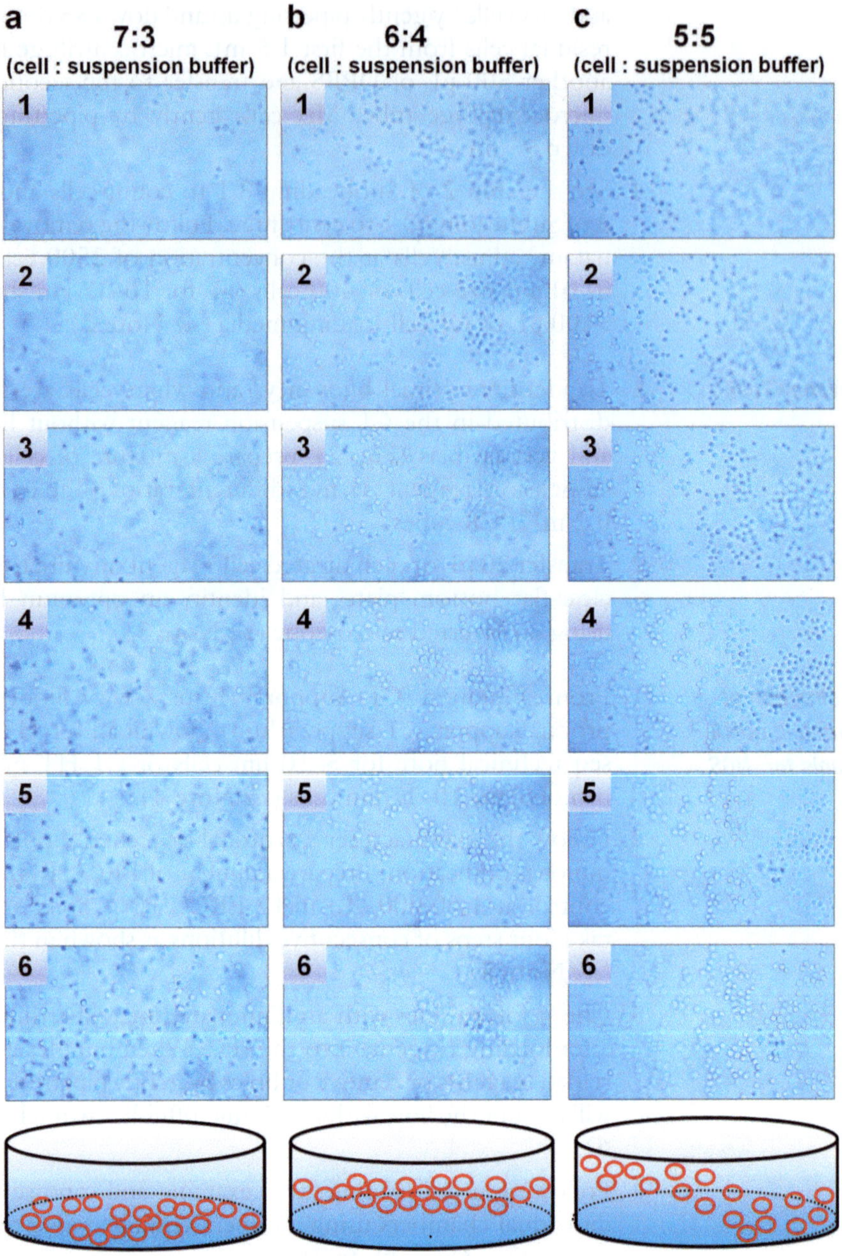

Fig. 2 Buoyancy test conducted using H1299 lung adenocarcinoma epithelial cell line. (**a**) Buoyancy at 7:3 ratio of cells to suspension reagent. The majority of cells settle in the lower layers (picture number 5 and 6). (**b**) Buoyancy at 6:4 ratio of cells to suspension reagent. Independent of cell size, the majority of cells settled horizontally in the middle layer (picture number 3–5). This 6:4 buoyancy ratio is the default recommended by C1 HT mRNA protocol. (**c**) Buoyancy at 5:5 ratio of cells to suspension reagent. The majority of cells are horizontally biased showing inverse-linear distribution. Picture number 1 presents the highest layer, while picture number 6 indicates the lowest layer. Note: prior to visualization/imaging, ensure that the inverted microscope is leveled

Table 1
Buoyancy test

Cell: C1 suspension reagent	1:9	9:1	7:3	6:4	5:5
Cells[a] (μL)	3	27	21	18	15
C1 suspension reagent[b] (μL)	27	3	9	12	15
Total volume (μL)	30				

[a]Adjusted at the concentration of 400 cells/μL in C1 cell-loading media
[b]Suspension reagent from C1 Single-Cell mRNA Seq HT Reagent Kit v2 (Fluidigm, Inc.)

Table 2
Preparation of ERCC RNA Spike-in diluent

Serial dilution	1:10	1:100	1:600	1:60,000
Previous diluent (μL)	1[a]	2	2	2
Nuclease-free H_2O (μL)	9	18	10	198

[a]ERCC RNA Spike-in in an original vial

quantification systems. Following this, adjust the concentration of individual cDNA samples to 0.2 ng/μL with 1× TE buffer.

6. Construct 3′-end enriched cDNA libraries from a diluent (0.2 ng/μL) of cDNA samples, then clean up cDNA libraries using a bead purification method as described in the C1 HT mRNA protocol or technical note. Quantify cDNA libraries using Agilent High Sensitivity DNA kits in Bioanalyzer 2100 systems or gel-electrophoresis-based DNA quantification systems to obtain molar concentration (Fig. 4) (*see* **Note 8**).
7. Perform NGS using Illumina sequencing systems, such as NextSeq or NovaSeq instruments, at the mode of "Paired-end read" as sequence dataset from "Read1" are used only to demultiplex cDNA libraries and Read2 dataset will be actually used for read mapping (*see* **Note 9**).
8. Set a read depth in NGS to obtain at least >50,000 reads per single cell library when sequencing >1600 libraries (*see* **Note 10**).

3.4 Bioinformatic Analysis

1. Linux operating system is recommended for informatic analysis. Download the C1 mRNA Sequencing High Throughput Demultiplexer Script from Fluidigm (https://www.fluidigm.com/software) and demultiplex NGS dataset to 20 read sets of

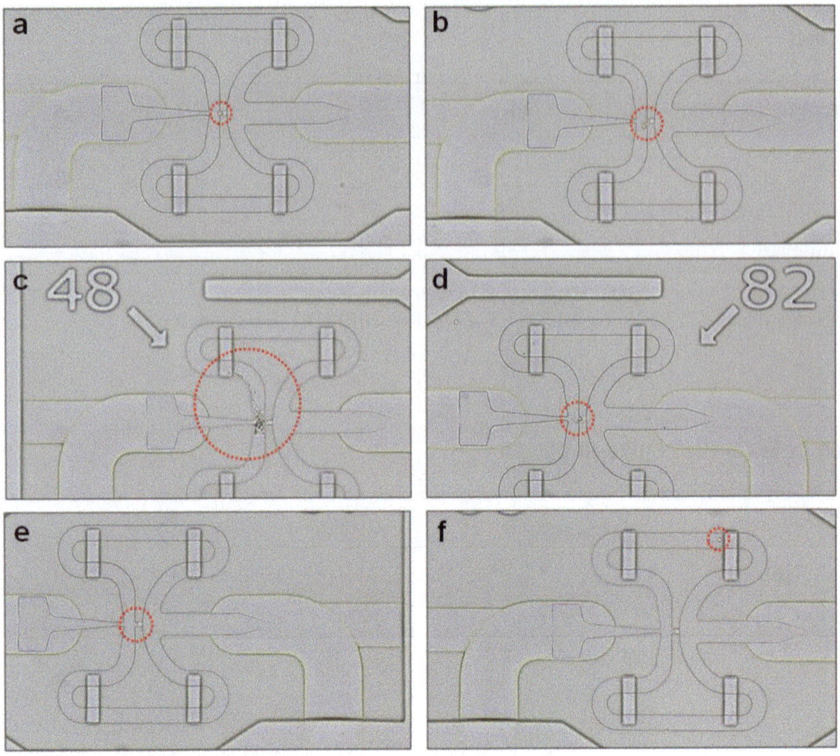

Fig. 3 Single cell-capturing chamber of IFC. (**a**) An intact single cell captured within the chamber's cell-capturing site. (**b**) Double cells captured within the chamber's cell-capturing site. (**c**) A clump of cell debris blocked the cell path. (**d**) A single cell with cell debris captured within the chamber's cell-capturing site. (**e**) Empty chamber where no cells were captured. (**f**) Single cell captured within the cell path of the chamber's cell-capturing site. Highlighted red-dotted lines indicate cell-capturing sites (**a–e**) or cell (**f**) in a chamber. The number at top left/right indicates chamber ID

40 prepooled libraries following manufacturer's instructions in C1 HT mRNA-seq technical note or C1 HT mRNA-seq protocol.

2. Download a 2nd demultiplexing tool, sabre, from https://github.com/najoshi/sabre and install the tool in Linux operating system. Demultiplex 800 individual 3′-end enriched single cell cDNA libraries from the 20 raw read sets preliminarily demultiplexed in the Subheading 3.4, **step 1**. When demultiplexing, use "`sabre se`" command that indicates to demultiplex read sets at "single-end read" mode (*see* **Note 11** and **12**).

3. Download BBDuk Trimmer from https://sourceforge.net/projects/bbmap/ and install the tool in Linux, Macintoch, or Windows operating system. When demultiplexing, set read length >30 bases and quality score >Q30 (equivalent to Phred score 30 indicating a potential sequencing-error rate

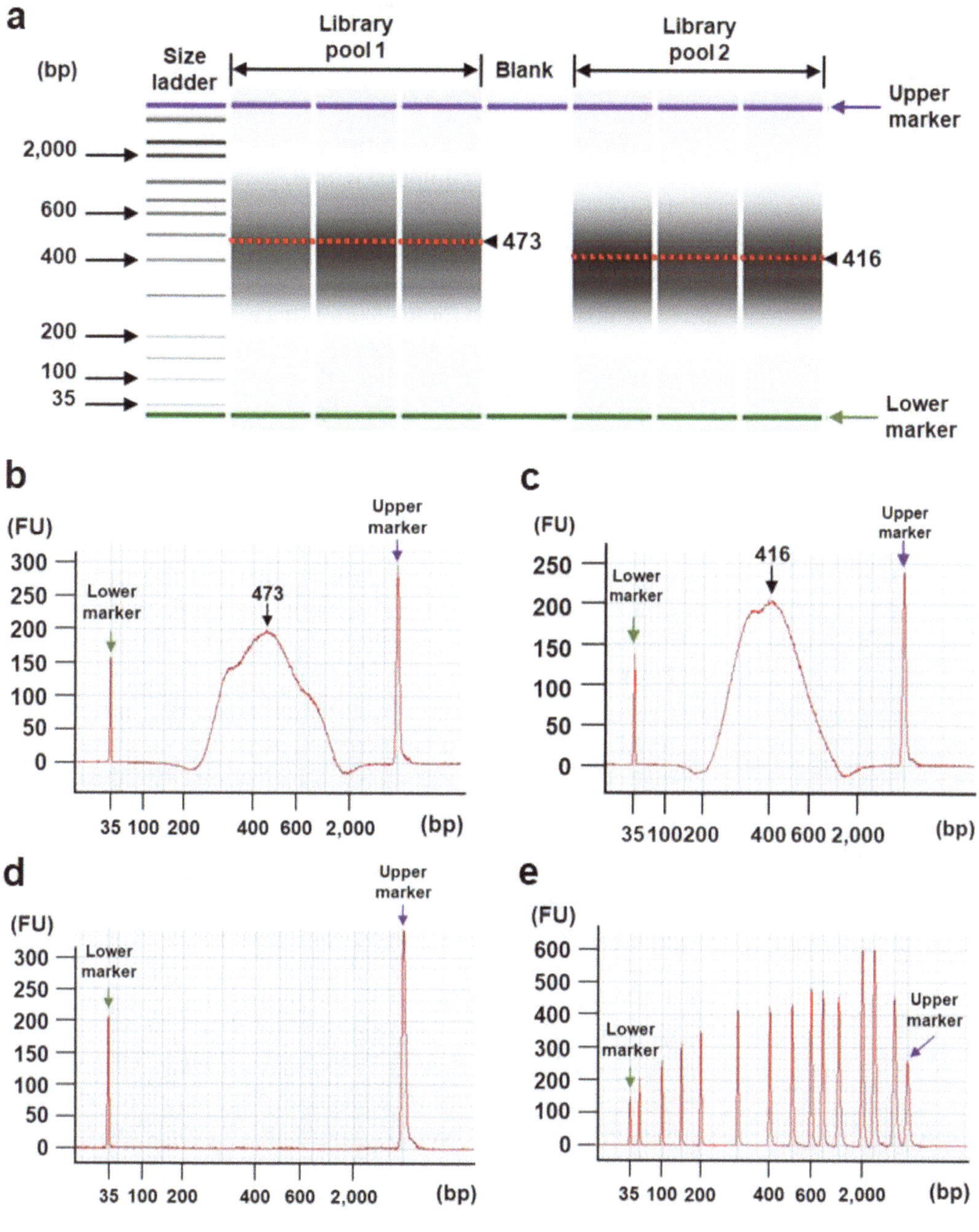

Fig. 4 Quality control for cDNA library pools using microfluidic electrophoresis systems. (**a**) Triplicate of each library pool was loaded on a microfluidic gel for electrophoresis. Red dotted lines in the center of gel image present an averaged size of each library pool indicated by black arrows with the numbers. Purple and green arrows highlight upper and lower markers at 10,380 bp and 35 bp, respectively. A size standard ladder was loaded followed by library pools. (**b–e**) Electropherogram images corresponding to the samples loaded on the microfluidic gel (**a**). Library pool 1 (**b**), library pool 2 (**c**), blank (**d**), and size ladder (**e**). Blank included only upper and lower markers

on 1 of 1000 bases). This trimming tool will also remove residual adapter/primer sequence-contaminated reads.

4. Download HISAT2 from http://daehwankimlab.github.io/hisat2/download/ and install the tool on Linux operating system. The tool is a memory-efficient sequence aligner [5], so that a computer equipped with RAM >8 giga bytes (GB) and 4-core CPU equal to or higher than Intel® Core™ i5 processors (or equivalent to AMD processors) will be enough for read mapping on human genome (*see* **Note 13**).
5. Download FeatureCounts from https://sourceforge.net/projects/subread/files/ and install the tool in Linux operating system. Count the number of reads aligned to individual genes using FeatureCounts. Refer to some useful example commands available in the following link; http://bioinf.wehi.edu.au/featureCounts/.
6. According to read-counting results, prepare a dataset matrix organized by read counts per gene with cell in rows and columns, respectively. Save the matrix in a text file format. Then create a second text file containing batch information of cells indicating the source of cells, such as cell line or patient ID (*see* **Note 14**).
7. Create an account in a web-based single-cell analysis tool, ASAP. Upload the two prepared text files (the read-count matrix dataset and the batch information) to ASAP. When processing dataset in ASAP, it is recommended to (1) set filtering criteria at >1 count per million (CPM) and >4 reads per gene, (2) select a normalizing tool, voom, (3) use normalized expression values for clustering analysis, (4) reduce highly variable dataset of normalized expression values using t-distributed stochastic neighbor embedding (t-SNE) analysis at default configuration (values of 30 perplexity and 0.5 theta) and (5) rearrange single cells using a clustering tool, SC3, according to a difference in gene expression (*see* **Note 15**).
8. To detect DEGs in ASAP, select one of the generated clusters as a control group and compare normalized expression values per gene in cells of the control group with those in cells of other individual groups (or clusters) (*see* **Note 16**).
9. When needed, gene set enrichment analyses can be performed using gene ontology (GO), KEGG pathway, or MSigDB Oncogenic Signatures database through ASAP at FDR-corrected *P* value <0.05 (*see* **Note 17**).
10. To further profile DEGs on heatmaps, import the read-count matrix to Morpheus, Broad Institute (*see* **Note 18**). Adjust the read-count matrix using Z-score, and perform a hierarchical clustering analysis at the following parameters: (1) Metric-One minus Pearson correlation; (2) Linkage method-Average; and

(3) Cluster-rows (by genes). Once after clustering is done, adjust color scheme with preferred color matching, for example, blue for low expression and red for high expression. If needed, adjust values of minimum and maximum color scheme in the range from −1.5 (low expression) to 1.5 (high expression).

11. Venn diagrams can present a difference in the number of DEGs among intersample, intrasample, and scRNA-seq clusters where individual cells were reorganized from different samples. To obtain intersample DEGs, set a sample as a control and individually compare normalized expression values per gene in cells of the selected control sample with those in cells in other samples. To identify intrasample DEGs, select cells of a sample in two different scRNA-seq clusters and compare normalized expression values per gene in cells of a sample in a scRNA-seq cluster with those in cells of the sample in a different scRNA-seq cluster (*see* **Note 19**).

12. To identify statistical significance on DEG detection, create volcano plots composed of $\log_2$-scaled fold change value per gene in x-axis and corresponding FDR-corrected P value in y-axis using GraphPad.

3.5 DEG Validation at Gene Level

1. To validate fold change values of scRNA-seq DEGs at the bulk-cell level, isolate total RNA from the same samples that were used for scRNA-seq analysis, following the manufacturer's manual of TRIzol™ Reagent (invitrogen) or commercially available total RNA purification kits (*see* **Note 20**).

2. Remove coisolated gDNA using DNase I (1 unit per 1~5 μg of total RNA) or TURBO™ DNA-free kits (invitrogen) following manufacturer's protocol. If coisolated gDNAs are removed by DNase I, it is required to clean up total RNA for removal of residual DNase I and enzyme-reaction buffer components using user's preferred methods, such as ethanol precipitation or commercially available total RNA clean-up kits. Quantify RNA molecules using a UV spectrophotometer, and confirm purity of RNA samples in the range of A260/A280 absorbance ratios from 1.8 to 2.1.

3. Prepare RNA diluents adjusted at the concentration of 200 ng/μL in a total volume >30 μL to validate fold change values of scRNA-seq DEGs at bulk-cell level. Use >5 μL (1 μg) of RNA diluent to synthesize single-strand cDNA per sample for two-step qRT-PCR method that first prepares cDNA samples from total RNAs and then performs qPCR analysis. Repeat qRT-PCR validation at least three times per RNA sample and set at least two replicates per gene of interest per assay (*see* **Note 21**).

4. Design primer pairs for genes of interest in qPCR analyses referring to already existing qPCR primer databases, such as "PrimerBank" or "RTPrimerDB-The Real-Time PCR and Probe" Databases, for human or mouse genes. When needed, custom primer pairs can be designed using open-source software, such as Primer3.
5. To obtain normalized fold change values, use a relative quantification method, such as the Pfaffl's method [6], considering primer amplification efficiency or delta-delta ($2^{-\Delta\Delta Ct}$) method [7] using a difference in fluorescent cycle threshold (Ct) between reference and genes of interest as well as control and experimental samples (*see* **Note 22**).

3.6 DEG Validation at Protein Level

1. For validation of DEGs at protein level, lyse cells with cell-lysis buffer in a 1.5 mL ultracentrifuge tube.
2. Centrifugate lysed cells at 4 °C and 125,000 × g for 10 min, and transfer supernatant to a new 1.5 mL microcentrifuge tube (used for a bench-top centrifuge).
3. Quantify protein concentration using users' preferred method, for example, Bradford Assay.
4. Prepare preferred lysate concentration (e.g., 10 μg) in 5× sample loading buffer, and denature at >95 °C for 5 min in heating block.
5. Prepare preferred SDS-PAGE gel for electrophoresis method.
6. Transfer proteins to membrane using users' preferred transfer systems.
7. Probe membrane for protein of interest with selected primary antibody at 4 °C overnight while rocking gently. Optimization of antibody concentration may be required.
8. Remove primary antibody followed by membrane washes in TBST for 5 min at room temperature on swing rocker systems at 80 rpm. Repeat up to three times.
9. Probe membranes with optimized secondary antibody.
10. Remove secondary antibody followed by membrane washes in TBST for 5 min at room temperature on swing rocker systems at 80 rpm. Repeat up to three times.
11. Visualize protein expression using preferred chemiluminescence or fluorescence methods in imaging systems.

4 Notes

1. We listed only those reagents and equipment that require additional information due to modification and/or a demand on more detailed descriptions.
2. If cells fall within a wide range for cell size, it is recommended to calculate cell concentration at the two different cell size ranges, (1) 5–10 μm and (2) 10–17 μm. This will help the user choose the proper IFC and increase single-cell capturing rate.
3. It is recommended to find an optimum ratio (v/v) between cells and C1 suspension reagent (Fluidigm, Inc.) as cells from various populations may have different buoyancies, consequently affecting the rate at which single cells are captured. It is important to mix cells and suspension by gently pipetting, ensuring no bubbles are generated in 0.2 mL PCR tubes. Bubbles will interfere with proper observation of cell-suspension mixtures in 384-well microplate. If bubbles are found and buoyancy observation is interrupted due to bubbles, remove bubbles by aspirating with a pipette or re-make the cell-suspension reagent mixture in new 0.2 mL PCR tubes. Also, when examining cells at the five titrations, it can be difficult to find a clear difference in buoyancy among the following ratios 7:3, 6:4, and 5:5 of cells to suspension reagents. Refer to the two polarized ratios, 1:9 and 9:1, of cells to suspension reagent as those two ratios can help refocus the microscope and find a difference in buoyancy among the three tested ratios. If a best buoyance is not found within 30 min, prepare a new mixture of cells and C1 suspension reagent and return to microscope to identify optimal cell buoyancy.
4. At this time, scRNA-seq HT analysis for cell size >17 μm is not available.
5. Upon arrival of a new vial of ERCC RNA Spike-in, it is recommended to prepare aliquots of 1.5 μL of ERCC RNA Spike-in in low DNA/Protein-binding 0.5 mL or 1.5 mL microcentrifuge tubes to avoid degradation of RNA Spike-in control transcripts by repeated thaw-freeze cycles. After sealing individual aliquots with parafilm, store aliquots at −80 °C for future use.
6. Technical tips to minimize the generation of bubbles are as follows: (1) Gently pipette up and down >3 times immersing the end of tips in a certain depth (0.5–1 cm) of reagents to wet with aspirating reagent. It may help prevent reagent from evaporation and bubble generation, and also minimize retention of reagent due to a difference in viscosity among reagents when expelling reagent from tips. (2) Aspirate 20–30% higher volume of reagents than the one indicated by protocol and,

when expelling reagent, always stop at the first stop of pipettes ensuring that a tiny volume of reagents still remain in tips (usually 1–5 mm from the end of tips). It is not necessary to load an exact volume of reagent to IFCs as only an already set volume of reagents per well of an IFC finally enters microfluidic tunnels of IFCs. (3) Use an illuminated magnifier with 3× enlarging capability.

7. We recommend that this step be conducted as quickly as possible within 30 min to prevent loaded reagents from evaporating and to ensure cell viability. If microscope does not have image-capture capability, visually examine each IFC chamber and record single-cell capturing rate (scoring) in the C1 HT IFC Workbook. This step can be done easily with two people, where one person scores each chamber while the second person records the results. If image capture is available, photograph the IFC, and then score cell-capturing rates while C1 is operating. In addition, the established functions in the 3rd spreadsheet of the C1 HT IFC Workbook provide an accurate volume of 1× TE buffer to adjust cDNA samples at 0.2 ng/μL.
8. To obtain more accurate fold change values of DEGs, it is recommended to perform equimolarly pooling cDNA libraries.
9. Name the 20 prepooled cDNA samples as per the C1 HT IFC Workbook.
10. It is required to sequence libraries for Read1 > 26 bases of read length and for Read2 > 75 bases of read length as Read1 is used to demultiplex raw NGS dataset to 20 read sets of 40 prepooled libraries, while Read2 is used for sequence alignment, finally contributing to gene expression profiling among single cells in downstream bioinformatic analysis.
11. The preliminarily demultiplexed read sets are composed of read dataset sequenced from only "Read2" as read dataset sequenced from "Read1" was removed during the pre-demultiplexing step.
12. We recommend to numerically name 800 individual libraries per IFC such as 001, 002, 003, and so on, as it is more convenient for scripting to run multiple commands on a single line with a looping command, for, in downstream procedures.
13. It may be helpful to use main or optional arguments as follows: (1) Main arguments: `-k` (nonspecific match), `-p` (multiple processors/cores), `-x` (basename of indexed reference), `-u` (single-end read) and `-s` (SAM output); and (2) Optional arguments: -`q` (fastq file format), and `-phred33` (Phred score 33).
14. If individual read sets were numerically named, the 1st column includes the numbers starting from 001 and batch information is provided in the 2nd column.

15. When "Batch file groups" is selected on created clusters following a t-SNE analysis, individual batches are highlighted in different colors. To visualize a specific gene, use Ensembl gene ID (https://useast.ensembl.org/index.html). It is also available to visualize clusters in three dimensions that may help identify subclusters.
16. It is suggested to choose a cluster located as a control group in the center of cluster plot, where perplexity values are close to "0" in two dimensions. Then, individually compare gene expression dataset in cells of the control group with those in cells of other clusters to detect DEGs generally showing fold change differences $\geq |2|$ at the statistical significance of FDR-corrected P value <0.05.
17. From one clustered group comparison, two gene sets can be created, for example, an upregulated and downregulated gene sets from the comparison of Group 1 with Group 2, and the two gene sets can generate two separate enriched functional annotation datasets.
18. Set the range of z-score from -1.50 (low expression) to $+1.50$ (high expression) after selecting a favorable color match.
19. The Venn diagram among intersample, intrasample, and t-SNE cluster DEGs (scRNA-seq DEGs) presents how many genes potentially associated with transcriptomic heterogeneity are composed of scRNA-seq DEGs overlapping with intrasample DEGs. Also, genes commonly found in the three DEG sets indicate that scRNA-seq overcomes the limit of traditional bulk RNA-seq in detecting intrasample DEGs.
20. Users can use their own total RNA isolation method.
21. In general, 2 μL of cDNA template is used for qPCR per low- or high-profile assay. To repeat qPCR at least three times per gene with two replates per sample, users need >12 μL of cDNA template per gene per sample. It is also required to include one reverse transcriptase-free reaction per sample per assay for identification of amplification on targeting genes from undigested gDNAs using another 2 μL of cDNA template, that is, another >6 μL of cDNA template per three times repeated assays. Therefore, it is recommended to prepare cDNA template in total volume >20 μL for fold change validations of a scRNA-seq DEG using qRT-PCR. In addition, an extra volume of cDNA template may need for optimization procedures, such as primer screening or identification of a best annealing temperature. A cDNA template-free reaction per gene per assay is also required as a negative control to evaluate contamination of qPCR master mix.

22. If a house-keeping gene used as an endogenous reference gene presents inconsistent Ct values per sample along with repeating qPCR assays, users can employ an absolute quantification method that applies a standard curve approach to generation of serial dilutions of a gene of interest. To apply the absolute quantification method for qPCR in unknown (or experimental) samples, it is required to know the quantity (or copy numbers) of genes of interest in the prepared serial dilutions.

Acknowledgment

The authors thank members of the Marignani Discovery Research Laboratory for their ongoing support. PAM is supported by the Cancer Research Society, The Canadian Cancer Society Diane Campbell designated research fund (grant #706202), and the Dalhousie Medical Research Foundation. The authors acknowledge that Dalhousie University campuses are located on original lands of the Mi'kma'ki, the ancestral and unceded territory of the Mi'kmaq People.

Financial Support Cancer Research Society, The Canadian Cancer Society Diane Campbell designated research fund (grant #706202), Dalhousie Medical Research Foundation.

References

1. Choi YH, Kim JK (2019) Dissecting cellular heterogeneity using single-cell RNA sequencing. Mol Cells 42(3):189–199
2. Valdes-Mora F, Handler K, Law AMK, Salomon R, Oakes SR, Ormandy CJ, Gallego-Ortega D (2018) Single-cell transcriptomics in cancer immunobiology: the future of precision oncology. Front Immunol 9(2582):1–30
3. Sharma A, Cao EY, Kumar V, Zhang X, Leong HS, Wong AML, Ramakrishnan N, Hakimullah M, Teo HMV, Chong FT, Chia S, Thangavelu MT, Kwang XL, Gupta R, Clark JR, Periyasamy G, Iyer NG, DasGupta R (2018) Longitudinal single-cell RNA sequencing of patient-derivedprimary cells reveals drug-induced infidelity in stem cell hierarchy. Nat Commun 9(1):4931
4. Kim J-H, Xu Z, Marignani PA (2021) Identifying unique markers of lung cancer using single-cell RNA sequencing. NPJ Genom Med. 15:6 (1):87
5. Kim D, Langmead B, Salzberg SL (2015) HISAT: a fast spliced aligner with low memory requirements. Nat Methods 12(4): 357–360
6. Pfaffl MW (2001) A new mathematical model for relative quantification in real-time RT-PCR. Nucleic Acids Res 29(9):e45
7. Livak KJ, Schmittgen TD (2001) Analysis of relative gene expression data using real-time quantitative PCR and the 2(-Delta Delta C(T)) method. Methods 25(4):402–408

Chapter 20

Sequences to Differences in Gene Expression: Analysis of RNA-Seq Data

Polina V. Pavlovich and Pierre Cauchy

Abstract

RNA-Seq is now a routinely employed assay to measure gene expression. As the technique matured over the last decade, so have dedicated analytic tools. In this chapter, we first describe the mainstream as well as the most up-to-date protocols and their implications on downstream analysis. We then detail the steps entailing RNA-Seq analysis in three main stages: (i) preprocessing and data preparation, (ii) upstream processing, and (iii) high-level analyses. We review the most recent and relevant tools as one workflow following a stepwise order. The chapter further encompasses in-depth features of these tools. Details of the required code are made available throughout the chapter, as well as of the underlying statistics. We illustrate these steps with analysis of publicly available RNA-Seq data.

Key words RNA-Seq, Bioinformatics, Protocols, Pipeline, Alignment, Differential expression, Statistics

1 Introduction

Transcription is a mechanism central to cell homeostasis that is tightly regulated at several levels. Its most basic expression consists in the production of mRNA from a DNA template following recruitment of RNA polymerase II and the elongating factors by the preinitiation complex [1]. The main readout toward its quantification consists in measuring mRNA in bulk cells. With the advent of high-throughput sequencing, these assays have shifted from amplicon-, concatemer-, or probe-based (reverse transcription/RT-PCR, microarrays and SAGE), to truly genome-wide in the case of RNA-Seq. RNA-Seq combines the advantages of each of those techniques in that it allows unbiased quantification over the entire genome, is sensitive to low expression levels, and enables

Supplementary Information The online version contains supplementary material available at [https://doi.org/10.1007/978-1-0716-2376-3_20].

Sherri L. Christian (ed.), *Cancer Cell Biology: Methods and Protocols*,
Methods in Molecular Biology, vol. 2508, https://doi.org/10.1007/978-1-0716-2376-3_20,

discovery of both novel genes and transcripts [2] (*see* **Note 1**). Thanks to the relative simplicity of its underlying protocol and the resulting high coverage of the transcriptome per its high-throughput sequencing nature, RNA-Seq has rapidly risen to being one of the most popular OMICS assays. In this chapter, we will be covering the bioinformatics underlying RNA-Seq analysis.

To date, most RNA-Seq assays are performed using Illumina sequencing, which consists of bridge amplification followed by sequencing by synthesis using a reversible dye terminator [3]. Variants of sequencing by synthesis include ion-conductor sequencing (Ion Proton) and single molecule fluorescent sequencing (Helicos). Alternative sequencing methods comprise pyrosequencing (454), sequencing by ligation (SOLiD), and nanopore sequencing (Oxford Nanopore). While downstream analysis is similar for all these techniques, the initial processing steps can diverge greatly; we will thus be covering analysis based on Illumina sequencing in this chapter. We will illustrate each stage step by step using previously published RNA-Seq data from undifferentiated and differentiating mouse embryonic cells (mESC) grown using the leukemia inhibitory factor (LIF) and retinoic acid (RA) [4].

2 Materials

2.1 Experimental Setups

Several steps can impact on RNA-Seq readout and downstream analyses. A key step in RNA-Seq conduction, as with other gene expression techniques, is the choice of the cell compartment; either total RNA, cytoplasmic or nuclear RNA. By and large, the standard method is total RNA, which is the one we will be detailing. Of note, cytoplasmic and nuclear RNAs are more used for special applications and will obviously produce different results due to differential mRNA splicing. Another important choice is the method of purification/RT, as mRNA only represents 10% of total RNA, the rest consisting mostly of rRNA [5]. Poly(A) capture using oligo-d(T) primers effectively selects against rRNA; however, previous studies have shown substantially higher amounts of reads concentrated at the 3′ end as a result. In the case of rRNA depletion, this method will generally be accompanied by random priming for the RT reaction [6].

2.2 Single and Paired-End Sequencing

While early RNA-Seq experiments employed single-end sequencing methods, most modern RNA-Seq assays are performed in paired-end mode. In this mode, sequencing of the insert is carried out from both ends, in the forward and reverse directions, respectively. Read 1 and read 2 are sequenced from their respective primers, which are located 3′ of their respective adapters, P5 and P7 (Fig. 1). Historically, high-throughput sequencing length started from 20 bases [7], which was barely above the required length to

a

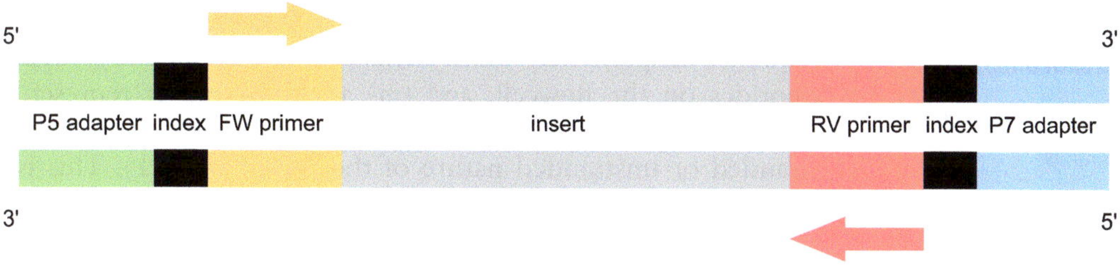

b

```
@HWUSI-EAS1692_42:1:115:16472:11830 length=76
CCTGCATGATGACGAACAAATACTGACTATCTGAAGATCTTATTAGCTCTATCTGATGTCGTATGCCGTCTTCTGC
+
GGDGGGDEGGEBBGBGGDGGDGDEGGGGGE:>>,.>5218B=CCBGGGEGEFBFE-@A<>267555:@;8EEACG#
```

c

```
@J00182:144:H3NTYBBXY:5:2103:8978:11425 1:N:0:ATATGGAT+TAATACAG
GCAGCCTTTGCTTGTGAGGTAAGAGACCAAGCATTGACTGTGGGGAGGGGATGGCGAGGAAGGAGGCAGACTTGG
+
AAFFFJJJJJJJJJJJJJJJJJJJJJJJJJJJJJJJJJJJJJJJJJJJJJJFJJJJJJJJJJJJJJJJJJJJJJJ

@J00182:144:H3NTYBBXY:5:2103:8978:11425 2:N:0:ATATGGAT+TAATACAG
GGAAGACAGCTGGATGGCAGTGCTGCTCCCTTTGGGCGGCTTGAGGGGACAAGGCTGGGGGGCCGAGTGGAAAGC
+
AAFFFJJJJJJJJJJJJJJJJJJJJJJJJJJJJJJJJJJJJJJJJJJJJJJJJJJJJJJJJJJJJJJJJJJJJJJ
```

Fig. 1 Structure of sequencing reads and fastq files. (**a**) Structure of a paired-end read with indexes. (**b**) Example read in fastq format from a single-end, unindexed RNA-Seq assay [28]. (**c**) Example read pair in fastq format from a paired-end, indexed RNA-Seq assay [4]

uniquely identify a sequence in the human genome (4^16). This gradually climbed to 25, 36, 50, 75 and now commonly 100, 150, or 200 bases in each direction. Note that there can also be a read 3, which corresponds to the library index in the case of barcoded samples. This sequence, called the index, consists of a short read (typically 6 or 8 bases) with its own primer in the Illumina adapter. This read is important when dealing with direct output from the Illumina sequencing software. In this case, the index will be used for demultiplexing base calls. With paired-end sequencing, the probability of covering an exon junction is thus highly increased, since this will allow determining whether two sequenced ends of a fragment span a known or unknown junction as they will map to genomic regions located far apart. The effective insert length also increases, further increasing effective mapping of individual chimeric reads should one span a junction. Thus, we will be focusing on paired-end sequencing in this chapter.

2.3 Library Preparation Strandedness

2.3.1 Unstranded Libraries

In RNA-Seq, the library preparation stage can be summarized by (i) reverse transcription (RT) to produce cDNA, (ii) the ligation of sequence adapters that will hybridize to complementary oligonucleotides on the flowcell, and (iii) amplification. It represents a critical step that is achieved by several methods that dictate the stranded or unstranded nature of the assay (Table 1). This is an important parameter throughout RNA-Seq analyses. Generally, stranded libraries will be of better quality as they will allow to ascertain without doubt whether a fragment corresponds to a transcript, since the strand can be matched. In unstranded libraries, signal can thus also originate from background on the reverse strand of a transcript of interest, or even from a long noncoding RNA or miRNA on the other strand. In all cases, sequencing adapters are ligated following RNA fragmentation. Libraries are prepared either from single- or double-stranded cDNA reverse-transcribed from RNA. Sequencing libraries originating from double-stranded cDNA will be unstranded as the strand information from which the transcript was produced is not preserved.

2.3.2 Stranded Libraries

The case is more nuanced for other library types. While standard protocols will produce unstranded libraries, numerous techniques have been devised to result in stranded RNA-Seq libraries (Fig. 2). Most methods consist in labelling the first strand before second strand synthesis takes place or in removing the second strand. A key difference at the analysis step between them stems from whether the forward strand of the library corresponds to the first or second cDNA strand. This will ultimately be dictated by the location of P5 and P7 on the first strand. First strand ligation can only result in 5′ ligation of P5, and 3′ ligation of P7, as adapters are in 5′-3′ order and thus have free 3′ and 5′ ends only, respectively. The original Illumina direct RNA ligation method uses sequential ligation of a 3′ preadenylated adapter and subsequently of a 5′ adapter [8]. The dUTP method is now employed by the TruSeq stranded and NEB kits. dUTPs are incorporated during second strand synthesis, followed by adapter ligation; the strand containing dUTPs is degraded by uracil-DNA-glycosylase [9]. A downside for this method is the selection against GC-rich sequences. In the DLAF method, double-stranded sequencing adaptors are directly ligated to the first strand with overhangs, which prevents second strand synthesis [10]. The forward strand adapters are then subsequently removed via uracil-specific excision as they specifically contain dUTP residues. In the case of low-input assays, the Smart-seq technique was devised with in situ amplification to recover full-length material in the nanogram range [11]. Following cell lysis, an oligo(dT) primer is hybridized onto the polyA tail with subsequent first-strand cDNA synthesis. A polyC overhang is ligated onto the 3′ end of the newly synthesized strand to ensure that only full length cDNAs are produced. Strand specificity is obtained via template switching

Table 1
Illumina-derived library protocols and strandedness

Library kit name	Manufacturer	Stranded	Stranded method	sequenced strand	featureCounts strand
TruSeq mRNA	Illumina	No	N/A	N/A	0
KAPA hyperprep	Roche	No	N/A	N/A	0
NEB Ultra II RNA	NEB	No	N/A	N/A	0
Nextera XT	Illumina	No	N/A	N/A	0
SMART-Seq	Takara	No	N/A	N/A	0
TruSeq stranded mRNA	Illumina	Yes	dUTP	First	1
KAPA hyperprep stranded	Roche	Yes	dUTP	First	1
NEB Ultra II directional RNA	NEB	No	dUTP	First	1
SMART-Seq stranded	Takara	Yes	Template switch tagging	Second	2
ScriptSeq	Illumina	Yes	Terminal tagging	Second	2
SMARTer-Seq	Takara	Yes	Template switch tagging	Second	2
Illumina direct ligation	Illumina	Yes	Direct ligation	Second	2
DLAF	N/A	Yes	ds adapter with dUTP	Second	2

to the overhang of the newly synthesized first strand of cDNA, which, in the presence of the Smart adapter, results in tagging of the mRNA with the subsequent second-strand synthesis using template-switch.

2.4 Mate Orientation

In paired-end Illumina TruSeq sequencing, the forward and reverse reads will represent the complementary and sense strands, respectively, due to the synthesizing nature of the sequencing method. However, during RNA-Seq analysis, the orientation of mates also plays a role to determine whether read pairs align concordantly (in the expected orientation) or discordantly (in an orientation differing from that expected). Importantly, for Illumina sequencing, mates always map inward, thus the orientation will be forward reverse (fr) (Fig. 3). Other platforms, including SOLiD, allow ff or rf. For direct ligation and tagging methods, this situation is reversed and the forward read will represent the sense strand, and conversely, the reverse read will represent the complementary strand. For second strand synthesis (e.g., SOLiD), the read

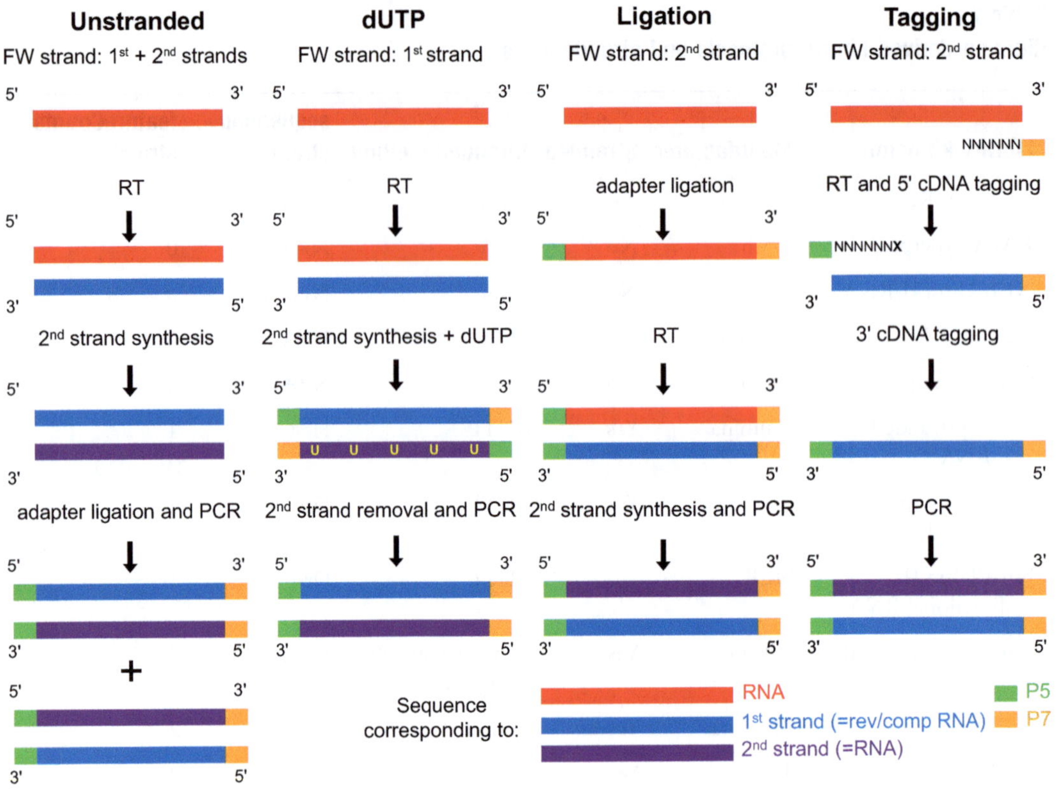

Fig. 2 Stranded and unstranded RNA-Seq protocols. Methodologies to produce unstranded and stranded libraries. Adapter ligation is always in 5′-3′ order and double stranded for DNA (adapters are Y shaped), with P5 and P7 having free 3′ and 5′ ends only. Hence the dUTP method results in the first strand (reverse complement of the RNA) being the forward strand in the library as P5 and P7 will be 5′ and 3′ of it, respectively. In contrast, RNA ligation and first-strand tagging methods lead to the first strand having P7 and P5 at the 5′ and 3′ ends, respectively, which results in the second strand (same orientation as the RNA) being the forward strand of the library. *U* dUTP, *NNNNNN* random hexamer, *X* terminal block [70]

orientation will be the same as direct ligation methods. Of note, while modifications of single-end sequencing methods allow strand-specific sequencing, we will not be covering those.

2.5 Resource Considerations and Workflows

In this chapter, we will help you build pipelines toward RNA-Seq analysis using standard tools (Table 2). Generally, processing and analysis of high-throughput sequencing data is performed via a (collection of) script(s) that contain all the necessary code (Annex 1). Good practice dictates creating and maintaining automated pipelines that take file or directory names and conditions as parameters. When piecing those together, an important step to keep in mind is also to implement the creation of log files that are useful for debugging or troubleshooting purposes. This is usually done using redirection, for example,

```
<command> 1> <command.stdout> 2> <command.stderr>
```

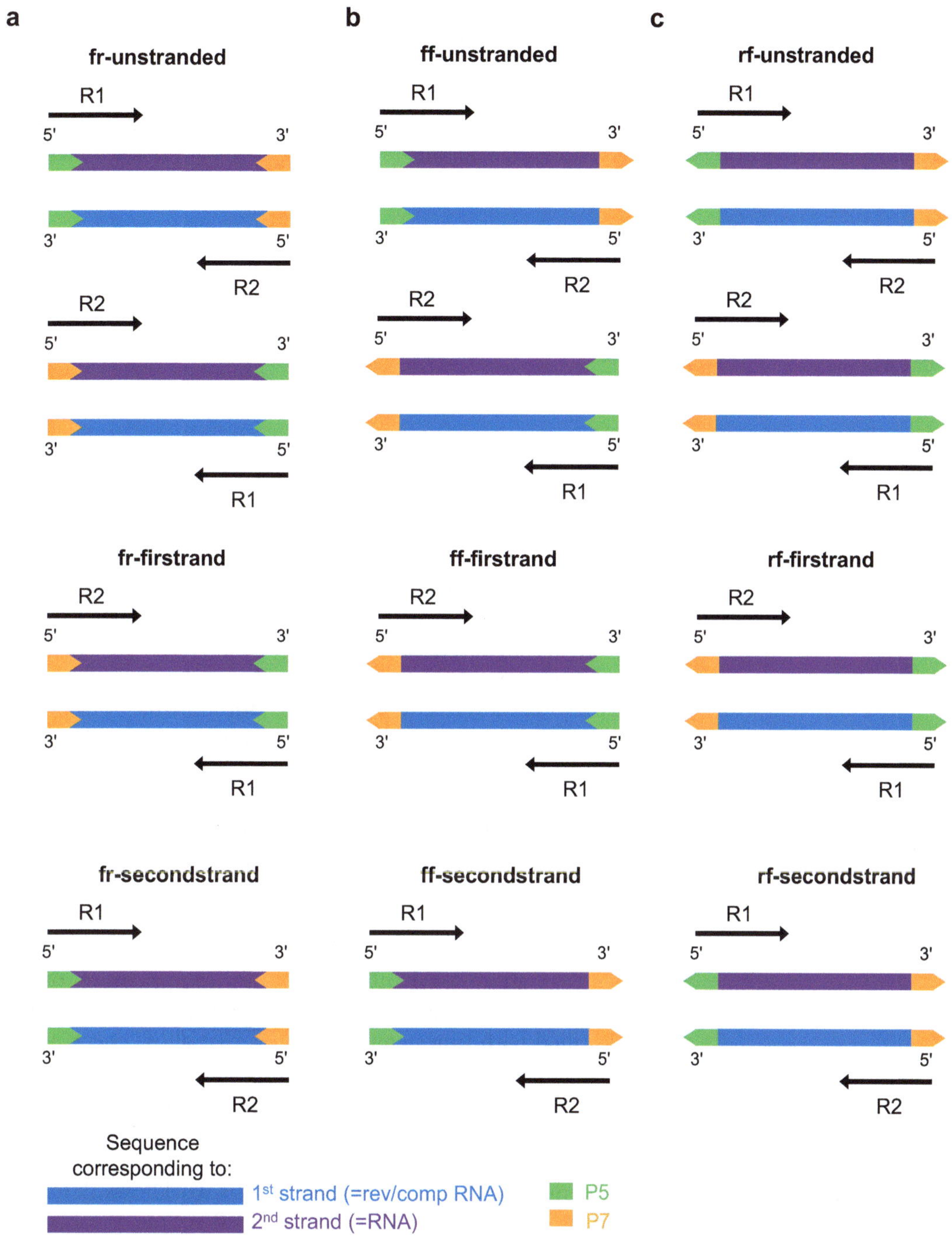

Fig. 3 Read orientation in stranded and unstranded RNA-Seq protocols. Orientation of mates with adapters in forward-reverse (fr), forward-forward (ff) and reverse-forward (**a–c**) with corresponding unstranded, first-strand and second-strand libraries. Arrows show reads and their associated sequence in the library, corresponding to a strand matching either the first cDNA strand or the second cDNA strand (amounting to the RNA strand). Resulting titles correspond to settings to be used in RNA-Seq analyses

Table 2
Sequence of RNA-Seq processing and analysis tools covered in this chapter

Function	Program name
Quality control	fastQC/multiQC
trimming	Trimmomatic
alignment	STAR
duplicate removal	samtools
MAPQ filtering	samtools
depth coverage	bamCoverage
quantification	featureCounts
Differential analysis	DESeq2
Gene ontology	homer

In terms of resources, the workflows that we will be describing here use STAR, which, for mammalian genomes, requires 32 Gb RAM and, on top of space requirements dictated by fastq and output files, requires >25 Gb disk space for each index. See **Note 1** for cases where a reference genome is not available. The higher the number of threads, the faster the processing will be; so, in the case of local processing, a high-end machine will be preferred. Access to a computer cluster will greatly increase analysis time; in this case, all commands shown in this chapter can be encapsulated with grid commands to submit jobs, for example, slurm or sbatch. Alternatively, user-parametered, modular, installable pipelines can readily be used, such as snakePipes, DolphinNext, or wg-blimp [12–14] (*see* **Note 2**). These pipelines usually implement install functions that allow local installation and encapsulation of the required programs and annotation files upon deployment. These tools are called with timestamps and rules that allow checking for correct execution as well as debugging and troubleshooting. Finally, all steps described here can be performed using Galaxy [15], which will require upload of fastq files via ftp.

2.6 Implementation

The scripts making up the RNA-Seq analysis workflows described here are Unix-based and R-based bash and R scripts, respectively. Thus, the requirements are a Unix-based environment (Linux, Unix, MacOS, BSD, Cygwin, Windows Subsystem for Linux, etc.) as well as an R installation ($\geq$3.6.3) with associated library dependencies for DESeq2. The programs to be additionally added can be found in Table 2. The sequencing data should be under the form of fastq.gz or fastq files. The pipelines consist of three parts:

(i) *Preprocessing*: collection of genomic annotations, workflow initialization, STAR index building [16], quality control of fastq files with FastQC [17] and multiQC [18], adapter trimming with Trimmomatic [19].

(ii) *Upstream processing*: alignment with STAR [16], read coverage generation with DeepTools [20] and count generation with featureCounts [21].

(iii) *Downstream analyses*: normalization and differential gene expression analysis with DESeq2 [22], optional batch correction with limma [23], and further analyses such as pathway/ gene ontology analyses, and gene set enrichment analyses (GSEA) [24].

We will also be covering alternative programs used for each of those steps, where applicable.

3 Methods

3.1 Preprocessing and Data Preparation

3.1.1 Annotation Files Checklist: Adapter Sequences

This part lists the required annotation files for the upstream processing and downstream analyses. For read trimming, while Trimmomatic [19] includes Illumina and Nextera adapter sequences, alternative programs such as cutadapt [25] or Trimgalore [26] will require adapter sequences as either direct arguments passed to those programs or as fasta adapter files. These sequences can be obtained from Illumina or Nextera or alternatively from Trimmomatic. Trimming programs search for adapters in both forward and reverse directions; so, the read 2 (reverse read) primer need not be reverse complemented, but can be given in the forward direction (Fig. 1a).

3.1.2 Annotation Files Checklist: Annotations Required to Build Indexes and Processing

Importantly, in order to build genome indexes, most RNA-Seq dedicated aligners, including STAR, will require two files: a gene transfer file (GTF) describing all gene features in the genome of interest, and a complete genome sequence file as fasta format. The GTF file itself represents all gene features known in the genome of interest. These include gene information with transcript annotation including coordinates of transcription start sites (TSSs), transcription termination sites (TTSs), known exons, coding sequences, start codons, stop codons, 5′ and 3′ UTRs. Importantly, these GTF files will also be used by featureCounts. Currently, two parallel annotation systems are in use: the UCSC Genome Browser and ENSEMBL systems. These use different coordinate systems and slightly different genome revisions. To illustrate these differences, a same feature is shown in GTF format using both nomenclatures in Tables 3 and 4 for a same exon from the mouse gene *Xkr4*. While it is possible to convert one from another using chromosome

Table 3
An exon of the Xkr4 gene in GTF format using the UCSC Genome Browser nomenclature, in the mm10 mouse genome

chr1	refGene	exon	3214482	3216968	.
-	.	gene_id "Xkr4";		transcript_id "NM_001011874";	
exon_number "1";		exon_id "NM_001011874.1";		gene_name	"Xkr4";

mapping files, in this chapter, we will be using the UCSC Genome Browser coordinate system, which is more directly compatible with external tools (e.g., Deeptools) and allows genome browsing with several public datasets including ENCODE ones. Further, when outputting genome-wide depth coverage files, an important file is what is known as the blacklist file. This is typically a bed file that contains positions of rRNA, satellite, and simple repeats that would otherwise incorrectly yield very high read depths as mapping to multiple places in the genome [27]. These regions should thus be removed. The download links to GTF annotation files, complete genome fasta sequences, and blacklist files for the UCSC Genome Browser and ENSEMBL can be found in Table 5.

Finally, when using snakePipes, the GTF and fasta files will be required to build indices. Read coverage files will also be masked against the given blacklist file. Importantly, the automatically generated yaml file will contain all paths to indices and annotation files required by snakePipes. The snakePipes samplesheet will also be used to identify samples during the differential gene expression analysis in the downstream analyses. This file should be in text-separated format and should have the structure shown in Table 6 for each sample.

3.1.3 Fastq Files

High-throughput sequencing data is generally delivered under the form of fastq files. They represent the starting input for most high-throughput sequencing analyses. Fastq files are derived from base calls, using bcl2fastq for Illumina sequencing (*see* **Note 3**). Also *see* **Note 4** to obtain fastq files from previously published RNA-Seq. The fastq format includes the following information: (i) flowcell and read identifier, (ii) read DNA sequence, and (iii) quality scores, as seen in a single-end RNA-Seq read example (Fig. 1b) [28]. Paired-end fastq files will include the same information, however, spanning two files, one for read 1 and the other one for its mate read 2 (Fig. 1c) [4]. Quality scores are embedded in fastq files and represent base-calling scores derived from dye fluorescence intensities. They are known as Phred scores and range on a scale from 0 to 42, encoded in ASCII. The main scale in use is the Phred33 scale (ASCII base 33), and some older files (from, e.g.,

Table 4
An exon of the Xkr4 gene in GTF format using the ENSEMBL nomenclature, in the GCRm38 mouse genome

1	ensembl_havana	exon	3214482	3216968	.
-	.	gene_id	"ENSMUSG 00000051951";	gene_version	"5";
transcript_id	"ENSMUST 00000070533";	transcript_version	"4";	exon_number	"3";
gene_name	"Xkr4";	gene_source	"ensembl_havana";	gene_biotype	"protein_ coding";
havana_gene	"OTTMUSG 00000026353";	havana_gene_ version	"2";	transcript_name	"Xkr4--001";
transcript_ source	"ensembl_havana";	transcript_biotype	"protein_coding";	tag	"CCDS";
ccds_id	"CCDS14803";	havana_transcript	"OTTMUST 00000065166";	havana_transcript_ version	"1";
exon_id	"ENSMUSE 00000448840";	exon_version	"2";	tag	"basic";
transcript_ support_level	"1";				

Table 5
URL models at which UCSC Genome Browser and ENSEMBL complete fasta genome sequences, GTF annotation files, and blacklists can be found

	UCSC Genome Browser	ENSEMBL
Genome fasta	https://hgdownload.soe.ucsc.edu/goldenPath/<genome>/bigZips	http://ftp.ensembl.org/pub/<release_version>/fasta/<organism>/dna
GTF file	https://hgdownload.soe.ucsc.edu/goldenPath/<genome>/bigZips/genes	http://ftp.ensembl.org/pub/<release_version>/gtf/<organism>
Black list	https://github.com/Boyle-Lab/Blacklist/tree/master/lists	

Table 6
Structure of an mRNA sample sheet for a sample set of two conditions with three biological replicates each

name	condition	Merge	Read1	Read2	Batch
<name1>	<condition1>	<merge1>	<name1_R1.fq.gz>	<name1_R2.fq.gz>	<batch1>
<name2>	<condition1>	<merge1>	<name2_R1.fq.gz>	<name2_R2.fq.gz>	<batch2>
<name3>	<condition1>	<merge1>	<name3_R1.fq.gz>	<name3_R2.fq.gz>	<batch3>
<name4>	<condition2>	<merge2>	<name4_R1.fq.gz>	<name4_R2.fq.gz>	<batch4>
<name5>	<condition2>	<merge2>	<name5_R1.fq.gz>	<name5_R2.fq.gz>	<batch5>
<name6>	<condition2>	<merge2>	<name6_R1.fq.gz>	<name6_R2.fq.gz>	<batch6>

The samplesheet is compatible with snakePipes

older Illumina GAII Instruments) use the Phred 64 scale (ASCII base 64). Phred scores (Q) correspond to

$$Q = -\log_{10} E,$$

where E is the sequencing error probability [29]. These values were originally computed from peak intensities and shapes from known sequences using Sanger sequencing. A good cutoff for Q is 20 for RNA-Seq data, which corresponds to 99% predicted accuracy [30]. Table 7 illustrates qualities encoded using the current Phred33 scale.

Table 7
Scale of STAR MAPQ alignment scores and corresponding number of alignments

Phred score	ASCII33	Error (%)	Accuracy (%)
0–10	!"#$&'()*+	10	90
11–20	,-./0123456	1	99
21–30	789:;<=> ?@A	0.1	99.9
31–40	BCDEFGHI	0.01	99.99
41–42	JK	0.001	99.999

3.1.4 Workflow Initialization

The bash pipeline that we describe here, rna-align.sh (Annex 1), is used in the following way:

```
rna-align.sh -f|--fastq <fastq_folder> -g|--genomedir
<genomedir> -o|--outdir <outdir> [--trim] -t|--threads
<threads> -a|--annotation <annotationGTF> -s|--samplesheet
<samplesheet>
```

where <fastq_folder> represents the path of the directory that contains the fastq files (either in the directory root or in subfolders), <genomedir> the genome index directory, <outdir> the path of the output directory, <threads> the number of threads, <annotationGTF> the path to the genome GTF file, <samplesheet> the samplesheet file. The presence of the --trim flag will indicate whether trimming should be performed or not.

The very first step consists in argument parsing (Annex 1). The next step is argument checking and informational messages. If a genome index directory is not passed or not already present, then, if a GTF and fasta file are provided, a flag for index building will be created, if not the workflow will exit after QC and adapter trimming. (Annex 1). Finally, for the sake of path name convenience, all fastq files are aliased locally (Annex 1).

The mRNA-Seq pipeline from snakePipes is started using the command:

```
mRNA-seq --DAG -j <threads> -i <fastq_folder> -o <outdir>
<genome.yaml> --sampleSheet <samplesheet>
```

3.1.5 Index Building

A reference index consists in a collection of k-mers that short reads are aligned against. For RNA-Seq aligners, these indexes are built from both the genome and transcriptome, which they use to map spliced reads [31]. STAR uses uncompressed suffix array indexes (SA), which increases performance. This is in contrast with other short read aligners (e.g., bowtie2, hisat and bwa) that use compressed Burrows-Wheeler transformed indexes. In this workflow, a genome index must be available or built for alignment with STAR. Index building is achieved by running STAR in genomeGenerate mode. This step will output the required SA and index files. For mammalian genomes, these files have sizes in the 25 Gb range, so an accumulation of indexes must account for sufficient disk space. Allow roughly the same time as alignment of a single dataset for this process, for example, 1 h on a 64-thread configuration for mammalian genomes. This is achieved via the following command from the workflow:

```
STAR --runThreadN <threads> --runMode genomeGenerate --genomeDir <genomedir> --genomeFastaFiles <genomefasta> --sjdbGTFfile <genomegtf> --sjdbOverhang 100
```

where `--sjdbOverhang` corresponds to the length within a read that can be used to call a junction. For example, for 100 bp reads, a parameter of 100 allows, for example, 75 bp to map to one end of a junction, and 25 on the other side [16].

When using snakePipes, genome indices will be built as well as other annotation files required by snakePipes (genome yaml configuration file, gene annotation bed file). This is achieved using the createIndices command:

```
createIndices -o <outdir> --genome <genome.fa> --gtf <genomegtf> --blacklist <blacklist>
```

3.1.6 FastQC Generation and Inspection: FastQC and MultiQC

FastQC [17] is an architecture-compiled java application that can be run either in GUI or in command line mode. In order to be integrated in RNA-Seq workflows, it is recommended to use it in command line mode; FastQC can be run with the following command:

```
fastqc *.fastq.gz -t <threads> -o <outdir/fastqc>
```

Thereafter, the program multiQC [18] can be run to produce a combined report for all FastQC reports:

```
multiqc <outdir>
```

3.1.7 FastQC Generation and Inspection: Assessment of Base Qualities

This step is essential in assessing the usability of the downstream analyses. FastQC produces reports that include several metrics. The first one is the assessment of sequencing quality scores for each read. As sequencing length increases, sequencing quality decreases. According to the FastQC documentation, median quality values that fall below 25 or lower quartile quality values below 10 warrant trimming. Figure 4a, b shows example outputs for base quality control with FastQC and multiQC.

3.1.8 FastQC Generation and Inspection: Adapter Content

The second critical value is the adapter content, which is often linked with the overrepresented sequences value. A short insert size, and thus sequencing in the $3'$ direction past the insert, can explain the presence of adapters in reads (Fig. 4c). According to the FastQC documentation, values above 5% result in warnings being issued and warrant trimming.

3.1.9 FastQC Generation and Inspection: G/C Content

The third critical value is the GC content of reads. Generally, one expects a distribution that matches the genome of interest. Overrepresentation of adapter sequences or of adapter dimers will also result in variations from the normal distribution, and values above 15% divergence will warrant trimming.

3.1.10 Adapter Trimming

The purpose of read trimming is to remove sequences, which will impair alignment, resulting in lower alignment scores. Reads can be trimmed using two metrics: (i) sequencing quality and (ii) adapter matches. For sequence quality, Trimmomatic performs sliding-window trimming, whereby a sliding window showing an average quality lower than the specified value represents a cut-off point. A sliding window of 4 bases with a mean quality of 15 represents a good threshold. This is specified using the SLIDINGWINDOW:< window size>:<quality> argument. For adapter trimming, reads are trimmed for adapters at the $3'$ end as this is where they are likely to be encountered. This is enabled with the ILLUMINACLIP:< fastaWithAdaptersEtc>:<seed mismatches>:<palindrome clip threshold>:<simple clip threshold> switch. Good results are obtained using ILLUMINACLIP:<adapters.fa> :2:30:10. Trimmomatic will produce "_paired" and "_unpaired" fastq.gz files, depending on whether both mates have survived the trimming process or not. Trimmomatic is thus started using the command:

```
java -jar Trimmomatic.jar <read1.fastq.gz> <read1_unpaired.
fastq.gz> <read1_unpaired.fastq.gz> <read2.fastq.gz>
<read2_unpaired.fastq.gz> <read2_unpaired.fastq.gz> IL-
LUMINACLIP:<adapters.fa> :2:30:10 LEADING:3 TRAILING:3 SLI-
DINGWINDOW:4:15 MINLEN:36
```

3.2 Upstream Processing

This part comprises the bulk of the resource-intensive tasks. The first step is to align the data to a reference genome and transcriptome. This is accomplished using splice-aware short read aligners. They are splice-aware in that they can detect junctions, either via the presence of chimeric reads, or if two mates map far apart, and rely on both the genome and transcriptome for mapping. They belong to a class of special alignment algorithms and are optimized for short reads and speed in that they use seed-and-extend strategies that combine index matching and standard alignment algorithms (e.g. Smith-Waterman) for the final alignment [32]. The seed-and-extend strategy involves chopping a read in shorter

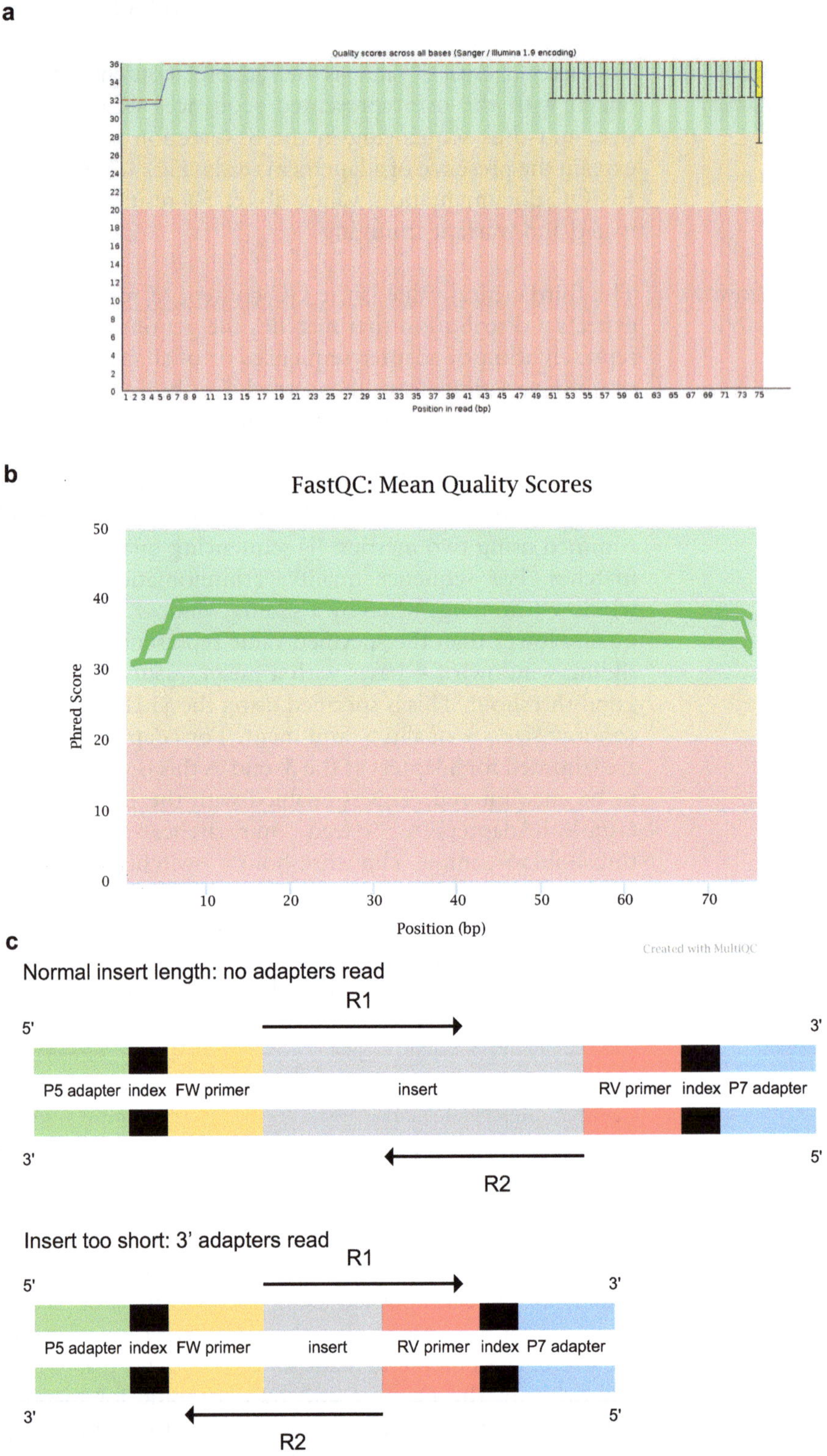

Fig. 4 Upstream processing quality control outputs. (**a**) Output of base qualities for one run using FastQC. Lower quartile quality values should mostly be in the green. (**b**) Output of base qualities for one run using multiQC. (**c**) Example of adapter contamination due to short insert length. In typical conditions, insert length is greater than read length (top). However, RNA fragmentation can lead to insert length being smaller than read length, resulting in sequencing through adapters in the 3′ direction (bottom)

Table 8
Scale of STAR MAPQ alignment scores and corresponding number of alignments

MAPQ	Alignments
0	≥10
1	4–9
2	3
3	2
255	1

k-mers along a sliding window. These seeds are then matched against the reference genome using an index method. Seed extension is then carried out using standard alignment algorithms. STAR uses a maximal mappable prefix (MMP) approach to match reads. This approach entails finding the longest substring of a read that matches one or more substrings of a genome [16]. Alternative splice-aware aligners to STAR include hisat2 [33], tophat2 [34], subread/Rsubread [35, 36], SOAPsplice [37], and GSNAP [38] (Table 8), benchmarked in [31].

3.2.1 Alignment

STAR can be run in single- or two-pass mode. In two-pass mode, the first pass is run in high stringency mode. Junctions discovered from this pass are used during a second pass with lower stringency, which allows higher alignment rates [39]. Unlike legacy aligners such as Tophat2 and other aligners (e.g., hisat2), STAR does not require the library type to be input when aligning. STAR will by default output only concordant reads. Importantly, STAR usually requires a higher limit of the number of open files than is typically set on Unix/Unix-like systems. This can be increased to a value compatible with STAR with the command:

```
ulimit -n 4096
```

This value corresponds to the maximum number of open files under Ubuntu, and is supported by MacOS. STAR alignment is then started using the command:

```
STAR --genomeDir <genome> --runThreadN <threads> --readFilesIn <read1.fastq.gz> <read2.fastq.gz> --outFileNamePrefix <STAR_output> --outSAMtype BAM SortedByCoordinate --outSAMunmapped Within --outSAMattributes Standard --readFilesCommand <zcat | gunzip -c>
```

This will produce a number of output files, the main one being a binary alignment map (bam) file, which corresponds to a binary representation of sequence alignment map files (sam). The bam file will be coordinate-sorted, which can then be used for further processing. Importantly, in paired-end mode, each mate will reference its counterparts, which allows the pair to be used as a fragment in subsequent analyses. The distance to the other mate will be indicated in column 9. Here, the final.out output logs should be checked, as these will indicate the percentage of aligned reads. A good cut-off point here is 50%, with lower mapping rates indicating a problem in the library [40].

The other important output of STAR consists in a tab-separated file called <outFileNamePrefixSJ.out.tab>, which represents the splice junctions. To increase the sensitivity and specificity to detect splice junctions, STAR can then optionally be run in two-pass mode using the `--twopassMode` switch. This can be useful when performing downstream splicing analyses (*see* **Note 5**).

3.2.2 Bam File Filtering

Another important metric in RNA-Seq alignment is the number of PCR-duplicated reads. PCR duplicates are identified by reads having the same start coordinate and orientation. In paired-end mode, PCR duplicates are identified by both mates having start coordinates and orientation found in one or more other read pairs. As of 2021, the consensus is now not to remove these for high-complexity libraries such as standard bulk RNA-Seq, as these often correspond to highly expressed genes [41]. However, when dealing with low-input starting material, such as that used in Smart-seq, library complexity is often reduced due to the need for a higher number of PCR cycles during amplification. Thus, we only recommend to use de-duplication in low input protocols. This is achieved using:

```
samtools markdup -r <bam>
```

Another important metric following alignment is the mapping quality scores, or MAPQ values. According to its original, sam and STAR specifications [16, 42, 43], an MAPQ score is defined as follows:

$$\text{MAPQ} = \text{int}\left(-10 \log_{10} P\right),$$

where P is the probability that the mapping position is wrong.

P itself is expressed as

$$P = \left(1 - \frac{1}{N}\right),$$

where N is the total number of alignments for a read. A factor influencing the probability that the mapping position is wrong is the difference in scores between a primary alignment and a secondary alignment. The resulting MAPQ scale is thus shown in Table 9.

Table 9
Comparison of features of popular splice-aware short read aligners

Aligner	STAR	hisat2	tophat2	GSNAP	SOAPsplice	(R)subread
Index type	Suffix array	Burrows-Wheeler Ferragina-Manzini	Burrows-Wheeler	Hash table Suffix array	Burrows-Wheeler	Hash table gapped
Method	Maximal mappable prefix	Seed-and-extend	Seed-and-extend	Oligomer chaining	Seed-and-extend	Seed-and-vote
Two-pass	Yes	Yes	No	No	No	Yes
Strand arg.	No	Yes	Yes	Yes	Yes	Yes
Discordant	Not included	Included	Included	Included	Included	Not included
Staged align	No	No	Yes	Yes	No	No
Out junctions	Yes	Yes	Yes	Yes	Yes	Yes
Runtime GTF	No	No	Yes	No	No	Yes

arg: argument; discordant: discordantly aligned mates included in alignment

A recommended setting to use is a cutoff of 10 [44]. For other high-throughput applications, such as variant calling, higher MAPQ values are often required [45]. Note that the MAPQ filtering can also be performed when running featureCounts (see below), running it directly after alignment ensures that data viewed following bamCoverage [20] (see below) mirrors expression values obtained with DESeq2 (see below). MAPQ filtering at this stage is run using the command

```
samtools view -bq 10 <bam>
```

Please note that this can also be accomplished using the `--bamRemoveDuplicatesType` switch when aligning with STAR.

3.2.3 Visualization and Coverage File Generation

One way to directly inspect the data is to use the Integrative Genomics Viewer (IGV) [46]. Rather than viewing the data in raw aligned reads, one can visualize the data using the coverage depth, that is, the number of reads per base or genomic window (Fig. 5a, b). To be able to open bam files in IGV and for depth coverage file generation, bam files must first be indexed. This is achieved using the command

```
samtools index -@ <threads> <bam>
```

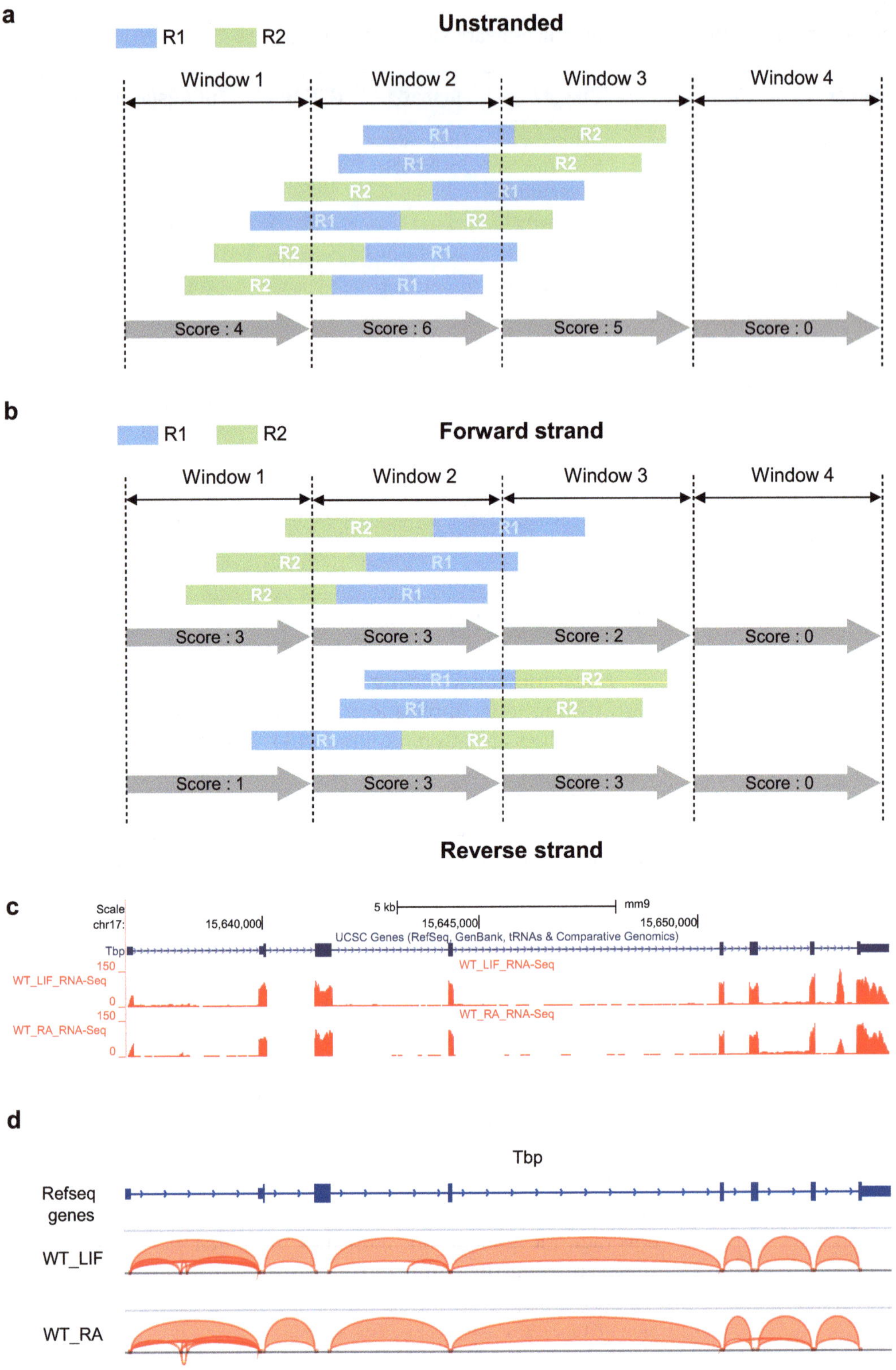

Fig. 5 Read depth coverage generation and example genome browser outputs. (**a**, **b**) Read depth coverage generation. In unstranded assays (**a**), the number of fragments made up of both mates in any orientation is

IGV and the UCSC Genome Browser will allow viewing depth coverages and individual read alignments (Fig. 5c). A good RNA-Seq experiment should have most of its signal concentrated on exons, 5′ and 3′ UTRs. RNA depth coverage tracks are also very important to help determine whether 5′-3′ RNA decay has occurred, as in this case, signal will be weaker toward the 5′ ends [47]. Importantly, IGV will also allow visualizing junctions (Fig. 5d), either directly computed from bam files, or by loading the SJout.tab file from STAR.

A more permanent way to inspect depth coverages is to generate bigwig files. These represent binary versions of wiggle or bedgraph files, which are in turn successions of read coverages along a window. Conveniently, these files can be linked to the UCSC Genome Browser. Normalization is performed using the Reads Per Kilobase of transcript per Million mapped reads (RPKM) method. In paired-end mode, pairs will be treated as one fragment with a size given by the region covered between the leftmost and rightmost coordinates from their mates. For unstranded protocols (Fig. 5a), coverage generation is achieved using the bamCoverage [20] command:

```
bamCoverage --bam <bam> -o <bigwig> --binSize 10 --normal-
izeUsing RPKM --effectiveGenomeSize <genome_size> --black-
ListFileName <blacklist> -p <threads>
```

For stranded protocols, it is possible to create strand-specific coverage tracks (Fig. 5b). This is useful when ascertaining that signal originates from a transcript, as most of the reads in this case should mirror the strand that a transcript of interest is in. This is obtained adding the `--filterRNAstrand` switch to the bamCoverage command:

```
bamCoverage --bam <bam> -o <bigwig> --binSize 10 --normal-
izeUsing RPKM --effectiveGenomeSize <genome_size> --black-
ListFileName <blacklist> -p <threads> --filterRNAstrand
forward
```

```
bamCoverage --bam <bam> -o <bigwig> --binSize 10 --normal-
izeUsing RPKM --effectiveGenomeSize <genome_size> --black-
ListFileName <blacklist> -p <threads> --filterRNAstrand
```

Fig. 5 (continued) collated per genomic window. In stranded assays, here in first-strand assays such as the dUTP method where the forward strand is given by the direction of read 2 (**b**), the number of fragments made up of both mates is collated per genomic window separately in the forward and reverse orientations. This would be reversed in second-strand assays (**c**) UCSC Genome Browser of the *Tbp* locus screenshot of two RNA-Seq experiments from mouse ES cells under LIF and RA conditions [4]. (**d**) IGV Browser screenshot of RNA junctions at the *Tbp* detected by STAR in the same conditions

```
reverse
```

This will create two tracks each for the forward and reverse strands. Directionality is based on read 2. For first-strand directional methods (reverse-stranded from the RNA sequence), this will result in coverage tracks corresponding to the strandedness of transcripts. Note that for second-strand directional, this will correspond to the opposite strands and will need to be swapped to reflect orientation of transcripts.

To account for variability in signal-to-noise ratios, even in assays performed in batch, an optional step consists in scaling the datasets, especially if the results of the previous bamCoverage commands result in assays on different scales. This is accomplished using multiBamSummary [20]:

```
multiBamSummary bins --bamfiles <bam> -o <npz> --scalingFac-
tor <scaling_factors> -p <threads>
```

This is followed by the bamCoverage command again, the using `--scaleFactor` switch:

```
bamCoverage --bam <bam> -o <bigwig> --binSize 10 --normal-
izeUsing RPKM --effectiveGenomeSize <genome_size> --black-
ListFileName <blacklist> -p <threads> -scaleFactor
<scaling_factors>
```

Please note that it also possible to produce bedGraph coverage files (the nonbinary representation of bigWig files) when running STAR, using --outWigType bedGraph. For stranded libraries, the additional parameter `--outWigStrand "Stranded"` would be used to produce strand-specific coverage files.

3.2.4 Quantification

The final step toward downstream analyses consists in counting the number of reads that overlap given features of the transcriptome. This is performed using the featureCounts program from the subread package for further processing [21]. featureCounts is a read summarization program that assigns aligned reads to genomic features. featureCounts uses the input GTF file as annotation to compute counts over all transcripts within. All counts on genomic features of a transcript (default: exons). featureCounts is entirely written in the C language. Therefore, while the preferred input of featureCounts is name-sorted bam files, its high speed of execution implies that on-the-fly reordering of input bam files is performed very quickly. Unsorted bam files are automatically detected; thus, the input can be in this format. When running, featureCounts uses a hash table to match chromosomes between bam and GTF files. Features subsequently use a hierarchical structure to match coordinates, whereby both datasets are divided into 128 kb bins and further divided into blocks. This initial matching of bins, and then of individual features from blocks, results in counts being computed rapidly.

Important considerations when running featureCounts include (i) the strandedness of the library (Table 1), (ii) annotation of primary hits, and (iii) whether reads should map to junctions only to reduce background noise in very low input samples [48]. By default, featureCounts will only count reads in exons; however, this can be set to other features, for example, CDS, 5′ and 3′ UTRs, using the -t switch. Other options include the minimum MAPQ, for which a value of 10, representing 10% error rate, is used in snakePipes [12]. featureCounts is run via the following command:

```
featureCounts -p -B -C -Q 10 --primary -T <threads> -s
<strand> -a <annotationGTF> -o <outdir/featureCounts>
```

3.3 High-Level Analysis

3.3.1 Underlying Statistics and Principles

In this part, we will discuss the postprocessing of bulk RNA-seq data (Annex 2). We primarily will use the package DESeq2, which applies negative binomial (NB) generalized linear models (GLMs) for calculating differential expression (DE) [22]. Reads are count-based and cannot have a value of −5 or 10.5. For this reason, the normal distribution is not suitable, especially at the low number of replicates seen in RNA-Seq. The data can be well approximated by a Poisson distribution when the number of observations is large and the proportions of the total count for each gene are small. However, when the data contains a low number of replicates, NB distribution suits better (Fig. 6a). A reason for this is that the Poisson distribution estimates that the mean and variance should be equal, and this is not the case (Fig. 6b). The commonly used design of the experiment includes 3–4 replicates, which are not enough to accurately estimate variance by looking at each gene.

For this reason, there are two per-gene estimates. The first one looks only at the data for a single gene (gene-est), and a final assessment includes both information from each gene and sharing information across genes. Biologically changing genes vary more than expected by the statistical model. To explain them, the NB distribution, unlike the Poisson distribution, includes an additional parameter called overdispersion or, in other words, quasi-likelihood dispersion. Fitting the mean-dispersion trend across all genes allows estimating the NB dispersion, whereas the Bayesian shrinkage approach estimates the quasi-likelihood dispersion [49].

Before differential expression analysis, a common practice is to filter lowly expressed genes. For example, keep genes with counts of ten or more in at least three samples. It removes more than half of the genes as not informative. And it reduces the time needed to fit per-gene parameters, such as dispersion and fold change estimates.

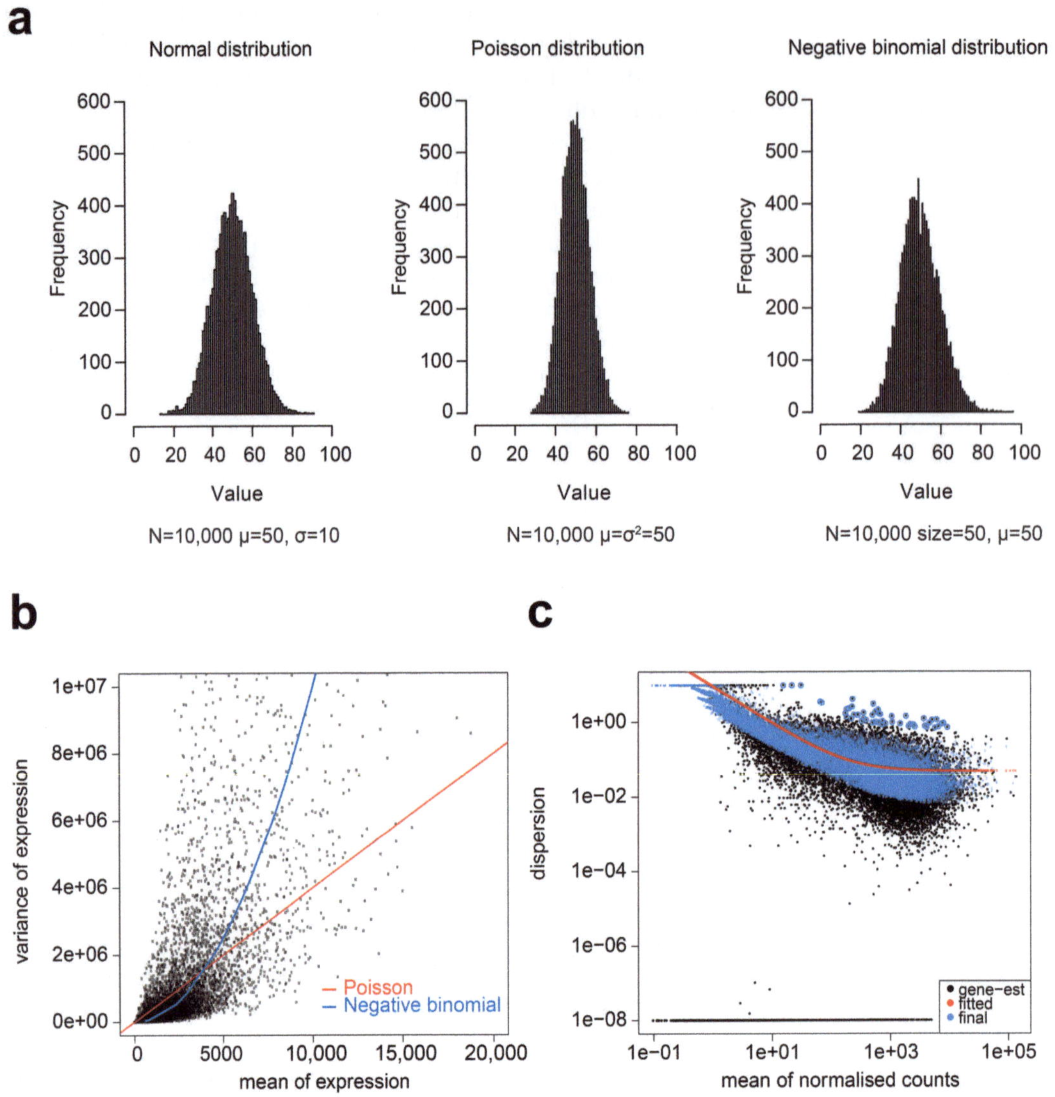

Fig. 6 Underlying statistics of the negative binomial distribution fitting in DESeq2. (**a**) Comparison of simulated normal, Poisson and negative binomial distributions at equivalent parameters. In the Poisson distribution, the mean equals the variance, which results in the negative binomial distribution being a better fit for the discrete normal distribution. (**b**) Regression fitting in count variances (*y*-axis) versus count averages (*x*-axis) using the Poisson (red) and negative binomial distribution (blue). Variances increase with averages, which results in the negative binomial distribution being a better fit. (**c**) Dispersions and model fit using DESeq2 on count data from [4]

3.3.2 Differential Expression Analysis: Principles

DESeq2 takes a table of raw counts and associated metadata as input. It performs an internal normalization called the median of ratios method, which consists of three steps. The first step is to create a pseudo-reference sample calculated as the geometric mean across all replicates. In the second step, calculate the ratios for every gene in a sample (sample/ref). Since most genes are not

Table 10
Features of three mainstream differential expression analysis packages

Package	DESeq2	edgeR	cuffdiff
Distribution	NB	Poisson/NB	Pooled/Poisson
Normalization	Median of ratios	TMM	Geometric/FPKM
Output	Counts/rlog/vst	TMM	FPKM
Test	Wald	Exact test	T-statistic
FDR	Benjamini-Hochberg	Benjamini-Hochberg	Benjamini-Hochberg

NB negative binomial, *TMM* trimmed mean of *M* values, *FPKM* Fragments Per Kilobase of transcript per Million mapped reads

biologically variable, most genes should have similar ratios within the sample. And the third step is to receive a size factor as the median value per column. Note that differentially expressed genes should not affect the median. This procedure corrects library size and RNA composition bias, for example, when only a tiny part of genes have noticeably more counts in one experiment condition than in the other. There are other tools for DE analysis, such as edgeR [50], which uses the weighted trimmed mean of *M*-values method (TMM) for normalization and cuffdiff [51], which outputs Fragments Per Kilobase of transcript per Million mapped reads (FPKM) (Table 10). The featureCounts object is read as follows in R:

```
countdata=read.table(<featureCounts>,head=T,row.names=1)
```

When creating a DESeq2 object, one needs to set a design formula. A design tells the DESeq2 the unwanted known sources of variation and the factor of interest during DE testing. The standard design (=~ condition) will model the samples where one condition's replicates will be considered a group (Annex 2). Additionally, if sex is a significant source of variation in the data, it should be included the model. All factors in the metadata that account for major sources of variation in the dataset should be present in the design. The last factor in the formula means the condition of interest (e.g., =~ sex + condition, or =~ batch + condition). This is accomplished bw, with one source of variation:

```
(condition <- factor(<conditions>))
(coldata <- data.frame(row.names=colnames(countdata), condi-
tion))
dds <- DESeqDataSetFromMatrix(countData=countdata, colData=-
coldata, design=~condition)
```

with two sources of variation (two conditions, condition and batch):

```
(condition <- data.frame(sample=colnames(countdata),condi-
tion=factor(<conditions>), batch=factor(<batch>)
(coldata <- data.frame(row.names=colnames(countdata), condi-
tion))
dds <- DESeqDataSetFromMatrix(countData=countdata, colData=-
coldata, design=~batch+condition)
```

where <conditions> is a character vector with the conditions, and <batches> is a numeric vector with the batch numbers.

DESeq2 is then run with the following commands:

```
dds <- DESeq(dds)
```

The function DESeq() estimates the size factors, dispersions and performs negative binomial GLM fitting and DE statistics. By default a, Wald test is used by DESeq2 to identify DE genes between two conditions. One can extract the final result table with the results() function:

```
res <- results(dds)
```

As input, this function takes a DESeq object and contrast information about which two states to be compared. One can specify the contrast in two ways. The first approach takes a character vector with three elements: the name of the factor (of interest) in the design formula, for example, "condition," and the name of the two factors levels to compare. The second way is to provide a list with two character vectors, which are elements of resultsNames (object). The results table consists of the following columns:

- baseMean: mean of normalized counts for all samples
- log2FoldChange: log2 fold change
- lfcSE: standard error
- stat: Wald statistic
- pvalue: Wald test *p*-value
- padj: BH adjusted *p*-values

The padj column is *p*-value adjusted by Benjamini-Hochberg false discovery rate (FDR). It is the most critical column of the results. The padj threshold, for example, 0.1, controls a set by its FDR, meaning no more than 10% of the genes in such a set on average should be false positives.

3.3.3 Differential Expression Analysis: Gene-Level Filtering

For selected genes, log2FC, *p*-value or padj can be NA. It can happen in one of three scenarios. In the first one, DESeq excludes from the statistics genes with zero counts in all samples to increase the power to detect DE genes. Note that DESeq2 does not remove genes from the dataset. They will all be present at the table. In the second scenario, genes with an extreme count outlier are detected with the Cook's distance. For this, one needs to have at least three replicates, because it is hard to decide on an outlier from two dots. And the third one, DESeq() function does not include genes with low mean normalized counts. They are less likely to be DE due to high dispersion.

3.3.4 Differential Expression Analysis: Fold Change

Fold change threshold can help to detect the most biologically relevant genes from the extensive list. For this, the results table contains another substantial column, log2FoldChange (LFC). Note that a threshold log2FoldChange >1 would translate into actual fold change (FC) more than 2. The results() function has an option lfcThrehsold to add a threshold. It performs a statistical two-tailed test for the alternative hypothesis that the log2FoldChange is greater than the specified absolute value. Note that if needed, one-tail tests are optionally also available (*see* **Note 6**). Genes with low counts and high dispersion values tend to have high FC. However, the information about those genes is insufficient. For such genes, DESeq2 allows for the shrinkage of the LFC estimates toward zero to highlight more relevant genes by log2-FoldChange. Shrinkage does not change the total number of significantly DE genes. But one most likely wants to use the shrunken LFC for ranking genes for Gene Set Enrichment Analysis (GSEA).

3.3.5 Normalization

Further to its internal count normalization, DESeq2 offers two additional normalization methods not used for DESeq() statistics but for visualization, clustering, and other machine learning tasks (Annex 2). The first method is regularized-logarithm transformation (rlog) transformation [22]. It does not require prior estimation of size factors and dispersions. Rlog normalizes data using library size and transforms the count data to the log2 scale minimizing differences between samples for genes with small counts. It is important because, due to the discrete nature of the data, low counts genes will have high variance on a log scale, and they will drive the information in principal component analysis (PCA) and clustering. Another tool is variance stabilizing transformation (VST) [52]. It is faster than rlog on medium and big datasets (>30 samples). The rlog works well on small datasets (<30 groups), potentially outperforming the VST when the range of sequencing depth across samples is wide, for example, an order of magnitude difference. An R package limma offers a normalization by columns to have the same quantiles. One can call it with the normalizeQuantiles() function [23]. DESeq normalizations are run in R using the commands:

```
rld <- rlogTransformation(dds)
vsd <- vst(dds,blind=F)
```

3.3.6 Batch Effect Removal

Sometimes some samples can differ by treatment, purification, sorting (if applicable), library preparation, sequencing date or environment. It is essential to eliminate these differences before driving biological conclusions from the dataset. DESeq2 does not offer batch removal techniques. Instead, one can run batch effect correction with the limma::removeBatchEffects method [23] (Annex 2). It takes a matrix with log-transformed values (after rlog or VST) and batch information and returns a corrected matrix. One can use the result of this function via PCA, multidimensional scaling (MDS), or heatmaps. Figure 7 shows PCA of a set of two conditions (LIF and RA) with four replicates that were run in three batches, before and after batch correction. Another function for batch removal is combat from the sva package [53]. Behind it, there are parametric or nonparametric empirical Bayes frameworks. Combat also returns a matrix with the batch effect removed. It is crucial to have so-called bridging populations in the dataset, meaning at least one common population between the batches. This group will allow the model to distinguish between biological and technical differences and remove the second. Limma batch removal is obtained using:

```
assay(vsd)=limma::removeBatchEffect(assay(vsd),batch=vsd
$batch)
```

3.3.7 Quality Control

DESeq2 provides several options to control the quality of the datasets (Annex 2). For example, one can visualize samples by PCA (Fig. 7). It allows reducing dimensions of the dataset by performing linear transformations. The function in the package is called plotPCA(). Next, one can apply, for example, Pearson's correlation coefficient to measure the similarity between rows or columns and then use hierarchical clustering to identify groups of correlating genes or samples. This approach is called correlation clustering. A similar approach consists in plotting the sample distances (Fig. 8a).

To show how genes in individual replicates change between different conditions, it is suitable to use heatmaps with row z-scores (Fig. 8b). The row z-score is calculated by using the following formula:

$$z = \frac{X - \mu}{\sigma},$$

a

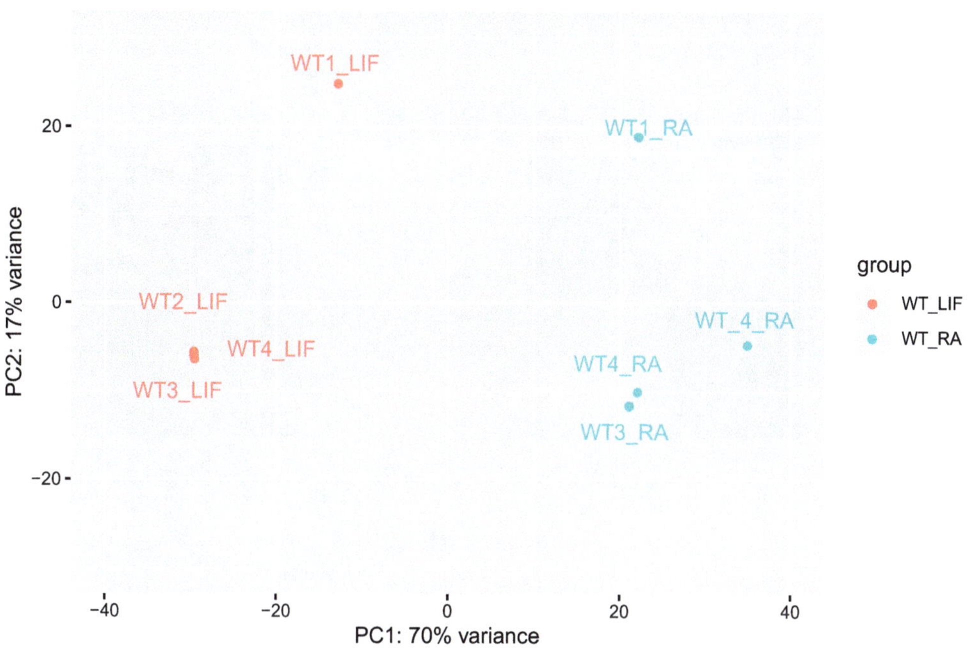

b

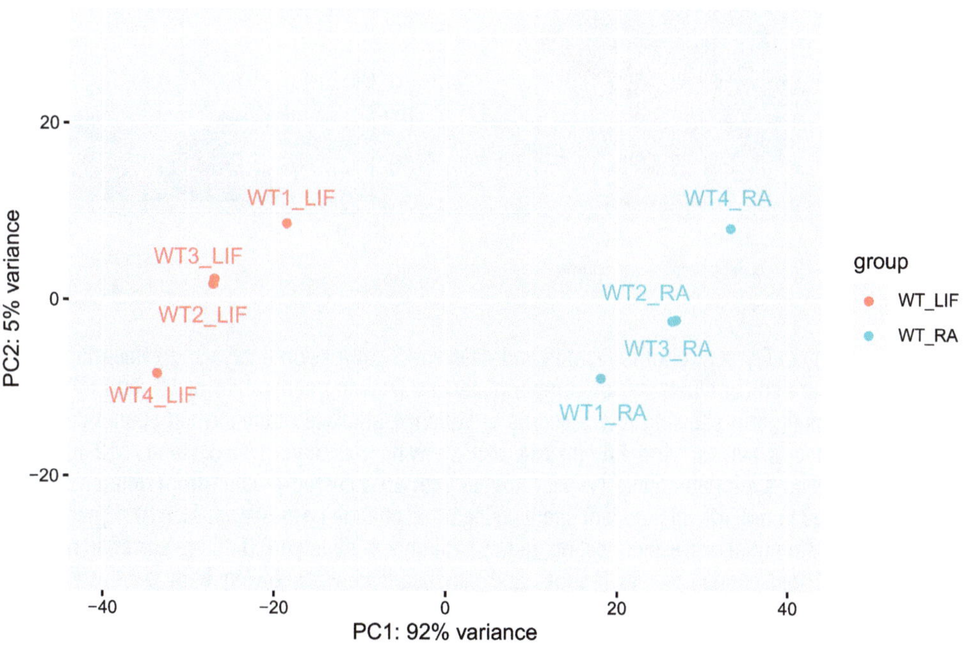

Fig. 7 Correlation quality control of replicates and batch correction. Principal component analyses of RNA-Seq replicates before (**a**) and after (**b**) batch correction. Samples WT1_LIF and WT1_RA are from batch 1, WT2_LIF, WT3_LIF, WT2_RA, WT3_RA from batch 2, WT4_LIF and WT4_RA from batch 3. Batch correction results in shorter distances to WT1_LIF and WT1_RA on the PC2 axis with the percentage of variance explained by PC2 decreasing, while that explained by PC1 increases

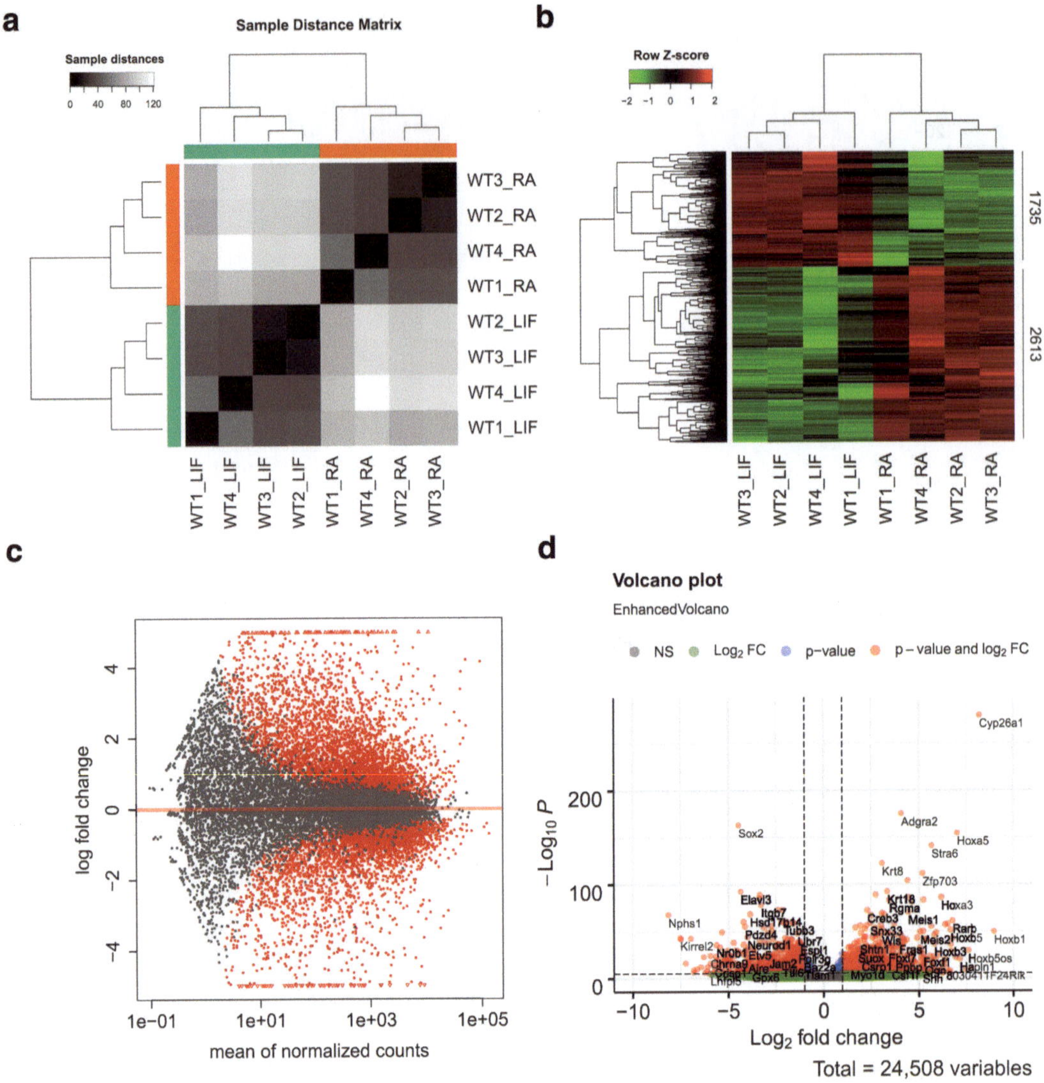

Fig. 8 Correlation quality control of replicates and differential expression analysis. (**a**) Heatmap and dendrogram showing hierarchical clustering of sample distances showing separation of LIF and RA datasets. (**b**) Heatmap and dendrogram showing row *z*-scores of variance-stabilized transformed counts on differentially expressed genes between LIF and RA. (**c**) MA plot showing the mean (*x*-axis) and log2 fold change of normalized condition-averaged counts (*y*-axis). Genes with an adjusted *p*-value from differential expression testing are labelled in red. (**d**) Volcano plot showing log2 fold change (*x*-axis) and −log10 adjusted *p*-value (*y*-axis). Following differential expression testing, genes with only *p*-adjusted <0.05 are shown in red, those with an absolute log2 fold change >1 in orange, and significant genes showing both *p*-adjusted <0.05 and absolute log2 fold change >1 in green

where X is the gene expression value in a sample of interest, μ the mean expression across all samples, and σ the standard deviation across all samples. It removes information about how high was the expression of each gene but shows relative changes across samples.

Another popular plot is the MA plot (Fig. 8c), plotMA(), which displays a log-ratio versus average:

$$M = \log_2 \text{treat} - \log_2 \text{control}$$

$$A = \frac{\log_2 \text{treat} + \log_2 \text{control}}{2}$$

In order to visualize the differences between two conditions. It shows log2FoldChange on the y-axis and the mean of normalized counts on the x-axis. One can use it to estimate how well the LFC shrinkage worked. For highlighting the top significant genes, it is useful to apply a volcano plot (Fig. 8d). It shows significance, −log10 *p*-value, versus log2FoldChange. See **Note 7** to obtain quality control plots in user-friendly web interfaces. The quality control plots (Fig. 8a–d) are obtained in R by the commands:

```
plotMA(res)
EnhancedVolcano(res,lab = rownames(res),x = 'log2FoldChange',y
= 'pvalue')
plotPCA(vsd, intgroup="condition")+ geom_text(aes(label =
name), position = nudge)
dev.off()
sampleDists <- dist( t( assay(vsd) ) )
heatmap.2(as.matrix(sampleDists), trace="none", col=co-
lorpanel(100, "black", "white"),ColSideColors=mycols[condi-
tion$condition], RowSideColors=mycols[condition$condition],
margin=c(10, 10), main="Sample Distance Matrix")
drgs=resdata[which((resdata$padj<0.05)&(abs(resdata
$log2FoldChange)>=1)),]
rownames(drgs)=drgs$Gene
drgs[,1:7]=NULL
drgs_log2_rowz=t(scale(t(log(drgs+1,2)),center=T,scale=T))
heatmap.2(drgs_log2_rowz,trace="none",density.info="none",
col=greenred(300), margins=c(10,10))
heatmap.2(cor((assay(vsd))),trace="none",density.info="none",
col=colorRampPalette(c("yellow","blue"))(300),margins=c
(10,10))
```

3.3.8 Pathway Analyses

After running the results() function, users have a list of differentially expressed genes. One can look at which Gene Ontology (GO) terms are enriched in the list. For this, DAVID or Homer or clusterProfiler can be used. Homer offers a Perl script findGO.pl, which takes the input file of Entrez Gene IDs, organism, and output directory. Homer assumes the cumulative hypergeometric distribution to calculate the enrichment [44]. The background usually consists of all genes of an organism, but one can specify the list of genes expressed in the dataset. Homer and the recent version of clusterProfiler also allow setting a custom background. clusterProfiler is an R package with multiple visualization options

[54]. The function enrichGO() performs a GO overrepresentation test. And the result can be visualized with barplot(), dotplot(), a gene-concept network cnetplot(), etc. DAVID is a web interface where one can upload the lists of DE genes and background genes [55]. Figure 9a shows the results of gene ontology analyses on the differentially regulated genes obtained from DESeq2 using Homer with Wikipathways gene ontologies [56]. Expectedly, pluripotent and cell cycle signatures are enriched in upregulated genes from undifferentiated ESCs, while differentiation signatures are enriched in upregulated genes from differentiated ESCs. To perform gene ontology analysis using Homer, the following commands are used:

```
findGO.pl <gene_list> <organism> <output> -cpu <threads>
[-bg <base.gene>]
```

In R, to plot the results, the following commands are run top plot the top 10 ontologies from Wikipathways:

```
library(gplots)
```

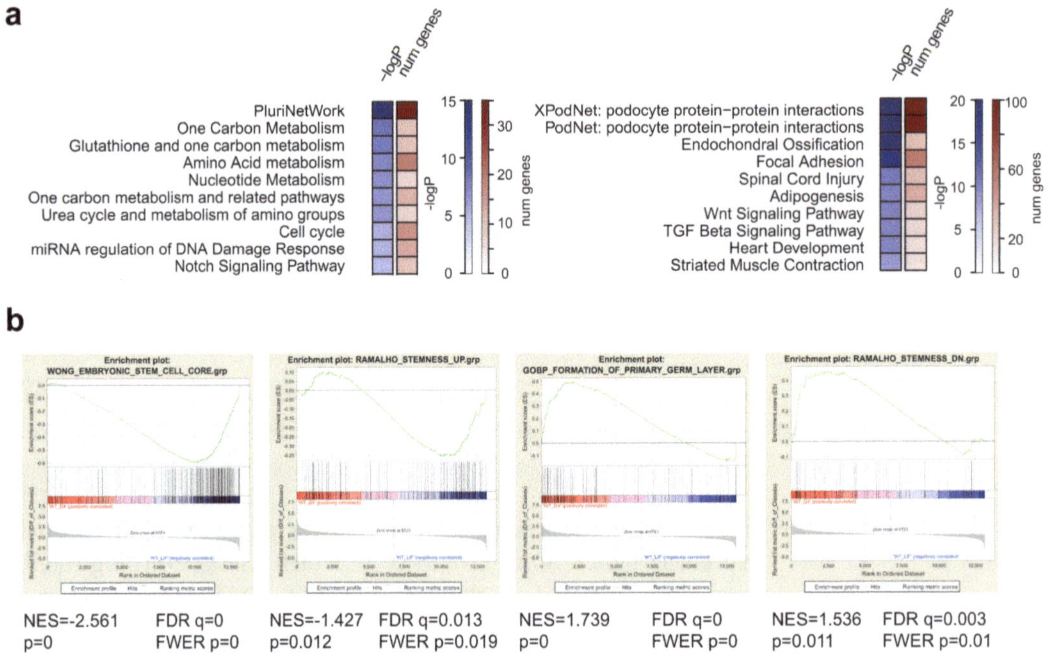

Fig. 9 Gene ontology and Gene set enrichment analyses from differentially regulated genes. (**a**) Heatmaps showing Wikipathways gene ontology analyses in differentially regulated genes from RNA-Seq of undifferentiated mESCs grown in LIF (left) versus differentiated mESCs grown in RA (right). –logP values are shown according to the blue scale, and the number of genes according to the red scale. (**b**) Gene set enrichment analyses of RNA-Seq of undifferentiated mESCs grown in LIF versus differentiated mESCs grown in RA using stemness and differentiation gene sets. Significance values are shown below each graph

```
wikipathways=read.table("wikipathways.txt",head=T,sep="\t",
fill=T)
x=cbind(head(-(wikipathways$logP),10),head(wikipathways$Tar-
get.Genes.in.Term,10))
rownames(x)=head(wikipathways$Term,10)
colnames(x)=c("-logP","num_genes")
pdf("wikipathways_heatmap_-logP.pdf")
heatmap.2(as.matrix(x[,1]), trace="none", density.info="-
none", dendrogram="none", Rowv=FALSE, Colv=FALSE, col=color-
RampPalette(c("white","darkblue"))(300), colsep=0:ncol(x),
rowsep=0:nrow(x), sepcolor="black", sepwidth=c(0.0001,
0.0001), margins=c(20,20))
dev.off()
pdf("wikipathways_heatmap_num_genes.pdf")
heatmap.2(as.matrix(x[,2]),trace="none",density.info="none",
dendrogram="none",Rowv=FALSE, Colv=FALSE, col=colorRampPal-
ette(c("white","darkred"))(300), colsep=0:ncol(x), rowsep=0:
nrow(x), sepcolor="black", sepwidth=c(0.0001, 0.0001), mar-
gins=c(20,20))
dev.off()
```

Another way to compare two biological conditions is to use Gene Set Enrichment Analysis (GSEA). GSEA preranked from Broad institute or the FGSEA R package takes gene names and corresponding logFCs [24, 57]. GSEA performs permutations to compute the significance of a signature (e.g., GO term or custom annotated gene set). The output is a table with Normalized Enrichment Score (NES) and adjusted *p*-value. One can visualize the result with the plotEnrichment() function of the FGSEA package. Performing GSEA is done using the following steps:

1. Prepare rlog/vst/pseudo $\log_2$ ($\log_2$+1) expression table (*see* Table 11 and Annex 2).
2. Prepare phenotype .cls file (*see* Table 12 and Annex 2).
3. Press "Load Data."
4. Optional: load annotations to be tested for enrichment in "Load Data" (*see* **Note 8**).
5. Press "Run GSEA."
6. Select expression dataset.
7. Select gene sets from "Gene sets database."
8. Select GENE_SYMBOL from "ChIP platform" (*see* **Note 9** if not available).
9. Select Diff_of_classes in "Metric for ranking genes."
10. Press "Run."

Table 11
Format of GSEA expression table

NAME	Description	<condition 1>	<condition 2>	<condition 3>
<gene1>	na	<$\log_2$ value 1>	<$\log_2$ value 1>	<$\log_2$ value 1>
<gene2>	na	<$\log_2$ value 2>	<$\log_2$ value 2>	<$\log_2$ value 2>
<gene3>	na	<$\log_2$ value 3>	<$\log_2$ value 3>	<$\log_2$ value 3>
<gene4>	na	<$\log_2$ value 4>	<$\log_2$ value 4>	<$\log_2$ value 4>
<gene5>	na	<$\log_2$ value 5>	<$\log_2$ value 5>	<$\log_2$ value 5>

Values are best rlog-, vst-, or pseudo $\log_2$- ($\log_2$+1) transformed to use the difference of classes metric

Table 12
Structure of the GSEA phenotype .cls file

<total sample number> <total conditions> 1
<condition 1> <condition 2> <condition 3> <...>
<condition 1 n times> <condition 2 n times> <condition 3 n times>

Figure 9b shows GSEA results using RNA-Seq data from WT ESC in LIF (undifferentiated) and RA conditions (differentiating), testing for enrichment of ESC and differentiation signatures [58, 59]. As expected, genes upregulated in undifferentiated ESCs show significant enrichment for stemness gene sets, and conversely those upregulated in differentiated ESCs show significant enrichment for differentiation gene sets.

3.3.9 Integration with Further RNA-Seq Datasets

Once the batch effect is corrected, one can perform integrative analysis to combine several bulk RNA-seq datasets (or RNA-Seq with other OMICS datasets, *see* **Note 10**). Having more than two conditions allows using hierarchical and k-means clustering of the union of DE genes (DEGs). Another approach to show how similar the two comparisons are is to use Venn diagrams of DEGs and calculate a hypergeometric *p*-value for the overlaps. One can also perform GSEA using DEG lists from two conditions as background and ranked expression from two others. An additional way to process this data is to perform metagene analysis [60]. These entail grouping coregulated genes into metagenes. When comparing different datasets, the ensuing dimensionality reduction yields better scatterplots and correlations. There are several hurdles to analyzing different experiments. Firstly, if one dataset has an unstranded library and another has a stranded one, both should be processed as unstranded. Secondly, if one library includes polyA-positive RNA and the second one has total RNA, it is recommended to disregard polyA-negative genes.

4 Notes

1. For de novo transcriptome assembly, if a genome is not yet available or for gene/isoform discovery, the Trinity package can be used for further analyses in this direction [61]. In this case, transcript abundances can be obtained from Trinity and passed directly to DESeq2 for differential expression analysis.
2. All steps shown here can be performed with snakePipes, with some variation from the examples shown here. For example, deeptools computes the correlation matrix based on genome bins of 10 kb per default. Alignment-free methods to perform quantification also exist, for example, Salmon, which is optionally used by snakePipes. Please visit https://snakepipes.readthedocs.io/en/latest/content/workflows/mRNA-seq.html for more information.
3. Demultiplexing and fastq generation: data is sometimes delivered directly from the sequencer software (e.g., Illumina HiSeq Analysis Sofware). In this case, it is possible to generate fastq files using bcl2fastq, in which case a sample sheet must be created or adapted from the experiment run. For more information, please consult https://support.illumina.com/content/dam/illumina-support/documents/documentation/software_documentation/bcl2fastq/bcl2fastq_letterbooklet_15038058brpmi.pdf
4. Public data: to analyze previously published data, all abovementioned aspects apply. This step can be integrated into workflows by using the sra-toolkit command fastq-dump:

```
fastq-dump -A <SRR_accession> --split-3 --gzip
```

where SRR_accession is the run number at the sequence read archive (SRA) [62]. For paired-end experiments, this will generate one gzip compressed fastq file per read, called <SRR_accession_1.fastq.gz> and <SRR_accession_2.fastq.gz>. If the data is single-end, this will simply create a one gzip compressed fastq file called <SRR_accession.fastq.gz>. Note that fastq-dump can also be used using standard output using the -Z switch. For single-end experiments, a STAR align command would thus use

```
<(fastq-dump -A <SRR_accession> -Z')
```

as the input read instead of a file. For paired-end files, the standard output of fastq-dump using --split-spot can be used and the first and second blocks of four lines every 8 lines can be used for read 1, and 2, respectively, as the input reads:

```
<(fastq-dump -A <SRR_accession> -Z --split-spot | sed '5~8d;6~8d;7~8d;8~8d') <(fastq-dump -A <SRR_accession> -Z --split-spot | sed '1~8d;2~8d;3~8d;4~8d')
```

5. Further analyses include splice-based analyses and differential splicing analyses. Two popular packages are MISO and DexSeq [63, 64].
6. The outputs of differential expression analysis and downstream quality controls (PCA, heatmaps, MA and volcano plots) can also conveniently be viewed and exported in a user-friendly manner using R shinyapps. For examples, please see https://kcvi.shinyapps.io/START/ or https://rnaseqvisualization.shinyapps.io/scatmat/.
7. For one-tailed tests, in this case, the alternative hypothesis will be that the actual log2FoldChange value, not absolute, will be greater than a given threshold. These can be used if changes are only expected in one direction and negligible or irrelevant on the other direction. Since it is a more conservative approach, a much smaller set of genes will be expected.
8. GSEA gene set annotations can be downloaded from the following webpage: https://www.gsea-msigdb.org/gsea/msigdb/index.jsp.
9. If not available, the GENE_SYMBOL.chip file can be obtained at https://github.com/zhezhangsh/DEGandMore/raw/master/examples/gsea/GENE_SYMBOL.chip.
10. It is possible and common to integrate RNA-Seq data with other types of OMICS data. Notably, other data types can be correlated at the gene feature level with RNA-Seq, for example, using the GSEA suite, for example, ATAC-Seq and whole-genome bisulfite sequencing (wGBS). In this case, the gene sets will consist of the genes to which peaks and CpGs are annotated to. Other popular methods include plotting expression data as boxplots per class of differential regions [28, 65], or plotting the corresponding expression fold change from two genomic assays sorted by fold change [66, 67]. Integration with metabolomics is also possible using metabolograms [68, 69].

Acknowledgment

This work was supported by the Max Planck Society and the German Cancer Research Centre (DKFZ).

References

1. Buratowski S (2009) Progression through the RNA polymerase II CTD cycle. Mol Cell 36(4):541–546. https://doi.org/10.1016/j.molcel.2009.10.019
2. Franzen O, Jerlstrom-Hultqvist J, Einarsson E, Ankarklev J, Ferella M, Andersson B, Svard SG (2013) Transcriptome profiling of Giardia intestinalis using strand-specific RNA-seq. PLoS Comput Biol 9(3):e1003000. https://doi.org/10.1371/journal.pcbi.1003000
3. Meyer M, Kircher M (2010) Illumina sequencing library preparation for highly multiplexed target capture and sequencing. Cold Spring Harb Protoc 2010(6):pdb.prot5448. https://doi.org/10.1101/pdb.prot5448
4. Antonio Urrutia G, Ramachandran H, Cauchy P, Boo K, Ramamoorthy S, Boller S, Dogan E, Clapes T, Trompouki E, Torres-Padilla ME, Palvimo JJ, Pichler A, Grosschedl R (2021) ZFP451-mediated SUMOylation of SATB2 drives embryonic stem cell differentiation. Genes Dev 35(15–16):1142–1160. https://doi.org/10.1101/gad.345843.120
5. Zhao S, Zhang Y, Gamini R, Zhang B, von Schack D (2018) Evaluation of two main RNA-seq approaches for gene quantification in clinical RNA sequencing: polyA+ selection versus rRNA depletion. Sci Rep 8(1):4781. https://doi.org/10.1038/s41598-018-23226-4
6. Zhao W, He X, Hoadley KA, Parker JS, Hayes DN, Perou CM (2014) Comparison of RNA-Seq by poly (A) capture, ribosomal RNA depletion, and DNA microarray for expression profiling. BMC Genomics 15:419. https://doi.org/10.1186/1471-2164-15-419
7. Wei C-L, Ruan Y (2008) Multiplex sequencing of paired end ditags for transcriptome and genome analysis. In: Next generation genome sequencing. Wiley, pp 165–182. https://doi.org/10.1002/9783527625130.ch13
8. Levin JZ, Yassour M, Adiconis X, Nusbaum C, Thompson DA, Friedman N, Gnirke A, Regev A (2010) Comprehensive comparative analysis of strand-specific RNA sequencing methods. Nat Methods 7(9):709–715. https://doi.org/10.1038/nmeth.1491
9. Zhao S, Zhang Y, Gordon W, Quan J, Xi H, Du S, von Schack D, Zhang B (2015) Comparison of stranded and non-stranded RNA-seq transcriptome profiling and investigation of gene overlap. BMC Genomics 16:675. https://doi.org/10.1186/s12864-015-1876-7
10. Agarwal S, Macfarlan TS, Sartor MA, Iwase S (2015) Sequencing of first-strand cDNA library reveals full-length transcriptomes. Nat Commun 6:6002. https://doi.org/10.1038/ncomms7002
11. Picelli S, Faridani OR, Bjorklund AK, Winberg G, Sagasser S, Sandberg R (2014) Full-length RNA-seq from single cells using Smart-seq2. Nat Protoc 9(1):171–181. https://doi.org/10.1038/nprot.2014.006
12. Bhardwaj V, Heyne S, Sikora K, Rabbani L, Rauer M, Kilpert F, Richter AS, Ryan DP, Manke T (2019) snakePipes: facilitating flexible, scalable and integrative epigenomic analysis. Bioinformatics 35(22):4757–4759. https://doi.org/10.1093/bioinformatics/btz436
13. Woste M, Leitao E, Laurentino S, Horsthemke B, Rahmann S, Schroder C (2020) wg-blimp: an end-to-end analysis pipeline for whole genome bisulfite sequencing data. BMC Bioinformatics 21(1):169. https://doi.org/10.1186/s12859-020-3470-5
14. Yukselen O, Turkyilmaz O, Ozturk AR, Garber M, Kucukural A (2020) DolphinNext: a distributed data processing platform for high throughput genomics. BMC Genomics 21(1):310. https://doi.org/10.1186/s12864-020-6714-x
15. Afgan E, Baker D, van den Beek M, Blankenberg D, Bouvier D, Cech M, Chilton J, Clements D, Coraor N, Eberhard C, Gruning B, Guerler A, Hillman-Jackson J, Von Kuster G, Rasche E, Soranzo N, Turaga N, Taylor J, Nekrutenko A, Goecks J (2016) The Galaxy platform for accessible, reproducible and collaborative biomedical analyses: 2016 update. Nucleic Acids Res 44(W1):W3–W10. https://doi.org/10.1093/nar/gkw343
16. Dobin A, Davis CA, Schlesinger F, Drenkow J, Zaleski C, Jha S, Batut P, Chaisson M, Gingeras TR (2013) STAR: ultrafast universal RNA-seq aligner. Bioinformatics 29(1):15–21. https://doi.org/10.1093/bioinformatics/bts635
17. Andrews S (2010) FastQC: a quality control tool for high throughput sequence data. Github
18. Ewels P, Magnusson M, Lundin S, Kaller M (2016) MultiQC: summarize analysis results for multiple tools and samples in a single report. Bioinformatics 32(19):3047–3048. https://doi.org/10.1093/bioinformatics/btw354
19. Bolger AM, Lohse M, Usadel B (2014) Trimmomatic: a flexible trimmer for Illumina sequence data. Bioinformatics 30(15):

2114–2120. https://doi.org/10.1093/bioinformatics/btu170
20. Ramirez F, Dundar F, Diehl S, Gruning BA, Manke T (2014) deepTools: a flexible platform for exploring deep-sequencing data. Nucleic Acids Res 42(Web Server issue):W187–W191. https://doi.org/10.1093/nar/gku365
21. Liao Y, Smyth GK, Shi W (2014) featureCounts: an efficient general purpose program for assigning sequence reads to genomic features. Bioinformatics 30(7):923–930. https://doi.org/10.1093/bioinformatics/btt656
22. Love MI, Huber W, Anders S (2014) Moderated estimation of fold change and dispersion for RNA-seq data with DESeq2. Genome Biol 15(12):550. https://doi.org/10.1186/s13059-014-0550-8
23. Smyth GK (2004) Linear models and empirical bayes methods for assessing differential expression in microarray experiments. Stat Appl Genet Mol Biol 3(1):3. https://doi.org/10.2202/1544-6115.1027
24. Subramanian A, Tamayo P, Mootha VK, Mukherjee S, Ebert BL, Gillette MA, Paulovich A, Pomeroy SL, Golub TR, Lander ES, Mesirov JP (2005) Gene set enrichment analysis: a knowledge-based approach for interpreting genome-wide expression profiles. Proc Natl Acad Sci U S A 102(43):15545–15550. https://doi.org/10.1073/pnas.0506580102
25. Martin M (2011) Cutadapt removes adapter sequences from high-throughput sequencing reads. EMBnet J 17(1):3. https://doi.org/10.14806/ej.17.1.200
26. Krueger F (2016) Trim Galore! Github. https://doi.org/10.5281/zenodo.5127898
27. Amemiya HM, Kundaje A, Boyle AP (2019) The ENCODE blacklist: identification of problematic regions of the genome. Sci Rep 9(1):9354. https://doi.org/10.1038/s41598-019-45839-z
28. Cauchy P, Maqbool MA, Zacarias-Cabeza J, Vanhille L, Koch F, Fenouil R, Gut M, Gut I, Santana MA, Griffon A, Imbert J, Moraes-Cabe C, Bories JC, Ferrier P, Spicuglia S, Andrau JC (2016) Dynamic recruitment of Ets1 to both nucleosome-occupied and -depleted enhancer regions mediates a transcriptional program switch during early T-cell differentiation. Nucleic Acids Res 44(8):3567–3585. https://doi.org/10.1093/nar/gkv1475
29. Ewing B, Green P (1998) Base-calling of automated sequencer traces using phred. II. Error probabilities. Genome Res 8(3):186–194
30. Benelli M, Pescucci C, Marseglia G, Severgnini M, Torricelli F, Magi A (2012) Discovering chimeric transcripts in paired-end RNA-seq data by using EricScript. Bioinformatics 28(24):3232–3239. https://doi.org/10.1093/bioinformatics/bts617
31. Engstrom PG, Steijger T, Sipos B, Grant GR, Kahles A, Ratsch G, Goldman N, Hubbard TJ, Harrow J, Guigo R, Bertone P (2013) Systematic evaluation of spliced alignment programs for RNA-seq data. Nat Methods 10(12):1185–1191. https://doi.org/10.1038/nmeth.2722
32. Ye H, Meehan J, Tong W, Hong H (2015) Alignment of short reads: a crucial step for application of next-generation sequencing data in precision medicine. Pharmaceutics 7(4):523–541. https://doi.org/10.3390/pharmaceutics7040523
33. Kim D, Langmead B, Salzberg SL (2015) HISAT: a fast spliced aligner with low memory requirements. Nat Methods 12(4):357–360. https://doi.org/10.1038/nmeth.3317
34. Kim D, Pertea G, Trapnell C, Pimentel H, Kelley R, Salzberg SL (2013) TopHat2: accurate alignment of transcriptomes in the presence of insertions, deletions and gene fusions. Genome Biol 14(4):R36. https://doi.org/10.1186/gb-2013-14-4-r36
35. Liao Y, Smyth GK, Shi W (2013) The Subread aligner: fast, accurate and scalable read mapping by seed-and-vote. Nucleic Acids Res 41(10):e108. https://doi.org/10.1093/nar/gkt214
36. Liao Y, Smyth GK, Shi W (2019) The R package Rsubread is easier, faster, cheaper and better for alignment and quantification of RNA sequencing reads. Nucleic Acids Res 47(8):e47. https://doi.org/10.1093/nar/gkz114
37. Huang S, Zhang J, Li R, Zhang W, He Z, Lam TW, Peng Z, Yiu SM (2011) SOAPsplice: genome-wide ab initio detection of splice junctions from RNA-Seq data. Front Genet 2:46. https://doi.org/10.3389/fgene.2011.00046
38. Wu TD, Nacu S (2010) Fast and SNP-tolerant detection of complex variants and splicing in short reads. Bioinformatics 26(7):873–881. https://doi.org/10.1093/bioinformatics/btq057
39. Veeneman BA, Shukla S, Dhanasekaran SM, Chinnaiyan AM, Nesvizhskii AI (2016) Two-pass alignment improves novel splice junction quantification. Bioinformatics 32(1):43–49. https://doi.org/10.1093/bioinformatics/btv642
40. Dobin A, Gingeras TR (2015) Mapping RNA-seq reads with STAR. Curr Protoc Bioinformatics 51:11.14.11–11.14.19. https://doi.org/10.1002/0471250953.bi1114s51
41. Klepikova AV, Kasianov AS, Chesnokov MS, Lazarevich NL, Penin AA, Logacheva M

(2017) Effect of method of deduplication on estimation of differential gene expression using RNA-seq. PeerJ 5:e3091. https://doi.org/10.7717/peerj.3091

42. Li H, Ruan J, Durbin R (2008) Mapping short DNA sequencing reads and calling variants using mapping quality scores. Genome Res 18(11):1851–1858. https://doi.org/10.1101/gr.078212.108
43. Li H, Handsaker B, Wysoker A, Fennell T, Ruan J, Homer N, Marth G, Abecasis G, Durbin R (2009) The sequence alignment/map format and SAMtools. Bioinformatics 25(16):2078–2079. https://doi.org/10.1093/bioinformatics/btp352
44. Heinz S, Benner C, Spann N, Bertolino E, Lin YC, Laslo P, Cheng JX, Murre C, Singh H, Glass CK (2010) Simple combinations of lineage-determining transcription factors prime cis-regulatory elements required for macrophage and B cell identities. Mol Cell 38(4):576–589. https://doi.org/10.1016/j.molcel.2010.05.004
45. Ribeiro A, Golicz A, Hackett CA, Milne I, Stephen G, Marshall D, Flavell AJ, Bayer M (2015) An investigation of causes of false positive single nucleotide polymorphisms using simulated reads from a small eukaryote genome. BMC Bioinformatics 16:382. https://doi.org/10.1186/s12859-015-0801-z
46. Robinson JT, Thorvaldsdottir H, Winckler W, Guttman M, Lander ES, Getz G, Mesirov JP (2011) Integrative genomics viewer. Nat Biotechnol 29(1):24–26. https://doi.org/10.1038/nbt.1754
47. Tourriere H, Chebli K, Tazi J (2002) mRNA degradation machines in eukaryotic cells. Biochimie 84(8):821–837. https://doi.org/10.1016/s0300-9084(02)01445-1
48. Edgington-White B, Cauchy P, Assi SA, Hartmann S, Riggs AG, Mathas S, Cockerill PN, Bonifer C (2019) Global long terminal repeat activation participates in establishing the unique gene expression programme of classical Hodgkin lymphoma. Leukemia 33(6):1463–1474. https://doi.org/10.1038/s41375-018-0311-x
49. Ren X, Kuan PF (2020) Negative binomial additive model for RNA-Seq data analysis. BMC Bioinformatics 21(1):171. https://doi.org/10.1186/s12859-020-3506-x
50. Robinson MD, McCarthy DJ, Smyth GK (2010) edgeR: a Bioconductor package for differential expression analysis of digital gene expression data. Bioinformatics 26(1):139–140. https://doi.org/10.1093/bioinformatics/btp616
51. Trapnell C, Hendrickson DG, Sauvageau M, Goff L, Rinn JL, Pachter L (2013) Differential analysis of gene regulation at transcript resolution with RNA-seq. Nat Biotechnol 31(1):46–53. https://doi.org/10.1038/nbt.2450
52. Durbin BP, Hardin JS, Hawkins DM, Rocke DM (2002) A variance-stabilizing transformation for gene-expression microarray data. Bioinformatics 18(suppl_1):S105–S110. https://doi.org/10.1093/bioinformatics/18.suppl_1.S105
53. Leek JT, Johnson WE, Parker HS, Jaffe AE, Storey JD (2012) The sva package for removing batch effects and other unwanted variation in high-throughput experiments. Bioinformatics 28(6):882–883. https://doi.org/10.1093/bioinformatics/bts034
54. Wu T, Hu E, Xu S, Chen M, Guo P, Dai Z, Feng T, Zhou L, Tang W, Zhan L, Fu X, Liu S, Bo X, Yu G (2021) clusterProfiler 4.0: a universal enrichment tool for interpreting omics data. Innovation (N Y) 2(3):100141. https://doi.org/10.1016/j.xinn.2021.100141
55. Huang DW, Sherman BT, Lempicki RA (2009) Systematic and integrative analysis of large gene lists using DAVID bioinformatics resources. Nat Protoc 4(1):44–57. https://doi.org/10.1038/nprot.2008.211
56. Slenter DN, Kutmon M, Hanspers K, Riutta A, Windsor J, Nunes N, Melius J, Cirillo E, Coort SL, Digles D, Ehrhart F, Giesbertz P, Kalafati M, Martens M, Miller R, Nishida K, Rieswijk L, Waagmeester A, Eijssen LMT, Evelo CT, Pico AR, Willighagen EL (2018) WikiPathways: a multifaceted pathway database bridging metabolomics to other omics research. Nucleic Acids Res 46(D1):D661–D667. https://doi.org/10.1093/nar/gkx1064
57. Korotkevich G, Sukhov V, Budin N, Shpak B, Artyomov MN, Sergushichev A (2021) Fast gene set enrichment analysis. bioRxiv:060012. https://doi.org/10.1101/060012
58. Wong DJ, Liu H, Ridky TW, Cassarino D, Segal E, Chang HY (2008) Module map of stem cell genes guides creation of epithelial cancer stem cells. Cell Stem Cell 2(4):333–344. https://doi.org/10.1016/j.stem.2008.02.009
59. Ramalho-Santos M, Yoon S, Matsuzaki Y, Mulligan RC, Melton DA (2002) "Stemness": transcriptional profiling of embryonic and adult stem cells. Science 298(5593):597–600. https://doi.org/10.1126/science.1072530
60. Tamayo P, Scanfeld D, Ebert BL, Gillette MA, Roberts CW, Mesirov JP (2007) Metagene projection for cross-platform, cross-species characterization of global transcriptional states.

Proc Natl Acad Sci U S A 104(14):5959–5964. https://doi.org/10.1073/pnas.0701068104

61. Grabherr MG, Haas BJ, Yassour M, Levin JZ, Thompson DA, Amit I, Adiconis X, Fan L, Raychowdhury R, Zeng Q, Chen Z, Mauceli E, Hacohen N, Gnirke A, Rhind N, di Palma F, Birren BW, Nusbaum C, Lindblad-Toh K, Friedman N, Regev A (2011) Full-length transcriptome assembly from RNA-Seq data without a reference genome. Nat Biotechnol 29(7):644–652. https://doi.org/10.1038/nbt.1883
62. Leinonen R, Sugawara H, Shumway M (2011) The sequence read archive. Nucleic Acids Res 39(Database issue):D19–D21. https://doi.org/10.1093/nar/gkq1019
63. Katz Y, Wang ET, Airoldi EM, Burge CB (2010) Analysis and design of RNA sequencing experiments for identifying isoform regulation. Nat Methods 7(12):1009–1015. https://doi.org/10.1038/nmeth.1528
64. Anders S, Reyes A, Huber W (2012) Detecting differential usage of exons from RNA-seq data. Genome Res 22(10):2008–2017. https://doi.org/10.1101/gr.133744.111
65. Cauchy P, James SR, Zacarias-Cabeza J, Ptasinska A, Imperato MR, Assi SA, Piper J, Canestraro M, Hoogenkamp M, Raghavan M, Loke J, Akiki S, Clokie SJ, Richards SJ, Westhead DR, Griffiths MJ, Ott S, Bonifer C, Cockerill PN (2015) Chronic FLT3-ITD signaling in acute myeloid leukemia is connected to a specific chromatin signature. Cell Rep 12(5):821–836. https://doi.org/10.1016/j.celrep.2015.06.069
66. Obier N, Cauchy P, Assi SA, Gilmour J, Lie ALM, Lichtinger M, Hoogenkamp M, Noailles L, Cockerill PN, Lacaud G, Kouskoff V, Bonifer C (2016) Cooperative binding of AP-1 and TEAD4 modulates the balance between vascular smooth muscle and hemogenic cell fate. Development 143(23):4324–4340. https://doi.org/10.1242/dev.139857
67. Kreher S, Bouhlel MA, Cauchy P, Lamprecht B, Li S, Grau M, Hummel F, Kochert K, Anagnostopoulos I, Johrens K, Hummel M, Hiscott J, Wenzel SS, Lenz P, Schneider M, Kuppers R, Scheidereit C, Giefing M, Siebert R, Rajewsky K, Lenz G, Cockerill PN, Janz M, Dorken B, Bonifer C, Mathas S (2014) Mapping of transcription factor motifs in active chromatin identifies IRF5 as key regulator in classical Hodgkin lymphoma. Proc Natl Acad Sci U S A 111(42):E4513–E4522. https://doi.org/10.1073/pnas.1406985111
68. Hakimi AA, Reznik E, Lee CH, Creighton CJ, Brannon AR, Luna A, Aksoy BA, Liu EM, Shen R, Lee W, Chen Y, Stirdivant SM, Russo P, Chen YB, Tickoo SK, Reuter VE, Cheng EH, Sander C, Hsieh JJ (2016) An integrated metabolic atlas of clear cell renal cell carcinoma. Cancer Cell 29(1):104–116. https://doi.org/10.1016/j.ccell.2015.12.004
69. Schönberger K, Obier N, Romero-Mulero MC, Cauchy P, Mess J, Pavlovich PV, Zhang YW, Mitterer M, Rettkowski J, Lalioti M-E, Jäcklein K, Curtis JD, Féret B, Sommerkamp P, Morganti C, Ito K, Ghyselinck NB, Trompouki E, Buescher JM, Pearce EL, Cabezas-Wallscheid N (2021) Multilayer omics analysis reveals a non-classical retinoic acid signaling axis that regulates hematopoietic stem cell identity. Cell Stem Cell 29:1–18. https://doi.org/10.1016/j.stem.2021.10.002
70. Pease J, Sooknanan R (2012) A rapid, directional RNA-seq library preparation workflow for Illumina sequencing. Nat Methods 9(3):i–ii. https://doi.org/10.1038/nmeth.f.355

Chapter 21

Reverse Transcription-Quantitative Real-Time Polymerase Chain Reaction (RT-qPCR) for Gene Expression Analyses

Albert Caballero-Solares, Jennifer R. Hall, Xi Xue, and Matthew L. Rise

Abstract

The reverse transcription-quantitative real-time polymerase chain reaction (RT-qPCR) is considered to be the gold standard for gene expression research. However, for this claim to be valid, RT-qPCR studies must test and optimize the quality of its RNA templates and assays. This chapter describes the experimental procedures required to generate reliable and reproducible gene expression results using RT-qPCR.

Key words Polymerase chain reaction, Reverse transcription, RNA, cDNA, Gene of interest, Normalizer gene selection, Primer pairs, Quality testing, MIQE guidelines

1 Introduction

Quantitative real-time polymerase chain reaction (qPCR) allows nucleic acid abundance quantification via monitoring the reaction's fluorescence [1]. Since its discovery in the early 1990s [1, 2], qPCR has rapidly become the preferred analytical tool to detect/quantify low levels of DNA. The technique's versatility expanded to include RNA detection/quantification with the addition of a reverse-transcription (RT) step preceding the qPCR. Due to its high sensitivity, specificity, and reproducibility, RT-qPCR has been extensively used for research (e.g., the study of gene expression responses) and diagnostic purposes (e.g., pathogen detection) [3].

Many research areas have applied RT-qPCR to analyze differences in the abundance of target mRNA among samples from different groups/treatments. The great popularization of RT-qPCR brought about substantial variations and inconsistencies in the laboratory procedures and data analysis methods [4]. The soundness and reproducibility of the RT-qPCR data rely on the use of an optimal RNA sample quality and amplification performance. In addition, the relative quantification of mRNA requires the selection of normalizer genes (also referred to as endogenous controls,

Sherri L. Christian (ed.), *Cancer Cell Biology: Methods and Protocols*,
Methods in Molecular Biology, vol. 2508, https://doi.org/10.1007/978-1-0716-2376-3_21,

reference genes, or housekeeping genes) stably expressed across biological replicates and groups/treatments. The need for standardization in the RT-qPCR methods and their reporting in research articles prompted the creation of the MIQE guidelines (i.e., Minimum Information for Publication of Quantitative Real-Time PCR Experiments) [4, 5].

Despite the existence of the MIQE guidelines and their rigorous enforcement by many scientific journals and reviewers, unfortunately, there are some recently published studies that still fail to provide the information required to ascertain the validity of RT-qPCR results and/or continue to use inadequate procedures (e.g., using one arbitrarily selected normalizer gene, failure to provide evidence of high RNA purity/integrity, and adequate performance of the qPCR primer pairs) [6]. This chapter will describe the MIQE-compliant procedures required to complete an RT-qPCR study, quality-check the RNA templates and the qPCR assays, and prevent contamination and bias. There are several types of fluorescence-based technologies for amplicon detection in RT-qPCR, such as hydrolysis probes (e.g., TaqMan®; Applied Biosystems/Thermo Fisher Scientific), hairpin primers (e.g., LUX™; Invitrogen), and intercalating dyes (e.g., SYBR® Green; Molecular Probes/Thermo Fisher Scientific) [7]. The analytical procedures explained below will be those for SYBR® Green-based qPCR, one of the most extensively used qPCR technologies for gene expression research.

In the interest of clarity, the text below will use 3 terms that are defined as follows: (1) Obtaining the quantification cycle (C_q) [also referred to as threshold cycle (C_t) or crossing point (C_p); that is, the PCR cycle at which fluorescence increases above the background] of a given gene of interest (GOI) or normalizer gene for all biological replicates is a *qPCR run*. (2) Calculating the transcript levels of a given GOI relative to 2—or more—normalizer genes is an *RT-qPCR experiment*. (3) Determining the relative quantification of all targeted GOIs is an *RT-qPCR study*. The description and figures of the Materials and Methods are based on a mock RT-qPCR study analyzing the mRNA levels of 6 GOIs and 6 candidate normalizer genes for a *Control* group and a *Treatment* group, comprised of 10 biological replicates (i.e., samples) each.

2 Materials

This section will list the reagents, consumables, and instruments required for quality-testing the total RNA templates and the qPCR primer pairs, and for cDNA synthesis and qPCR. Regarding the required personal protective equipment (PPE), RT-qPCR users should follow basic safety requirements for molecular biology laboratories such as wearing closed-toe shoes, clothing that fully covers their legs, a lab coat, disposable gloves, and safety goggles. Some procedures will require specific PPE, which will be indicated in the subheadings below.

2.1 Total RNA Sample Purity-Testing (see Note 1)

1. DNase-treated, column-purified RNA for each biological replicate (*see* **Note 2**).
2. Ultrapure nuclease-free water (UPNFW).
3. A single-channel P2 or P10 pipette—that is, capable of accurately pipetting volumes ranges of 0.2–2 μL or 1–10 μL, respectively.
4. Nuclease-free sterile filter pipette tips for a P2 or P10 pipette.
5. Ultraviolet (UV)/visible spectrophotometer designed to quantify and purity-test RNA using small solution volumes [e.g., <2 μL; NanoDrop™ spectrophotometer (Thermo Scientific)].

2.2 Total RNA Sample Integrity-Testing via Agarose Gel Electrophoresis (see Notes 2 and 3)

1. 1× Tris acetate-EDTA (TAE) buffer: 40 mM Tris, 20 mM acetic acid, 1 mM EDTA or 1× Tris borate-EDTA (TBE) buffer: 89 mM Tris, 89 mM boric acid, 2 mM EDTA.
2. Molecular biology-grade agarose.
3. RNA-binding fluorescent dye [e.g., ethidium bromide, SYBR™ Safe (Invitrogen), or GelRed® (Biotium)].
4. A loading or tracking dye buffer [e.g., BlueJuice gel loading buffer 10× (Invitrogen)].
5. DNA size marker [e.g., 1 Kb Plus DNA Ladder (Invitrogen)].
6. Single-channel pipettes to pipette different volume ranges (e.g., 1–10 μL, 2–20 μL, 10–100 μL, 100–1000 μL).
7. Nuclease-free sterile filter pipette tips for different volume ranges (e.g., 1–10 μL, 2–20 μL, 10–100 μL, 100–1000 μL).
8. PCR strip tubes.
9. An Erlenmeyer flask.
10. A microwave.
11. A horizontal gel caster with combs.
12. An electrophoresis chamber—connected to a power supply.
13. A gel imaging system equipped with a UV transilluminator and camera.
14. Specific PPE: heat-resistant gloves, a chemical fume hood, and, if handling a mutagenic RNA-binding fluorescent dye (e.g., ethidium bromide), dedicated and adequately labeled laboratory work surfaces must be used.

2.3 qPCR Primer Design

1. National Center for Biotechnology Information (NCBI)'s nucleotide [e.g., nonredundant nucleotide (nr/nt) collection, expressed sequence tags (ESTs), and sequence read archive

(SRA)] and protein [i.e., nonredundant (nr) protein sequences] databases.

2. NCBI's Basic Local Alignment Search Tool (BLAST; https://blast.ncbi.nlm.nih.gov/Blast.cgi) for nucleotide (i.e., BLASTn) and protein (i.e., BLASTx) searches.
3. A multiple sequence alignment tool [e.g., Clustal Omega (https://www.ebi.ac.uk/Tools/msa/clustalo/)].
4. A primer design tool [e.g., Primer3 (https://bioinfo.ut.ee/primer3/); Primer Express (Applied Biosystems/Thermo Fisher Scientific); NCBI Primer-BLAST (https://www.ncbi.nlm.nih.gov/tools/primer-blast/)].
5. An oligo analyzer tool for manually designed primers (e.g., from Integrated DNA Technologies; https://www.idtdna.com/calc/analyzer).

2.4 Reagents, Consumables, and Instruments for cDNA Synthesis (Reverse Transcription; RT)

1. Random primers and/or oligo(dT) solution: Dilute stock solution [e.g., random primers: 3 μg/μL; oligo(dT): 0.5 μg/μL] or mix and dilute—for random primers + oligo (dT) combinations—to a final concentration of 250 ng/μL. Store at −20 °C.
2. dNTPs solution: Mix and dilute dATP, dCTP, dGTP, and dTTP stock solutions (e.g., 100 mM; Invitrogen) to 10 mM each using UPNFW. Store at −20 °C.
3. Reverse transcriptase solution [e.g., Moloney Murine Leukemia Virus Reverse Transcriptase (M-MLV RT), Superscript™ II; Invitrogen]: 200 U/μL.
4. 5× RT buffer, included in most commercially available reverse transcriptase kits (e.g., M-MLV, Superscript™ II): 250 mM Tris-base (pH 8.3 at room temperature), 375 mM KCl, 15 mM $MgCl_2$.
5. Dithiothreitol (DTT) solution: 100 mM [Also included in most commercially available reverse transcriptase kits (e.g., M-MLV, Superscript™ II)].
6. DNase-treated, column-purified RNA (*see* **Note 4**).
7. PCR strip tubes.
8. Nuclease-free 1.5-mL tubes.
9. Single-channel (and multichannel, for higher throughput) pipettes to pipette different volume ranges (e.g., 1–10 μL, 2–20 μL, 10–100 μL, 100–1000 μL).
10. Nuclease-free sterile filter pipette tips for different volume ranges (e.g., 1–10 μL, 2–20 μL, 10–100 μL, 100–1000 μL).
11. A thermal cycler with a heated lid.

2.5 Reagents, Consumables, and Instruments for qPCR

1. 2× Power SYBR® Green Master Mix.
2. Working primer solution: Mix and dilute forward (F) primer and reverse (R) primer stock solutions (usually, 100 μM) to 1.25 μM each using UPNFW. An intermediate 10 μM F&R primer solution can be prepared to ensure an accurate dilution and reduce the frequency the primer stock solutions are freeze-thawed and used. It is recommended to prepare a new working primer solution for each RT-qPCR study using the 10 μM solution. Store stock and diluted primer solutions at −20 °C.
3. Nuclease-free 1.5-mL tubes.
4. 96-well or 384-well PCR plates and optical-grade sealing tapes.
5. Single-channel (and multichannel, for higher throughput) pipettes to pipette different volume ranges (e.g., 1–10 μL, 2–20 μL, 10–100 μL, 100–1000 μL).
6. Nuclease-free sterile filter pipette tips to pipette different volume ranges (e.g., 1–10 μL, 2–20 μL, 10–100 μL, 100–1000 μL).
7. A qPCR system.

3 Methods

3.1 Total RNA Sample Purity-Testing (see Note 1)

1. Measure the solution's absorbance at 260 nm to calculate its RNA concentration using a UV spectrophotometer [e.g., NanoDrop™ spectrophotometers (Thermo Scientific)].
2. Measure absorbance at 230 (e.g., phenol contamination) and 280 (e.g., protein contamination) nm to assess purity.
3. Total RNA samples with A260/A280 < 2.0 and/or A260/230 < 1.8 should undergo the purification procedure again.

3.2 Total RNA Sample Integrity-Testing via Agarose Gel Electrophoresis (see Notes 2 and 3)

1. Add the required quantity of agarose to prepare a 1% agarose/buffer gel into an Erlenmeyer flask (*see* **Note 5**).
2. Add the appropriate volume of 1× TAE or 1× TBE buffer into the Erlenmeyer flask (e.g., 100 mL of buffer for 1 g of agarose).
3. Heat the mixture contained in the Erlenmeyer flask in a microwave to boil (*see* **Note 6**).
4. Retrieve the Erlenmeyer flask from the microwave—using heat-resistant gloves—and swirl gently.
5. Repeat **steps 3** and **4** if the agarose is not fully dissolved.
6. Add the appropriate volume of RNA-binding fluorescent dye to the 1% agarose/buffer solution according to desired final concentration (e.g., 0.5 μg/mL for ethidium bromide; check vendor's recommendation; *see* **Note** 7) and swirl gently.

7. Let the 1% agarose/buffer/dye mixture cool down to 50–65 °C (approximately 10 min).
8. Prepare the gel caster and place the comb(s) to create wells where total RNA samples will be loaded.
9. Gently pour the 1% agarose/buffer/dye solution into the gel caster.
10. Immediately check for air bubbles in the surface of the poured 1% agarose/buffer/dye solution. Remove them by pushing against the gel caster's walls (e.g., using a pipette tip).
11. Once the gel has solidified, remove the comb(s) and transfer to an electrophoresis chamber—filled with enough 1× TAE or 1× TBE buffer to cover the gel.
12. For each biological replicate, add into a strip tube, tracking dye buffer for a 1× final concentration [e.g., BlueJuice gel loading buffer 10× (Invitrogen)], 0.5–1 μg of total RNA, and enough UPNFW to make up a final volume of 10 μL (e.g., 1 μL of 10× tracking dye buffer + 2 μL of 500 ng/μL total RNA sample + 7 μL of UPNFW).
13. For each row of lanes on the gel, add into a strip tube, tracking dye buffer (1× final concentration), 500 ng of DNA size marker [e.g., 1 Kb Plus DNA Ladder (Invitrogen)], and enough UPNFW to make up a final volume of 10 μL.
14. Carefully load 10 μL of DNA size marker/tracking dye buffer solution into the first lane of each row on the gel.
15. Carefully load each total RNA/tracking dye buffer into its corresponding lane on the gel.
16. Run the gel at 90–150 V (*see* **Note 8**) until the tracking dye has progressed ¾ of the path between rows, or the row and the gel's bottom edge.
17. Visualize the separated ribosomal RNA (rRNA) bands and mRNA smear on the gel under UV transillumination. Gel images of intact total RNA samples will show tight 18S, and 28S, rRNA bands around 1.8–1.9 and 4.1–5.0 Kb—they vary from one species to another, respectively (Fig. 1). Noticeable bands above the 28S rRNA band and close to the well could constitute evidence of genomic DNA (gDNA) contamination.

3.3 qPCR Primer Design

1. Obtain the mRNA sequences available for the study's GOIs, including paralogues/isoforms. Depending on the species/study, these could be de novo sequences [e.g., from RNA sequencing (RNA-seq) studies; suppression subtractive hybridization (SSH) cDNA libraries]; or sequences obtained directly from NCBI's (https://www.ncbi.nlm.nih.gov/) sequence databases (e.g., nucleotide) or from performing BLASTn searches of NCBI's sequence databases (*see* **Note 9**).

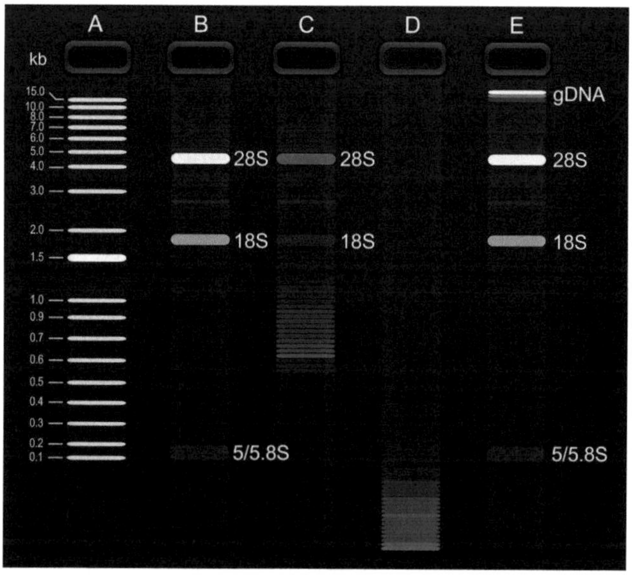

Fig. 1 Stylized examples of 1% agarose gel electrophoresis representing total RNA samples of different integrity levels, alongside a DNA size marker (**a**). Panel (**b**) represents a high integrity total RNA sample; the 28S and 18S rRNA bands are visible and tight, with 28S being approximately twice as intense as 18S. Depending on the purification protocol, gel images may show the 5/5.8S rRNA (between 120 and 150 nt). Panel (**c**) depicts a partially degraded total RNA sample; 28S and 18S rRNA bands are fading, and there is a smear of degraded RNA below the 18S band. Panel (**d**) represents a completely degraded total RNA sample; 28S and 18S rRNA bands are no longer visible—there is only a smear of degraded RNA at the bottom of the lane. Panel (**e**) shows a total RNA sample contaminated with gDNA

2. For each sequence, perform a BLASTx search of the NCBI's nonredundant (nr) protein sequence database to identify the coding sequence (CDS). This is particularly important for sequence data obtained from large-scale sequencing projects—which can sometimes be misassembled (e.g., chimeras)—or for immature transcripts (i.e., containing introns).
3. Align all of the available CDS for the GOI and potential paralogues (if applicable), using a multiple sequence alignment tool [e.g., Clustal Omega (https://www.ebi.ac.uk/Tools/msa/clustalo/)]. This alignment may identify the presence of single nucleotide polymorphisms (SNPs), or gene paralogues/isoforms with varying percentage identity. BLASTn searches against the NCBI genome database can also help identify paralogues (i.e., BLASTn hits in different regions of the genome).
4. Design the forward and reverse qPCR primers for each GOI using the compiled sequences and their alignment (*see* **Note 10**), and following these general guidelines: (A) Design

primers so that they are located in the CDS to ensure that the assay measures the GOI's transcript expression levels. The exception will be if splice variants are of interest. (B) Primers should not anneal to a sequence that contains an SNP, or generate an amplicon containing an SNP. (C) The DNA polymerase in the Power SYBR® Green Master Mix allows for a two-step thermal cycling protocol in that primer annealing and extension are combined in the same step (i.e., 60 °C for 1 min; *see* Subheading 3.8). Therefore, set the melting temperature (Tm) of the primers to 60 °C. (D) The desired GC content is 40–60%, and if possible, the last base pair (bp) at the 3′ end of the primer should be a G or C to promote binding. (E) Using a Tm of 60 °C and a 40–60% GC content, primer length will be ~20–22 bp. (F) Space the primers to generate an amplicon size between 90 and 200 bp. (G) In the case of gene paralogues, to ensure specificity, design primers to anneal in an area where sequence differences between the paralogues are ≥3 bp.

3.4 cDNA Synthesis (see Note 11)

Steps 4–8 and **10–14** of the cDNA synthesis protocol are represented in Fig. 2d.

1. Thaw a sufficient volume of the primer [250 ng/μL; random primers and/or oligo (dT)], dNTPs (10 mM), DTT (100 mM) and 5× RT buffer solutions to complete the cDNA synthesis run at room temperature. Thaw the total RNA samples on ice. The reverse transcriptase solution (200 U/μL) should remain at −20 °C until used.
2. RT priming mixture: Into a nuclease-free 1.5-mL tube, add 1 μL of random primers or oligo(dT) or random primers + oligo(dT) mixture solution (250 ng) and 1 μL of dNTPs solution (10 mM) for each RT reaction (Fig. 2b; *see* **Note 12**). Mix gently using a pipette, centrifuge briefly (i.e., >1000 × g for 5 s), and place the tube on ice.
3. The first reaction mixture will contain 1 μg of total RNA, 2 μL of RT priming mixture, and enough UPNFW to make a final total volume of 13 μL. Calculate the required volume of total RNA solution (i.e., V_{RNA}; equivalent to 1 μg) and UPNFW (i.e., $V_{UPNFW} = 13\ \mu L - 2\ \mu L - V_{RNA}$) for each biological replicate.
4. For each biological replicate, add its corresponding V_{UPNFW} into a PCR strip tube. Place the PCR strip tubes on ice.
5. Add 2 μL of RT priming mixture to each PCR strip tube.
6. For each biological replicate, add V_{RNA} into its corresponding PCR strip tube (final volume: 13 μL). Place the PCR strip tubes on ice. The biological replicates should be randomly distributed among the PCR strip tubes, as shown in Fig. 2a.

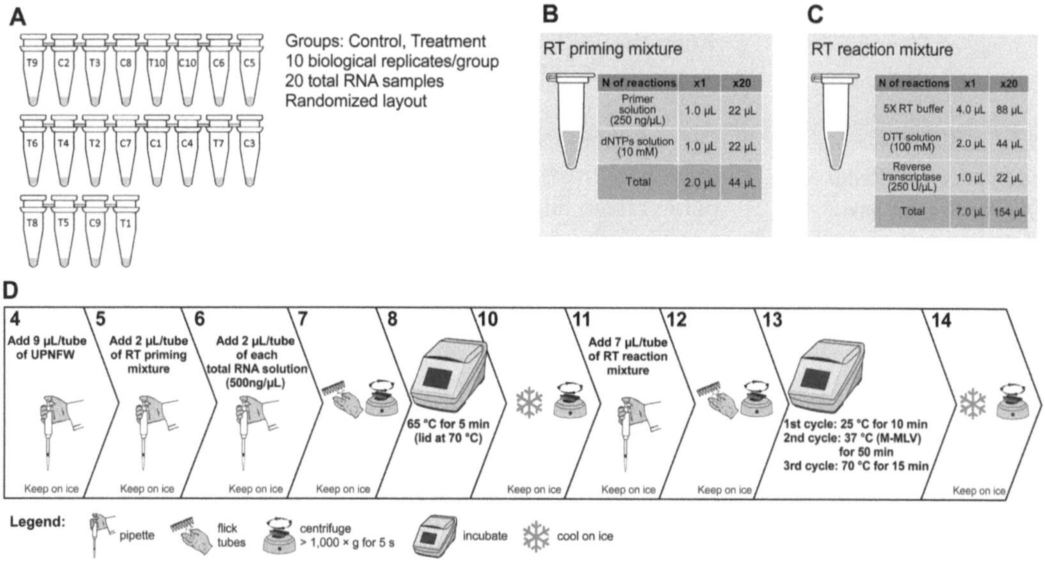

Fig. 2 cDNA synthesis. (**a**) Randomized biological replicate layout for the mock RT-qPCR study. (**b**) Reagent volumes to prepare enough RT priming mixture for 1 and 20 RT reactions. (**c**) Reagent volumes to prepare enough RT reaction mixture for 1 and 20 RT reactions. A 10% extra volume of each reagent has been added for the 20-RT reaction calculation (*see* **Note 12**). (**d**) Representation of **steps 4–8** and **10–14** of the cDNA synthesis protocol. Numbers in the top left corner of each workflow segment refer to the steps of Subheading 3.4

7. Mix by gently flicking the tubes, and centrifuge briefly. Proceed immediately to the next step.
8. Incubate the reaction mixtures at 65 °C for 5 min in a thermal cycler with the lid heated at 70 °C.
9. RT reaction mixture: While the reaction mixtures are being incubated at 65 °C, add 4 µL of 5× RT buffer, 2 µL of DTT solution (100 mM), and 1 µL of reverse transcriptase solution (200 U/µL) into a 1.5-mL tube for each RT reaction (Fig. 2c; *see* **Note 12**). Mix gently using a pipette, centrifuge briefly, and place the tube on ice.
10. Cool the reaction mixtures on ice for 2 min. Centrifuge briefly.
11. Add 7 µL of RT reaction mixture to each PCR strip tube (final volume: 20 µL).
12. Gently flick the tubes, and centrifuge briefly. Proceed immediately to the next step.
13. Incubate the reaction mixtures at 25 °C for 10 min. Then, incubate for 50 min at the optimal temperature of the selected reverse transcriptase (e.g., 37 °C if using M-MLV, 42 °C if using Superscript™ II). Finally, incubate at 70 °C for 15 min to stop the reaction.

14. Cool the cDNA solution on ice for 2 min and centrifuge briefly. Keep the PCR strip tubes on ice until proceeding to the next steps (i.e., Subheadings 3.5 or 3.6).

3.5 cDNA Pool Templates for Primer Quality Testing (see Note 13)

1. Prepare a cDNA pool for the Control and the Treatment groups by mixing 5 μL of RT product solution (cDNA templates) from all of the biological replicates for each group—or most, if the amount of RNA available is limited for some biological replicates. Each cDNA pool is then diluted 1:20 (i.e., to 2.5 ng/μL) using UPNFW (Fig. 3).
2. Mix well by pipetting and centrifuge briefly.
3. Divide the diluted cDNA pools into 300–400-μL aliquots (in 1.5-mL tubes). Store the aliquots at −20 °C until use.

3.6 Individual cDNA Templates for Normalizer Testing and RT-qPCR Experiments (see Note 13)

1. Dilute 15 μL of the RT product solution for each biological replicate with 585 μL of UPNFW (i.e., 1:40 dilution; final volume: 600 μL; final concentration: 1.25 ng/μL; Fig. 3).
2. Mix well by pipetting, and centrifuge briefly.
3. Divide the diluted cDNAs into three 200-μL aliquots (in strip tubes). Store the aliquots at −20 °C until use.

3.7 qPCR Mixture

Regardless of the purpose of the qPCR run, into a 1.5-mL tube, add 0.52 μL of working F&R primer solution (1.25 μM), 1.98 μL of UPNFW, and 6.5 μL of 2× Power SYBR® Green Master Mix for each reaction (final volume: 9 μL/reaction; Table 1; *see* **Note 12**). Mix thoroughly by finger flicking and centrifuge briefly. Place the tube containing the qPCR reaction mixture on ice.

3.8 qPCR Thermal Cycling Parameters

Using a calibrated qPCR system, for qPCR reaction mixtures prepared with Power SYBR® Green Master Mix, the real-time analysis program is: 1 cycle of 50 °C for 2 min, 1 cycle of 95 °C for 10 min, and 40 cycles of 95 °C for 15 s and 60 °C for 1 min. For other SYBR® reagents, check the thermal cycling conditions recommended by the manufacturer. The reaction mixture fluorescence is measured at the end of each 60 °C step. Ensure that the qPCR program includes a dissociation (also referred to as melt) curve analysis (*see* **Note 14**).

3.9 Primer Pair Quality Testing

1. Thaw on ice a sufficient volume of each cDNA pool to complete the primer pair quality-check qPCR run.
2. cDNA pool dilution series (for primer quality-check): Prepare a 5-point serial 1:3 dilution series using UPNFW, starting with the 2.5 ng/μL cDNA pool as the first point (Fig. 4). At each dilution step, mix 100 μL of the previous dilution with 200 μL of UPNFW: mix thoroughly by pipetting and centrifuge

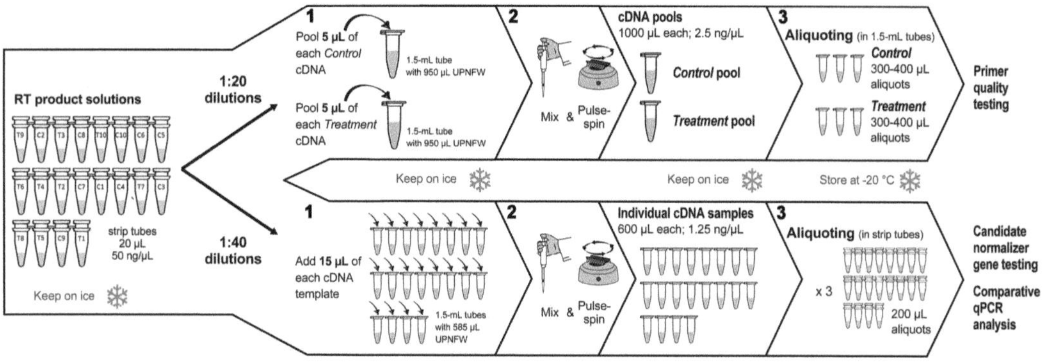

Fig. 3 cDNA templates (based on the proposed mock RT-qPCR study): Pooling, dilution, and aliquoting for primer quality testing (top workflow); dilution and aliquoting for candidate normalizer gene testing and comparative qPCR analysis (bottom workflow). Numbers in the top left corner of each workflow segment refer to the steps of Subheadings 3.5 and 3.6, respectively

Table 1
Reagents volumes needed to prepare enough qPCR mixture for 1 and 20 reactions

***N* of reactions**	**×1**	**×20[a]**
Working F&R primer solution (1.25 μM each)	0.52 μL	11.44 μL
UPNFW	1.98 μL	43.56 μL
2× Power SYBR® Green Master Mix	6.50 μL	143 μL
Total	9.0 μL	198 μL

[a]A 10% extra volume of each reagent has been added (*see* **Note 12**)

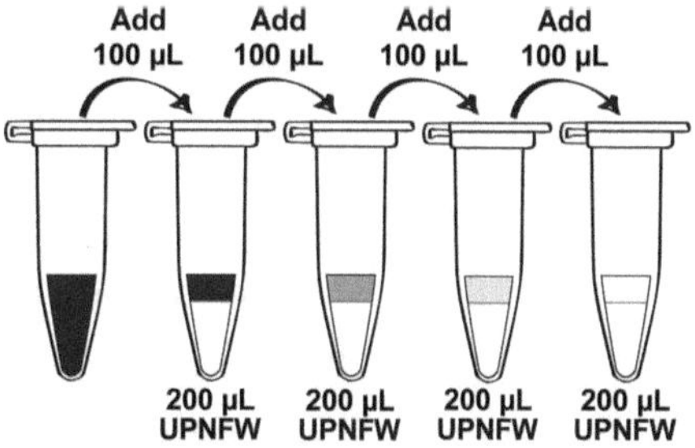

cDNA concentration (ng/μL)	2.50	0.83	0.28	0.09	0.03
cDNA input (ng)	10.00	3.33	1.11	0.37	0.12

Fig. 4 Preparation of the 5-point cDNA dilution series, cDNA concentration of each cDNA standard, and cDNA input in the PCR

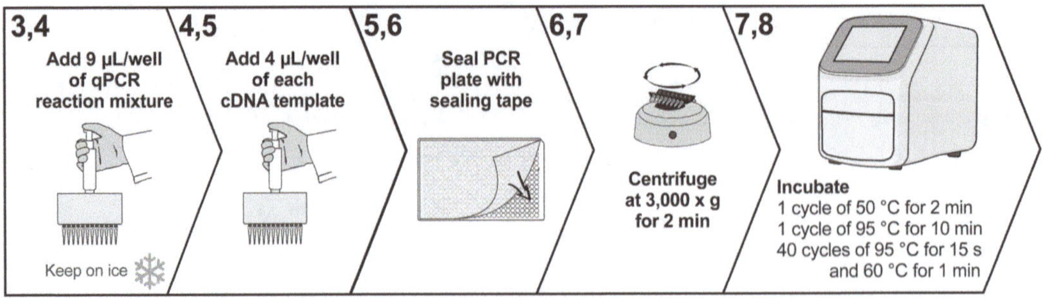

Fig. 5 Preparation of a qPCR run. The two numbers in the top left corner of each workflow segment refer to the steps of Subheadings 3.9 and 3.10 (the number on the right), and Subheading 3.11 (the number on the left)

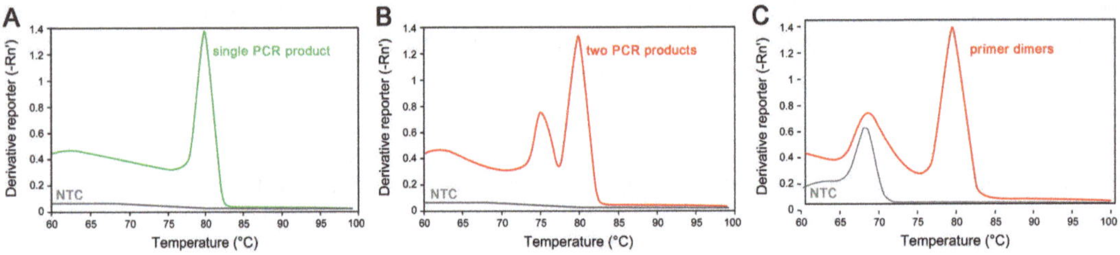

Fig. 6 Examples of different melt (dissociation) curve analysis outcomes. (**a**) Example of a single-product PCR with no signs of amplification in the NTC. (**b**) Example of a two-product PCR with no amplification in the NTC. (**c**) Example of primer dimer formation, as observed in the NTC's melt curve

briefly. Place the cDNA dilutions on ice. A no-template control (NTC; i.e., UPNFW only) is also included).

3. Prepare the qPCR mixture as indicated in Subheading 3.7.
4. Dispense 9 μL of qPCR reaction mixture into each well being used in the PCR plate, according to the previously designed qPCR run layout (*see* **Note 15**). Keep the PCR plate on ice (Fig. 5).
5. Dispense 4 μL of cDNA template for each technical replicate and standard solution (i.e., dilution point; Fig. 5) or NTC.
6. Seal the PCR plate using optical-grade sealing tape (Fig. 5).
7. Centrifuge the PCR plate at 3000 × g for 2 min (Fig. 5).
8. Insert the PCR plate in the qPCR system and proceed with the qPCR run (set qPCR thermal cycling parameters as described in Subheading 3.8; Fig. 5).
9. Check for a single amplification product: for a given primer pair, the melt curves corresponding to reactions with cDNA template should only show one sharp peak (Fig. 6a). Primer pairs with melt curves showing multiple peaks (Fig. 6b) should be discarded.

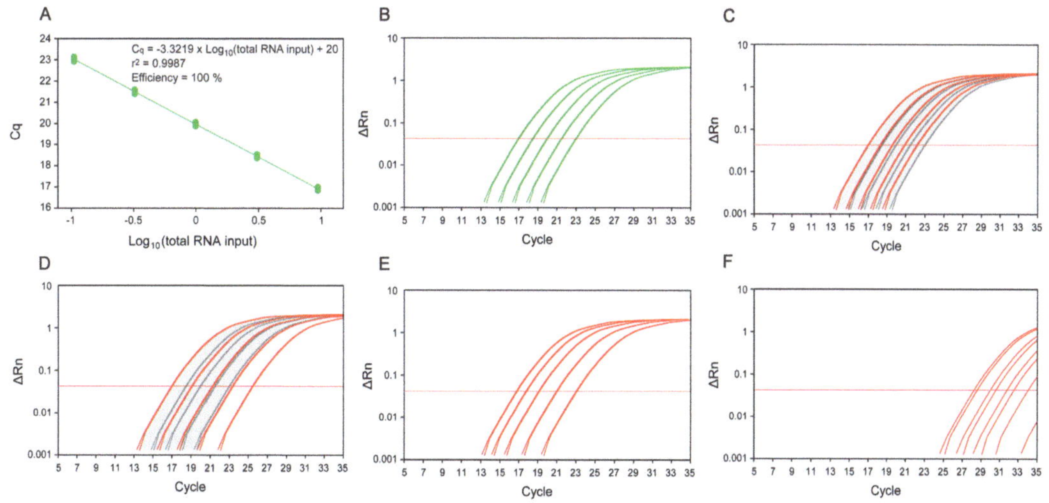

Fig. 7 Examples of primer pair amplification efficiency testing results. (**a**, **b**) Example of a primer pair with 100% amplification efficiency [as determined via linear regression analysis (**a**)], and evenly spaced amplification curves (**b**), obtained from a qPCR run with low variation among technical replicates [squared Pearson's correlation coefficient (r^2) > 0.995]. The horizontal red line represents the fluorescence threshold used for C_q determination. C_q corresponds to the cycle at which the fluorescence threshold level crosses the amplification curve. (**c**) Example of a primer pair with an amplification efficiency >100% (red amplification curves), in contrast with a 100% efficiency primer pair (gray amplification curves; gray area highlights the cycle range). The amplification curves of >100% efficiency primer pairs are closer to each other. (**d**) Example of a primer pair with an amplification efficiency <100% (red amplification curves), in contrast with a 100% efficiency primer pair (gray amplification curves; gray area highlights the cycle range). The amplification curves of <100% efficiency primer pairs are further from each other. (**e**) Unevenly spaced amplification curves may indicate a poorly prepared cDNA dilution series or the potential presence of PCR inhibitors (their effects are reduced as cDNA templates are diluted). (**f**) Low abundance transcripts may result in high C_q values and high technical replicate variability, limiting the user's capacity to evaluate the primer pair's amplification efficiency

10. Check for contamination or primer dimers: The melt curves of the NTCs should not show signs of amplification products (Fig. 6c).
11. For each dilution, check the C_q values of its corresponding technical replicates for pipetting imprecisions—any technical replicate deviating ≥ 0.5 C_q from the other two replicates should be removed.
12. Calculate the PCR amplification efficiency using the following formula [8]:

$$\mathrm{Eff} = 10^{\left(-\frac{1}{\mathrm{slope}}\right)},$$

which will provide the fold-change increase in amplicon levels in the reaction mixture—ideally, this number would be 2 (i.e., the amount of amplicons is doubled in each PCR cycle). For efficiency values expressed as % (e.g., 100% efficiency for an efficiency value of 2), use the following formula:

$$\text{Eff}\ (\%) = \left(10^{\left(-\frac{1}{\text{slope}}\right)} - 1\right) \times 100.$$

The slope is obtained from the linear regression between C_q and the $\log_{10}$-transformed quantity of cDNA input for each dilution (Fig. 7a).

13. Amplification plots should be evenly spaced between dilutions (Fig. 7b; *see* **Note 16**).

14. qPCR primer pairs showing amplification efficiencies above 110% (Fig. 7c) and below 85% (Fig. 7d) should be discarded; new primer pairs should be designed (*see* **Note 17**).

3.10 Candidate Normalizer Gene Testing

1. Select ≥4 candidate normalizer genes based on the specialized literature available or by mining large transcriptome profiling datasets (e.g., microarrays, RNA-seq) for genes with highly stable transcript levels across all biological replicates and experimental groups/treatments (*see* **Note 18**).
2. Thaw on ice a sufficient volume of each individual cDNA template to complete the normalizer testing qPCR run.
3. Prepare the qPCR mixture as indicated in Subheading 3.7 for each candidate normalizer gene included in the normalizer testing qPCR run.
4. Dispense 9 μL of qPCR reaction mixture in each well being used in the PCR plate, according to the previously designed qPCR run layout (*see* **Note 15**). Keep the PCR plate on ice (Fig. 5).
5. Dispense 4 μL of cDNA template (5 ng) for each technical and biological replicate, or UPNFW (i.e., NTC) according to the previously designed qPCR run layout (*see* **Note 15**; Fig. 5).
6. Seal the PCR plate using optical-grade sealing tape (Fig. 5).
7. Centrifuge the PCR plate at 3000 × g for 2 min (Fig. 5).
8. Insert the PCR plate in the qPCR system and proceed with the qPCR run (set qPCR thermal cycling parameters as described in Subheading 3.8; Fig. 5).
9. Check for contamination: The melt curves and qPCR amplification plots of the NTC should not show signs of amplification products.
10. For each biological replicate, check the C_q values of its corresponding technical replicates for pipetting imprecisions—any technical replicate deviating ≥0.5 C_q from the other two replicates should be removed.
11. Once the qPCR runs of all candidate normalizer genes have been completed, identify the 2 (sometimes more than 2 normalizer genes are recommended) most stably expressed across treatments using a widely accepted statistical method to rank

the expression stability of various genes (e.g., NormFinder [9], geNorm [10]; *see* **Note 19**).

3.11 Comparative qPCR Analysis

1. Thaw on ice a sufficient volume of each individual cDNA template to complete the comparative analysis qPCR run.
2. Proceed as indicated in "3.7 qPCR reaction mixture" for each gene included in the qPCR run.
3. Dispense 9 μL of qPCR reaction mixture in each well being used in the PCR plate, according to the previously designed qPCR run layout (*see* **Note 15**). Keep the PCR plate on ice (Fig. 5).
4. Dispense 4 μL of cDNA template (5 ng) for each technical and biological replicate, or UPNFW (i.e., NTC), according to the previously designed qPCR run layout (*see* **Note 15**; Fig. 5).
5. Seal the PCR plate using optical-grade sealing tape (Fig. 5).
6. Centrifuge the PCR plate at 3000 × g for 2 min (Fig. 5).
7. Insert the PCR plate in the qPCR system and proceed with the qPCR run (set qPCR thermal cycling parameters as described in Subheading 3.8; Fig. 5).
8. Check for contamination: The melt curves and qPCR amplification plots of the NTC should not show signs of amplification products.
9. For each biological replicate, check the C_q values of its corresponding technical replicates for pipetting imprecisions—any technical replicate deviating ≥ 0.5 C_q from the other two replicates should be removed.
10. Biological replicate mean C_q values will be used in the calculation of the normalized expression values (*see* Subheading 3.12).

3.12 RT-qPCR Experiment

A number of methods for analyzing relative gene expression data have been developed in the last couple of decades [11]. One common method is the "delta-delta-Ct method" by Livak and Schmittgen [12], which assumes 100% PCR efficiency and uses a single normalizer gene. Pfaffl [8] proposed an improved model that adjusts for differences in PCR efficiencies between the GOI and normalizer gene, but cannot deal with multiple normalizer genes. Here we will use a qBase relative quantification framework [11] to calculate the normalized expression values for each GOI. This method takes PCR efficiency and multiple stably expressed normalizer genes into account.

1. Choose a sample as the calibrator (i.e., assigned a relative expression value of 1.0). If you want to use the sample with the lowest expression as the calibrator, select a random sample here first, and then replace it later with your desired calibrator (see **step 9**).

2. Calculate the Cq deviation (i.e., ΔC_q value) of each sample versus the calibrator for the normalizers (independently):

$$\Delta C_{q\ \text{normalizer 1}} = C_{q\ \text{calibrator}} - C_{q\ \text{sample}}.$$

3. Calculate relative quantity (RQs) of the normalizers for each sample:

$$\text{RQ}_{\text{normalizer 1}} = \text{Eff}_{\text{normalizer 1}}{}^{\Delta C_{q\ \text{normalizer 1}}},$$

where $\text{Eff}_{\text{normalizer 1}}$ represent the normalizer's primer pair efficiency, expressed as a fold-change value (e.g., 2.0 for an amplification efficiency of 100%).

4. Calculate the Normalization Factor (NF; geometric mean of the normalizer RQs) for each biological replicate. NF will be used as the denominator in the formula to calculate the Normalized RQ (NRQ):

$$\text{NF} = \sqrt[n]{\text{RQ}_{\text{normalizer 1}} \times \ldots \times \text{RQ}_{\text{normalizer } n}}.$$

5. Calculate the C_q deviation (i.e., ΔC_q value) of each sample versus the calibrator for the GOI:

$$\Delta C_{q\ \text{GOI}} = C_{q\ \text{calibrator}} - C_{q\ \text{sample}}.$$

6. Calculate the RQ of the GOI for each biological replicate:

$$\text{RQ}_{\text{GOI}} = \text{Eff}_{\text{GOI}}{}^{\Delta C_{q\ \text{GOI}}}.$$

7. Calculate the NRQ of the GOI for each biological replicate:

$$\text{NRQ}_{\text{GOI}} = \frac{\text{RQ}_{\text{GOI}}}{\text{NF}}.$$

8. If you want the calibrator sample to be the sample with the lowest NRQ, locate that sample, use it as the calibrator, and repeat the analysis from **step 2**. Check to make sure there are no NRQ values <1.0.

4 Notes

1. The RT-qPCR users should be aware of the potential presence of endogenous (e.g., hemoglobin, fat, glycogen) and exogenous (e.g., heparin, phenol) RT and PCR inhibitors in their total RNA samples, as well as genomic DNA (gDNA) contamination [13, 14]. Total RNA samples should have been subjected to DNase treatment and purification (e.g.,

silica-membrane-based purification methods) to prevent gDNA, and RT and PCR inhibitors contamination.

2. RNA degradation by RNases is a major concern in gene expression studies, as it significantly reduces RT-qPCR performance (i.e., higher C_q [13]) and, consequently, the reliability of the gene expression data obtained. DNase contamination will compromise cDNA integrity and, thus, the reliability of the RT-qPCR results. Therefore, besides checking RNA integrity, the laboratory must have protocols for nuclease (i.e., RNases and DNases) decontamination. For example, the workspaces, pipettes, and instruments—if possible—should be cleaned with a nuclease decontamination solution [e.g., RNase AWAY™ (Invitrogen)]. Also, laboratories employing molecular biology techniques would preferably have an independent room/PCR workstation, and a set of pipettes and instruments (e.g., centrifuges) designated to prepare reaction mixtures and primer dilutions exclusively (i.e., RNA/DNA template-free spaces and equipment).
3. RNA integrity can also be assessed using a capillary-electrophoresis system [15] such as the Bioanalyzer (Agilent Technologies). This technology provides the user with an RNA integrity number (RIN; ranging from 1 to 10), calculated based on the entire electropherogram [16]. RIN > 5.0 has been proposed as the RNA integrity minimum for RT-qPCR [13]; however, it already implies a significant RNA degradation. RT-qPCR studies would preferably use total RNA samples with RIN > 8.0 (i.e., intact RNA samples) [13].
4. Adjusting all total RNA templates to the same concentration—100–500 ng/μL, preferably—will help to minimize the introduction of intersample variability during cDNA synthesis.
5. The Erlenmeyer flask's capacity should be at least 2.5 times larger than that of the 1% agarose/buffer solution (e.g., use a 250-mL flask if preparing a 100-mL gel) to prevent overflow.
6. Do not over boil the 1% agarose/buffer solution as water would evaporate, thus altering the agarose and buffer salts concentrations.
7. If ethidium bromide is used as the RNA-binding fluorescent dye, then it must be added to the 1% agarose/buffer solution in a chemical fume hood. All subsequent steps must be performed on dedicated work surfaces, adequately labeled as ethidium-bromide-contaminated.
8. Lower voltages will require a more extended run period but will favor higher RNA band resolution. Thicker gels (not recommended) will also require longer runs as the RNA will migrate more slowly.

9. When designing primer pairs for a GOI from a given species, the potential presence of gene paralogues/isoforms must be considered. Whole-genome duplication (WGD) and/or lineage-specific gene duplication events have resulted in some genes with paralogue numbers ranging from two to multigene or even superfamilies. The outcomes of gene duplication may include gene conservation, subfunctionalization, neofunctionalization, and pseudogenes [17, 18]. Therefore, if gene paralogues/isoforms are present, it is paramount that qPCR primers be designed in highly heterogeneous areas of the sequence to ensure specificity.

   For example, 13 aquaporin (AQP) subfamilies have been identified in mammals [19], while the zebrafish (*Danio rerio*) genome has revealed 18 putative members, present as single-copy genes or gene duplicates [20]. Two zebrafish *apq10* gene paralogues (*apq10a* and *aqp10b*) have been identified, which arose through a teleost-specific WGD event (~320–350 mya) [20, 21]. Furthermore, two *aqp10b* gene paralogues have been identified in rainbow smelt (*Osmerus mordax*), predicted to have arisen from a gene duplication event in the smelt lineage that occurred more recently than the teleost-specific WGD event [22].

   The common ancestor of salmonids experienced a more recent (~80 mya) WGD event, and it has been estimated that 30–70% of the arisen gene duplicates have been retained [23]. Consequently, gene paralogues are commonplace in salmonids; hence, salmonid nucleotide databases must be thoroughly mined to identify and obtain sequences for them. These paralogues are typically ~90% identical at the nucleotide level, making primers' placement in paralogue-specific areas difficult [24].

10. There are multiple free online primer design tools that are ideal for qPCR primer design [e.g., Primer3 (https://bioinfo.ut.ee/primer3/); Primer Express (Applied Biosystems/Thermo Fisher Scientific); NCBI Primer-BLAST (https://www.ncbi.nlm.nih.gov/tools/primer-blast/)]. However, for closely related gene paralogues, primers may have to be designed manually to ensure specificity. Manually designed primers should be assessed using an oligo analyzer tool (e.g., from Integrated DNA Technologies; https://www.idtdna.com/calc/analyzer) to check for the presence of hairpin loops, self- and hetero-dimers, and to verify the Tm. Furthermore, melt curve analysis can confirm primer specificity (i.e., single peak in the melt curve) and the absence of primer dimers. As it is common that primer pairs fail quality testing, it is recommended that at least 2 sets of primer pairs be designed and tested for each GOI.

11. In addition to preparing cDNA templates for qPCR-analyzing differences in the transcript levels of a given GOI between groups/treatments (Subheading 3.11), they are also prepared for two preparatory steps. The first one is testing the specificity and efficiency of the qPCR primer pairs for each GOI (Subheading 3.9). The second one is qPCR-analyzing the transcript expression levels of a set of candidate normalizer genes to identify the 2—sometimes more than 2—that are most stably expressed across biological replicates and groups/treatments (Subheading 3.10) [4].
12. It is recommended to add 10% extra volume of each reagent to ensure enough reagent mixture is available to complete all planned reactions.
13. The cDNA template dilutions (i.e., Subheadings 3.5 and 3.6) and qPCR (i.e., Subheadings 3.9, 3.10, and 3.11) procedures below will assume that the cDNA concentration corresponds to the input total RNA amount used in the cDNA synthesis step [i.e., if the total RNA input was 1 μg, then the final cDNA concentration in a 20 μL reaction volume is 50 ng/μL (**steps 3** and **11** of Subheading 3.4)].
14. At the end of the run, do not dispose of the PCR plate as the PCR products could be needed to verify the amplicon size—using agarose gel electrophoresis—and/or nucleotide sequence—via DNA sequencing analysis.
15. For any qPCR run (e.g., primer pair quality testing, comparative qPCR analysis), design a plate layout that (A) includes a no-template control (NTC) for each qPCR assay; (B) maximizes the number of genes that can be accommodated in each qPCR run (Fig. 8), considering the qPCR system available in the laboratory (e.g., adapted to 96-well or 384-well qPCR plate formats); and (C) adjusts to the RT-qPCR experimental design.

    RT-qPCR experiments can be designed to analyze the selected normalizer genes and GOIs independently (i.e., in different qPCR runs) or jointly (i.e., every qPCR run will also include the normalizer genes). However, the relative expression levels for each GOI (Subheading 3.12) must be calculated using GOI and normalizer C_q values generated from the same cDNA batch. If the number of reactions exceeds the qPCR plate capacity, then the biological replicates should be distributed across multiple plates. Each of the plates constituting the multiplate qPCR run should include triplicate technical replicates for a "linker" sample (i.e., the same biological replicate is included in all plates). The "linker" sample average C_q should not vary among plates >0.3 cycles; otherwise, the entire multiplate qPCR run must be repeated. Multiplate qPCR runs

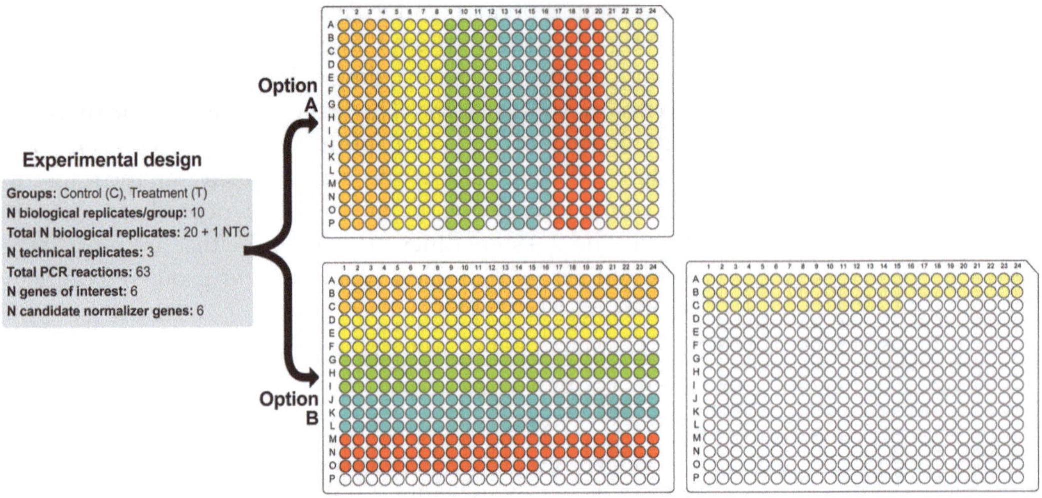

Fig. 8 Example of the importance of selecting a qPCR layout that maximizes the number of genes tested in each qPCR run. A mock RT-qPCR study comprised of 2 groups with 10 biological replicates each, intended to analyze the mRNA levels of 6 GOIs and 6 candidate normalizer genes using triplicate replicates, would translate into 63 PCR reactions per gene [(2 groups × 10 biological replicates/group × 3 technical replicates/biological replicate) + (1 NTC × 3 technical replicates)]. If arranged vertically (option A), a 384-well qPCR plate could accommodate 6 genes; that is, only 2 plates would be required to complete both the candidate normalizer gene testing and the comparative qPCR analysis. If arranged horizontally (option B), the study would require twice as many qPCR plates—and qPCR runs—to be completed

should be completed on the same day and using reagents from the same batch to prevent batch effects.

For normalizer gene testing and comparative qPCR analyses, design a plate layout that (A) distributes the biological replicates on the qPCR plate in a random fashion (i.e., not arranged based on the group/treatment to which they belong); (B) includes triplicate technical replicates (i.e., 3 reactions) for each biological replicate and the NTC.

For primer pair quality testing, design a plate layout that includes duplicate or triplicate technical replicates for each cDNA standard and the NTC.

16. Unevenly spaced amplification curves (Fig. 7e) may be indicative of a poorly prepared cDNA dilution series or the potential presence of PCR inhibitors (their effect is reduced as cDNA templates are diluted). Total RNA samples may be rechecked for contamination (Subheading 3.1), and the cDNA dilution series and primer testing qPCR run repeated.

17. Degraded cDNA templates may lead to low amplification efficiency. If cDNA integrity issues are suspected, a new batch of cDNA templates should be prepared. Primer amplification efficiency can be difficult to assess for GOIs with low transcript expression levels (i.e., $C_q > 30$ cycle; Fig. 7f). In this case, the

RT-qPCR user may consider increasing the cDNA template input or preparing a 1:2 dilution series—instead of 1:3.

18. The literature search for candidate normalizer genes can look for studies similar to the one undertaken—in terms of experimental conditions/treatments—using the same model species or others. In fact, several genes related to basic cellular functions have extensively been used as normalizer genes for RT-qPCR studies in multiple taxa; for example, elongation factor 1 alpha (*EF1A1*), beta-actin (*ACTB*), 18S ribosomal RNA (*18S*), TATA-box-binding protein (*TBP*), and glyceraldehyde-3-phosphate dehydrogenase (*GAPDH*).

19. Gene expression stability statistical approaches (e.g., geNorm) tend to disregard C_q differences among groups/treatments. Normalizer gene selection should also consider the comparison of groups/treatments mean C_q to detect significant differences (i.e., $\Delta C_q > 0.5$). When two normalizer genes are selected, it is also important to ensure they are not related to the same pathway or molecular function through their encoded proteins.

References

1. Higuchi R, Fockler C, Dollinger G, Watson R (1993) Kinetic PCR analysis: real-time monitoring of DNA amplification reactions. Bio/-Technology 11(9):1026–1030
2. Higuchi R, Dollinger G, Walsh PS, Griffith R (1992) Simultaneous amplification and detection of specific DNA sequences. Bio/Technology 10(4):413–417
3. Kralik P, Ricchi M (2017) A basic guide to real time PCR in microbial diagnostics: definitions, parameters, and everything. Front Microbiol 8: 108
4. Bustin SA, Benes V, Garson JA, Hellemans J, Huggett J, Kubista M, Mueller R, Nolan T, Pfaffl MW, Shipley GL (2009) The MIQE guidelines: minimum information for publication of quantitative real-time PCR experiments. Clin Chem 55(4):611–622
5. Bustin SA, Beaulieu J-F, Huggett J, Jaggi R, Kibenge FSB, Olsvik PA, Penning LC, Toegel S (2010) MIQE précis: practical implementation of minimum standard guidelines for fluorescence-based quantitative real-time PCR experiments. BMC Mol Biol 11(1):74
6. Abdel Nour AM, Nemer G, Khalil A (2020) The MIQE Guidelines' tenth anniversary: the good and bad students. Gene Rep 19:100630
7. VanGuilder HD, Vrana KE, Freeman WM (2008) Twenty-five years of quantitative PCR for gene expression analysis. BioTechniques 44(5):619–626
8. Pfaffl MW (2001) A new mathematical model for relative quantification in real-time RT–PCR. Nucleic Acids Res 29(9):e45
9. Andersen CL, Jensen JL, Ørntoft TF (2004) Normalization of real-time quantitative reverse transcription-PCR data: a model-based variance estimation approach to identify genes suited for normalization, applied to bladder and colon cancer data sets. Cancer Res 64(15):5245
10. Vandesompele J, De Preter K, Pattyn F, Poppe B, Van Roy N, De Paepe A, Speleman F (2002) Accurate normalization of real-time quantitative RT-PCR data by geometric averaging of multiple internal control genes. Genome Biol 3(7):1–12
11. Hellemans J, Mortier G, De Paepe A, Speleman F, Vandesompele J (2007) qBase relative quantification framework and software for management and automated analysis of real-time quantitative PCR data. Genome Biol 8(2):R19
12. Livak KJ, Schmittgen TD (2001) Analysis of relative gene expression data using real-time quantitative PCR and the 2−ΔΔCT method. Methods 25(4):402–408

13. Fleige S, Pfaffl MW (2006) RNA integrity and the effect on the real-time qRT-PCR performance. Mol Asp Med 27(2):126–139
14. Vermeulen J, De Preter K, Lefever S, Nuytens J, De Vloed F, Derveaux S, Hellemans J, Speleman F, Vandesompele J (2011) Measurable impact of RNA quality on gene expression results from quantitative PCR. Nucleic Acids Res 39(9):e63–e63
15. Mueller O, Hahnenberger K, Dittmann M, Yee H, Dubrow R, Nagle R, Ilsley D (2000) A microfluidic system for high-speed reproducible DNA sizing and quantitation. Electrophoresis 21(1):128–134
16. Schroeder A, Mueller O, Stocker S, Salowsky R, Leiber M, Gassmann M, Lightfoot S, Menzel W, Granzow M, Ragg T (2006) The RIN: an RNA integrity number for assigning integrity values to RNA measurements. BMC Mol Biol 7(1):3
17. Force A, Lynch M, Pickett FB, Amores A, Yan Y-L, Postlethwait J (1999) Preservation of duplicate genes by complementary, degenerative mutations. Genetics 151(4):1531–1545
18. Conant GC, Wolfe KH (2008) Turning a hobby into a job: how duplicated genes find new functions. Nat Rev Genet 9(12):938–950
19. Agre P, King LS, Yasui M, Guggino WB, Ottersen OP, Fujiyoshi Y, Engel A, Nielsen S (2002) Aquaporin water channels – from atomic structure to clinical medicine. J Physiol 542(1):3–16
20. Tingaud-Sequeira A, Calusinska M, Finn RN, Chauvigné F, Lozano J, Cerdà J (2010) The zebrafish genome encodes the largest vertebrate repertoire of functional aquaporins with dual paralogy and substrate specificities similar to mammals. BMC Evol Biol 10(1):38
21. Christoffels A, Koh EGL, Chia J-M, Brenner S, Aparicio S, Venkatesh B (2004) Fugu genome analysis provides evidence for a whole-genome duplication early during the evolution of ray-finned fishes. Mol Biol Evol 21(6):1146–1151
22. Hall JR, Clow KA, Rise ML, Driedzic WR (2015) Cloning and characterization of aquaglyceroporin genes from rainbow smelt (*Osmerus mordax*) and transcript expression in response to cold temperature. Comp Biochem Phys B 187:39–54
23. Lien S, Koop BF, Sandve SR, Miller JR, Kent MP, Nome T, Hvidsten TR, Leong JS, Minkley DR, Zimin A et al (2016) The Atlantic salmon genome provides insights into rediploidization. Nature 533:200
24. Caballero-Solares A, Hall JR, Xue X, Eslamloo K, Taylor RG, Parrish CC, Rise ML (2017) The dietary replacement of marine ingredients by terrestrial animal and plant alternatives modulates the antiviral immune response of Atlantic salmon (*Salmo salar*). Fish Shellfish Immunol 64:24–38

Chapter 22

Peptide-Affinity Isolation of Extracellular Vesicles and Cell-Free DNA From Human Plasma

Jeremy W. Roy, Catherine A. Taylor, Rodney J. Ouellette, and Stephen M. Lewis

Abstract

Extracellular vesicles (EVs) are membrane-bound nanoparticles that carry DNA, RNA, and protein cargoes and are found in a variety of biofluids. EVs, along with cell-free DNA (cfDNA), have attracted interest as a source of biomarker material for liquid biopsy, a process in which a sample of body fluid is used for the detection or monitoring of disease. The Vn96 synthetic peptide facilitates the isolation of both EVs and cfDNA from multiple body fluids, including human plasma, placing it as a versatile tool for the capture of multiple biomarker materials for disease detection and/or treatment monitoring. In this chapter, we describe an optimized protocol for Vn96-mediated isolation of EVs and cfDNA from human plasma samples, as well as downstream methods for EV enumeration and DNA, RNA, and protein extraction from the material captured by Vn96 for use in biomarker discovery or detection.

Key words Vn96, Extracellular vesicle, cfDNA, Plasma, Liquid biopsy, Nanoparticle tracking analysis

1 Introduction

Extracellular vesicles (EVs) are small, membrane-bound nanoparticles that are shed by cells [1]. EVs contain a sample of the molecules found within the parent cell, including DNA, RNA, and proteins [2], and are therefore a rich source of biomarkers for disease detection and/or monitoring [3]. EVs and their respective cargoes have become of significant interest for liquid biopsy, a process by which a sample of a body fluid (e.g., blood, urine, saliva) is interrogated for the presence of molecular markers of disease. Liquid biopsy has advantages compared to tissue biopsy, including the minimally invasive nature of sample collection, as well as the ability to longitudinally sample for disease markers over the course of disease progression or treatment regimen [4]. Therefore, simple, rapid, and clinically-amenable methods for EV isolation from

Sherri L. Christian (ed.), *Cancer Cell Biology: Methods and Protocols*,
Methods in Molecular Biology, vol. 2508, https://doi.org/10.1007/978-1-0716-2376-3_22,

patient biofluids (such as blood and urine) are valuable tools for collecting EVs for liquid biopsy-based disease biomarker analyses.

The Vn96 synthetic peptide binds to and captures EVs from a variety of biofluids [5], and its use has been optimized for the isolation of EVs from plasma [6]. Vn96 was designed to interact with members of the heat shock protein family (HSPs), which are present on the surface of a variety of EVs [7]. It has been hypothesized that Vn96's ability to interact with HSPs facilitates the capture and isolation of EVs from biofluids [5]. Moreover, the Vn96 synthetic peptide forms a cationic alpha helix [5], which could also interact with and/or insert into EV membranes [8], thereby causing EV aggregation to facilitate their capture. In addition to its ability to isolate EVs, Vn96 has recently been found to coisolate cell-free DNA (cfDNA) through an electrostatic interaction between the cationic Vn96 peptide and the negatively-charged DNA backbone [6, 9]. These observations have positioned Vn96 as a valuable and versatile tool for liquid-biopsy-based biomarker analyses that survey a broad spectrum of biomarker types (i.e., DNA, RNA, and protein) for disease detection and/or monitoring.

In this chapter, we will provide a detailed description of the optimized method for Vn96 synthetic peptide–mediated isolation of EVs and cfDNA from human plasma samples. We will also describe standard methods for the enumeration of Vn96-isolated EVs, as well as the isolation of DNA, RNA, and protein from the Vn96-EV/cfDNA pellet that can be used in a variety of downstream assays for biomarker discovery or detection, which can potentially be applied for the detection and/or monitoring of diseases, such as cancer.

2 Materials

Human plasma for use in EV and cfDNA isolation experiments can be purchased from commercial sources, retrieved from a biobank, or collected from volunteers under an approved ethics protocol.

2.1 Vn96 Peptide

1. Lyophilized Vn96 synthetic peptide (Vivitide; available in bulk, catalogue # ME-010-5000; or as part of the ME™ Kit [plasma], catalogue # ME-020p-kit).
2. 0.1 μm-filtered nuclease-free water (*see* **Note 1**).
3. 1× phosphate-buffered saline (PBS): 137 mM NaCl, 2.7 mM KCl, 10 mM Na_2HPO_4, 1.8 mM KH_2PO_4.
4. 0.625× PBS: 18.75 mL of 1× PBS diluted with 11.25 mL of nuclease-free water.
5. ProClin 300™ (Sigma-Aldrich, catalogue # 48912-U; *see* **Note 2**).

2.2 Reagents for Elution of EVs from Vn96

1. 2 M NaCl, 0.1 μm filter-sterilized (*see* **Note 3**).
2. 0.1 M glycine-NaOH, pH 10, 0.1 μm filter-sterilized (*see* **Note 4**).

2.3 Kits and Reagents for Nucleic Acid Extraction

1. Plasma/Serum Cell-Free Circulating DNA Purification Mini Kit from Norgen Biotek Corp. (catalogue # 55100).
2. *mir*Vana™ miRNA Isolation Kit without Phenol from Thermo Fisher Scientific (catalogue # AM1561; *see* **Note 5**).
3. Acid Phenol:Chloroform:Isoamyl alcohol (125:24:1 solution), pH 4.5.

2.4 Lysis of EVs for Western-Blot Analyses

1. SDS-Tris Lysis buffer: 2% SDS, 125 mM Tris–HCl, pH 6.8 (*see* **Note 6**).
2. Protease inhibitors: Protease Inhibitor Cocktail set III (Cedarlane, catalogue # 539134-1 mL). Store in aliquots at −20 °C.
3. 4× Laemmli sample buffer: 250 mM Tris base, 8% SDS, 40% glycerol, 0.002% bromophenol blue.
4. TCEP: Bond-Breaker TCEP solution (Fisher Scientific, catalogue # PI77720).

3 Methods

3.1 Isolation of EVs and cfDNA from Plasma Using the Vn96 Synthetic Peptide

The following protocol has been recently optimized in our laboratory for the isolation of EVs and cfDNA from human plasma for use in downstream biomarker discovery and/or detection assays; therefore, it differs somewhat from the protocol included with the Vn96 kit that is available from vivitide (The ME™ Kit [plasma]). The protocol described below represents a general method for EV/cfDNA isolation from plasma using the Vn96 synthetic peptide, and subsequent sections include instructions for downstream extraction of either DNA, or RNA, or protein for biomarker analyses. Simultaneous extraction of DNA, RNA, and protein from the same Vn96-isolated EV/cfDNA material is also possible (i.e., multiparametric extraction), and we have recently published a method for this purpose [6].

For isolation of EVs and cfDNA from human plasma, we ensure that plasma is prepared from whole blood within 30 min of collection in order to avoid platelet activation and release of platelet-derived EVs into the sample. Following preparation, the plasma may be used immediately for EV and cfDNA isolation or stored at −80 °C for future use.

1. Solubilize the lyophilized Vn96 synthetic peptide in 0.625× PBS with 0.05% ProClin 300™ (*see* **Note 7**) to a final concentration of 2.5 mg/mL. A typical stock solution of Vn96 peptide

requires the solubilization of 1 mg of lyophilized Vn96 into 0.4 mL of 0.625× PBS with 0.05% ProClin 300™. This solution *is not* 0.1 μm-filtered (*see* **Note 8**). The Vn96 stock solution may be stored at 4 °C for up to one year with little loss of activity.

2. Use fresh plasma or thaw plasma that was previously frozen at −80 °C slowly on ice. Fresh or thawed plasma is precleared of any residual cells, large particles, and debris by performing a 3000 × g centrifugation for 15 min prior to use for EV isolation (*see* **Note 9**). Transfer the supernatant to a new tube. Additional preclearing of plasma can be performed by centrifugation at 3000 × g to 17,000 × g for 15 min, or by filtration through a 0.8 μm filter. The optimal preclearing protocol should be empirically determined for each application (*see* **Note 10**).
3. Optional: Add 5 μL of protease inhibitor to each 1 mL plasma sample (*see* **Note 11**).
4. Dilute 1 mL of precleared plasma with 1 mL of 1× PBS (0.1 μm filtered) into a 2 mL microfuge tube; mix well by gentle inversion.
5. Add 40 μL of the Vn96 stock solution (2.5 mg/mL; 100 μg of Vn96 per mL plasma) and rotate end-over-end for 1 h at room temperature (*see* **Note 12**).
6. Centrifuge at 17,000 × g for 15 min at 4 °C (sedimentation coefficient = 41) to pellet Vn96-EV/cfDNA complexes. The Vn96-EV/cfDNA pellet should appear to be glassy or translucent and located on the bottom wall of the tube.
7. Quickly remove the supernatant from the pellet (*see* **Note 13**) and resuspend the pellet in 2 mL of 1× PBS (0.1 μm filtered); flick the tube to dislodge the pellet from the side of the tube.
8. Centrifuge at 17,000 × g for 10 min at 4 °C.
9. Remove the supernatant and wash the pellet with 2 mL of 1× PBS (0.1 μm filtered) two additional times; centrifuge as described in **step 8**.
10. Remove and discard all PBS from the pellet after the final wash. Use a quick spin (<10 s) to bring down any remaining PBS from sides of tube to facilitate removal. The resulting Vn96-EV/cfDNA pellet can be stored at −80 °C for future use or used immediately for the analyses described below.

3.2 Nanoparticle Tracking Analysis

Nanoparticle tracking analysis (NTA) permits the enumeration of the isolated EVs as well as determination of the size of the EVs that have been isolated. We routinely use the NanoSight NS300 (Malvern Panalytical Ltd) equipped with either a 405 nm or 488 nm

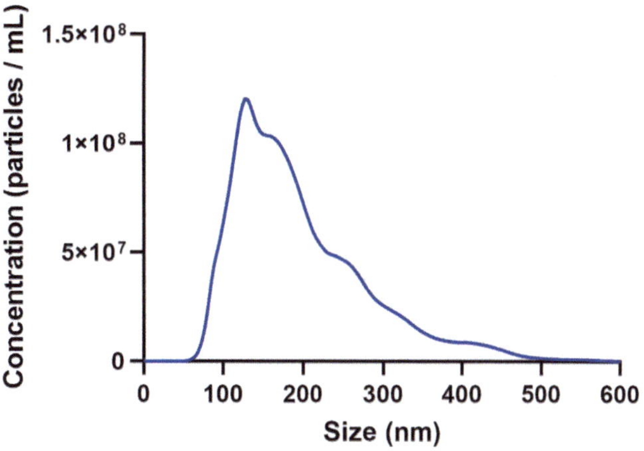

Fig. 1 Nanoparticle tracking analysis of EVs isolated from plasma using the Vn96 synthetic peptide. EVs were eluted from the Vn96-EV/cfDNA pellet with 2 M NaCl and measured using a NanoSight NS300 (Malvern Panalytical Ltd), followed by analysis with the Nanosight software (v3.2). The concentration of particles/mL versus particle size (nm) was determined for seven plasma samples ($n = 7$)

laser for this purpose and analyze the data using Nanosight software (v3.2).

1. Incubate the Vn96-EV/cfDNA pellet in 100 μL of 2 M NaCl (0.1 μm filtered) or 0.1 M glycine-NaOH, pH 10 (0.1 μm filtered) overnight at 4 °C to elute EVs from the pellet (*see* **Note 14**).
2. Following overnight incubation, vortex the tube well. Centrifuge at 17,000 × g for 10 min at room temperature to remove the undissolved pellet and transfer the supernatant containing the eluted EVs to a new tube.
3. Dilute the eluted EVs in water (0.1 μm filtered) to the appropriate concentration for NTA analysis (1×10^8 to 1×10^9 particles/mL) and analyze using a NanoSight instrument or similar according to manufacturer's instructions. An example of size distribution data obtained for EVs isolated from plasma using the Vn96 synthetic peptide is shown in Fig. 1.

3.3 DNA Extraction from the Vn96-EV/cfDNA Pellet

We recommend the use of the Plasma/Serum Cell-Free Circulating DNA Purification Mini Kit from Norgen Biotek Corp. for DNA isolation from the Vn96-EV/cfDNA pellet. Alternatively, the QIAamp Circulating Nucleic Acid Kit from Qiagen may also be used, although its requirement for a carrier RNA may interfere with accurate nucleic acid quantification following DNA extraction. The method detailed below uses the Plasma/Serum Cell-Free Circulating DNA Purification Mini Kit from Norgen Biotek Corp.

1. Resuspend the Vn96-EV/cfDNA pellet in 500 μL of 1× PBS; add 30 μL Proteinase K (from the kit) and vortex for 10 s. Incubate at 60 °C for 10 min (*see* **Note 15**).
2. Add 1 mL of Binding Buffer B and vortex for 10 s.
3. Transfer 750 μL of the mixture from **step 2** into a Mini Spin column that has been inserted into a collection tube. Centrifuge for 2 min at 3300 × g. Discard the flow-through.
4. Transfer the remaining ~750 μL of the mixture into the Mini Spin column and repeat the centrifugation described in **step 3**. Discard the flow-through.
5. Apply 500 μL of Solution WN to the column and centrifuge for 1 min at 3300 × g. Discard the flow-through.
6. Apply 500 μL of Wash Solution A to the column and centrifuge for 1 min at 3300 × g. Discard the flow-through.
7. Repeat **step 6** one more time (for a total of 2 washes with Wash Solution A).
8. Centrifuge the empty Mini Spin column for 2 min at 13,000 × g. Discard the collection tube and its contents.
9. Transfer the Mini Spin column to a fresh collection tube. Apply 100 μL of Elution Buffer B to the column and let stand at room temperature for 2 min. Centrifuge for 1 min at 400 × g, followed by centrifugation for 2 min at 5800 × g.
10. Add 300 μL Binding Buffer B to the 100 μL of eluted DNA, and mix well by vortexing for 10 s.
11. Transfer the entire mixture into a Micro Spin column assembled with one of the provided collection tubes. Centrifuge for 1 min at 3300 × g. Discard the flow-through.
12. Apply 500 μL of Wash Solution A to the Micro Spin column and centrifuge for 1 min at 3300 × g. Discard the flow-through.
13. Repeat **step 12** for a total of two washes.
14. Centrifuge the empty Micro Spin column for 2 min at 13,000 × g. Discard the collection tube and its contents.
15. Transfer the Micro Spin column to a 1.5 mL tube. Apply 30 μL of nuclease-free water to the Micro Spin column and incubate at room temperature for 2 min. Centrifuge for 1 min at 400 × g, followed by centrifugation for 2 min at 5800 × g.
16. Add another 25 μL of nuclease-free water to the Micro Spin column and incubate at room temperature for 2 min. Centrifuge for 1 min at 400 × g, followed by centrifugation for 2 min at 5800 × g. Discard the depleted Micro Spin column.
17. Concentrate the sample to a volume of ≤20 μL. (*see* **Note 16**).

18. After concentration, measure volume of sample and bring to ~20 μL with nuclease-free water, if required. Keep on ice for immediate use in a downstream assay or store at −20 °C.

3.4 RNA Extraction from the Vn96-EV/cfDNA Pellet

We recommend the use of the *mir*Vana™ miRNA Isolation kit from Thermo Fisher Scientific for the isolation of the small RNA species that are typically found in EVs (<200 nucleotides; e.g., miRNA, tRNA, snRNA, etc.) that will be used for downstream analyses, such as RNA sequencing or reverse transcription-polymerase chain reaction (RT-PCR).

1. In a fume hood, add 300 μL of Lysis/Binding buffer from the MirVana™ miRNA Isolation kit to the Vn96-EV/cfDNA pellet. Vortex vigorously for 30 s on high speed and ensure that the Vn96-EV/cfDNA pellet is completely disrupted.
2. In a fume hood, add 1/10th volume of miRNA Homogenate Additive to each tube (i.e., for 300 μL sample volume, add 30 μL miRNA Homogenate Additive). Mix well by inversion and incubate on ice for 10 min.
3. In a fume hood, add one volume of Acid Phenol:Chloroform: Isoamyl Alcohol equal to the original sample volume (i.e., for 300 μL sample add 300 μL Acid Phenol:Chloroform:Isoamyl Alcohol; *see* **Note 17**). Vortex for 1 min.
4. Centrifuge for 5 min at 10,000 × g at room temperature to separate the phases. If the interphase is not compact, repeat this centrifugation step.
5. In a fume hood, transfer the upper aqueous phase to a new tube; take note of the actual volume.
6. Prewarm RNase-free water to 95 °C for the RNA elution step; prepare 250 μL per sample.
7. In a fume hood, add 1.25 volumes of room temperature 100% ethanol to the aqueous phase (i.e., for a 300 μL sample, add 375 μL of ethanol). Invert several times to mix.
8. In a fume hood, place a *mir*Vana™ Filter Cartridge into a collection tube. Pipet up to 700 μL of the aqueous layer/ethanol mixture onto the Filter Cartridge.
9. Centrifuge for 15 s at 10,000 × g. Discard the flow-through. Repeat the centrifugation until the entire sample has been passed through the filter, if necessary.
10. Add 700 μL of Wash Solution 1 to the Filter Cartridge. Centrifuge for 10 s at 10,000 × g. Discard the flow-through.
11. Add 500 μL of Wash Solution 2/3 to the Filter Cartridge. Centrifuge for 15 s at 10, 000 × g, and discard the flow-through. Repeat this wash with an additional 500 μL Wash Solution 2/3. Discard the flow-through.

12. Centrifuge the Filter Cartridge for 1 min at 10,000 × g to remove residual fluid from the filter.
13. Transfer the Filter Cartridge to a new collection tube. Apply 100 μL of prewarmed (95 °C) RNase-free water to the center of the filter. Incubate for 2 min. Centrifuge for 30 s at 10,000 × g to elute the RNA.
14. Repeat the elution step using the same collection tube by adding an additional 100 μL of prewarmed RNase-free water to the center of the Filter Cartridge; incubate for 2 min and centrifuge as described in **step 13**.
15. Concentrate the sample to a volume of ≤20 μL (*see* **Note 18**).
16. After concentration, measure volume of sample and bring to ~20 μL with RNase-free water, if required. Keep on ice for immediate use in a downstream assay or store at −80 °C.

3.5 Protein Extraction from the Vn96-EV/cfDNA Pellet for Use in Western-Blot Analyses

The steps below are recommended for the isolation of proteins from the Vn96-EV/cfDNA pellet for use in Western-blot analyses for the detection of both canonical EV proteins, as well as specific EV protein cargoes. Steps for the preparation of protein samples under both reduced and nonreduced conditions are provided.

1. Prepare SDS-Tris Lysis buffer by supplementing with protease inhibitors immediately before use. For example, to 600 μL of SDS-Tris Lysis buffer, add 2 μL protease inhibitor cocktail III (Cedarlane).
2. Add 65 μL of SDS-Tris Lysis buffer supplemented with protease inhibitors to the Vn96-EV/cfDNA pellet.
3. Incubate at 70 °C for 15 min; mix by vortexing. If desired, a 5 μL sample may be taken at this step for protein quantification (e.g., BCA analysis; *see* **Note 19**). The remainder of the sample may be stored at −20 °C for later analysis or processed immediately.
4. Add 20 μL of 4× Laemmli Sample buffer to 60 μL of protein sample; incubate at 70 °C for 15 min. This is the nonreduced protein sample.
5. Load 20 μL per well on a 12% SDS-PAGE. Proceed with immunoblotting for antibodies that require nonreduced conditions (e.g., CD63, CD9, CD81). A sample blot for CD63 and CD9 is shown in Fig. 2a.
6. Aliquot 56 μL of the remaining sample prepared in **step 4** and add 2.8 μL of Bond-Breaker ™ TCEP Solution (*see* **Note 20**). Incubate at 95 °C for 5 min. This is the reduced protein sample.
7. Load 25 μL per well on a 10% or 12% SDS-PAGE. Perform immunoblotting with antibodies that require reduced

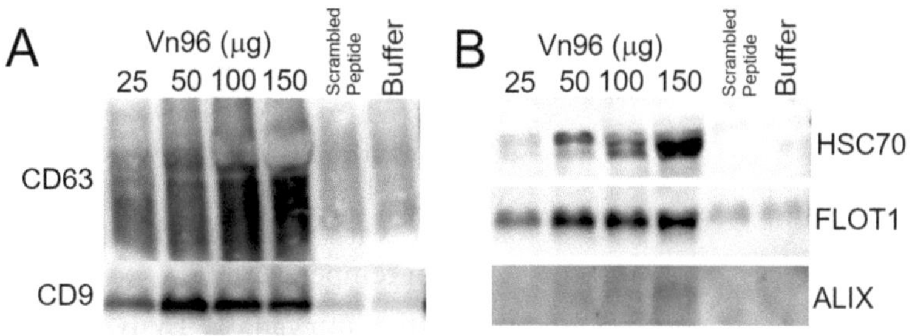

Fig. 2 Western-blot analyses of material isolated from plasma with the Vn96 synthetic peptide. (**a**) EVs were isolated from plasma using the indicated amounts of the Vn96 synthetic peptide, subjected to protein extraction, and 20 μL of nonreduced protein sample was loaded per lane on a 12% SDS-PAGE. Material isolated with a scrambled peptide as well as the addition of 0.625× PBS (buffer) were included as controls. Separated proteins were transferred to a PVDF membrane and immunoblotting for CD63 (Santa Cruz Biotechnology, catalogue # sc-5275) and CD9 (Santa Cruz Biotechnology, catalogue # sc-59140) was performed. (**b**) EVs were isolated from plasma using the indicated amounts of the Vn96 synthetic peptide, subjected to protein extraction, and 25 μL of reduced protein sample was loaded per lane on a 12% SDS-PAGE. Material isolated with a scrambled peptide as well as the addition of 0.625× PBS (buffer) were included as controls. Separated proteins were transferred to a PVDF membrane and immunoblotting for HSC70 (Santa Cruz Biotechnology, catalogue # sc-7298), flotillin-1 (FLOT1; New England Biolabs, catalogue # 18634S), and ALIX (Santa Cruz Biotechnology, catalogue # sc-53540) was performed

conditions (e.g., HSC70, HSC90, flotillin-1, TSG101, or ALIX). A sample blot for HSC70, flotillin-1, and ALIX is shown in Fig. 2b.

4 Notes

1. All reagents for Vn96-mediated EV/cfDNA isolation and elution are prepared using 0.1 μm-filtered water, in order to remove any nanoparticles that could contaminate the isolated EVs; moreover, prepared reagents are also 0.1 μm-filtered (unless otherwise noted) to remove nanoparticles that could have been introduced during preparation. We recommend the use of molecular-grade PBS and water that are certified RNase- and DNase-free.
2. ProClin 300™ is a bactericide that is added to prevent bacterial growth in the peptide solution during storage.
3. Add 11.69 g of NaCl to a 100 mL beaker or graduated cylinder, add 90 mL of water, and stir/mix until dissolved. Adjust

volume to 100 mL with water. Sterilize using a 0.1 μm filter and store at room temperature.

4. Add 0.751 g of glycine to a 100 mL beaker or graduated cylinder, add 80 mL of water, and stir/mix until dissolved. Adjust pH to 10 using NaOH and then adjust volume to 100 mL with water. Sterilize using a 0.1 μm filter, and store at room temperature.
5. Warm the mirVana™ Lysis/Binding buffer, Wash Solution 1 and Wash Solution 2/3 to room temperature before use. Ensure that the appropriate volume of ethanol has been added to each wash solution before using.
6. Combine 10 mL of 20% SDS, 25 mL of Tris–HCl buffer pH 6.8, and 65 mL of water and mix well. *Do not* filter-sterilize. Store at room temperature.
7. Add 15 μL of ProClin 300™ to 30 mL of 0.625× PBS and mix by inversion. This reagent may be stored at 4 °C for up to 1 year.
8. Solubilized Vn96 (2.5 mg/mL) will have a slightly opalescent appearance but should be relatively free of large visible particulates.
9. The effect of preclearing the plasma should be tested for your application. We routinely use 3000 × g for 15 min to preclear plasma that will be used for the isolation of EVs whose contents will be subjected to small RNA sequencing analyses.
10. In our experience, the use of a ≥16,000 × g centrifugation step can reduce EV and nucleic acid recovery [6].
11. Protease inhibitor cocktail can be added (we typically use Protease Inhibitor Cocktail set III from Cedarlane, catalogue # 539134-1 mL) in order to reduce degradation of the Vn96 peptide by plasma proteases during EV/cfDNA isolation procedures.
12. A range of 25–100 μg/mL of Vn96 synthetic peptide can be used to isolate cfDNA and EV-associated DNA from plasma. We have determined that Vn96 has a high affinity for DNA and that as little as 25 μg of Vn96 per mL of plasma is sufficient (add 10 μL of Vn96 stock solution per mL of plasma) for complete recovery of >95% of cfDNA and EV-associated DNA from plasma. For EVs that will be used for downstream RNA isolation, we typically use 150 μg of Vn96 per mL of plasma (add 60 μL of Vn96 stock solution per mL of plasma). A scrambled version of the Vn96 peptide that does not isolate EVs or cfDNA is available (vivitide, catalogue # ME-040-0100), which can be used as a negative control.
13. If samples are left at room temperature for too long, the Vn96-EV/cfDNA pellet will dislodge from the tube. If the pellet

dislodges before the supernatant has been removed, simply centrifuge at 17,000 × g for 2 min to bring the pellet back to bottom of tube before proceeding.

14. EVs that have been isolated with the Vn96 synthetic peptide can be eluted from the Vn96-EV/cfDNA pellet using either a 2 M NaCl solution or a 0.1 M glycine-NaOH, pH 10 solution; however, we recommend the use of 0.1 M glycine, pH 10 elution solution for salt-sensitive applications.
15. Longer incubation times at 60 °C (up to 1 h) can improve DNA recovery, but should be determined empirically for the intended application.
16. The typical yield of DNA for plasma collected from "healthy" individuals (e.g., control samples for disease biomarker detection) is 4.9 ± 4 ng per mL of plasma, but can be significantly higher in plasma samples from patients with diseases such as cancer (2–150 ng per mL of plasma). By concentrating the sample to ~20 μL, the final concentration of DNA should be sufficiently high to measure by automated gel electrophoresis. We use a vacuum concentrator from Eppendorf (Concentrator Plus/Vacufuge plus) for this purpose. Place open sample tubes in the vacuum concentrator and set to 45 °C, Mode V-AQ (vacuum-aqueous, concentrator). Run for ~15 min intervals, each time visually inspecting the residual sample volume; repeat until residual volume is ~20 μL. We typically use a Fragment Analyzer from Agilent to measure DNA concentration and size; the HS Small Fragment kit (Agilent, catalogue # DNF-477-0500) can accurately measure DNA concentration and size profile in samples with a DNA concentration as low as 100 pg/μL.
17. Be sure to withdraw from the bottom phase of the bottle of Acid Phenol:Chloroform:Isoamyl Alcohol (the upper phase contains the aqueous buffer).
18. The typical yield of RNA per mL of plasma can range from 2 to 150 ng. By concentrating the sample to ~20 μL, the final concentration should be sufficiently high to measure by an automated gel electrophoresis system. We typically use a vacuum concentrator from Eppendorf (Concentrator Plus/Vacufuge plus) for concentration of the RNA sample. Place open sample tubes in the vacuum concentrator and set to 45 °C, Mode V-AQ (vacuum-aqueous, concentrator). Run for ~15 min intervals, each time visually inspecting the residual sample volume; repeat until residual volume is ~20 μL. We recommend the use of the HS RNA kit (Agilent, catalogue # DNF-472-0500) and Fragment Analyzer from Agilent to accurately measure RNA concentration and size profile in samples with an RNA concentration as low as 50 pg/μL.

19. The Vn96 peptide present in the sample will contribute to the total protein concentration; therefore, the amount of input Vn96 synthetic peptide will be reflected in the values measured by the BCA assay.
20. Alternatively, dithiothreitol (DTT) or β-mercaptoethanol can be used as reducing agents at final concentrations of 0.1 M or 2.5% (v/v), respectively.

References

1. Théry C, Witwer KW, Aikawa E et al (2018) Minimal Information for Studies of Extracellular Vesicles 2018 (MISEV2018): a position statement of the International Society for Extracellular Vesicles and Update of the MISEV2014 guidelines. J Extracell Vesicles 7:1535750
2. Zaborowski MP, Balaj L, Breakefield XO et al (2015) Extracellular vesicles: composition, biological relevance, and methods of study. Bioscience 65:783–797
3. Properzi F, Logozzi M, Fais S (2013) Exosomes: the future of biomarkers in medicine. Biomark Med 7:769–778
4. Mathai R, Vidya R, Reddy B et al (2019) Potential utility of liquid biopsy as a diagnostic and prognostic tool for the assessment of solid tumors: implications in the precision oncology. J Clin Med 8:373
5. Ghosh A, Davey M, Chute IC et al (2014) Rapid isolation of extracellular vesicles from cell culture and biological fluids using a synthetic peptide with specific affinity for heat shock proteins. PLoS One 9:e110443
6. Roy JW, Taylor CA, Beauregard A-P et al (2021) A multiparametric extraction method for Vn96-isolated plasma extracellular vesicles and cell-free DNA that enables multi-omic profiling. Sci Rep 11:8085
7. Reddy VS, Madala SK, Trinath J et al (2018) Extracellular small heat shock proteins: Exosomal biogenesis and function. Cell Stress Chaperones 23:441–454
8. Su Y, Li S, Hong M (2013) Cationic membrane peptides: atomic-level insight of structure-activity relationships from solid-state NMR. Amino Acids 44:821–833
9. Zocco D, Bernardi S, Novelli M et al (2020) Isolation of extracellular vesicles improves the detection of mutant DNA from plasma of metastatic melanoma patients. Sci Rep 10:15745

Chapter 23

Characterizing Extracellular Vesicles Using Nanoparticle-Tracking Analysis

Modeline N. Longjohn and Sherri L. Christian

Abstract

Extracellular vesicles (EVs) are nanosized lipid bound particles secreted by cells. EVs transfer biologic material and can be found in all body fluids. Accurately characterizing the size and concentration of EVs is difficult because of the nanoscale size of EVs. An evolving solution to this problem is tracking the Brownian motion of EVs in suspension, a technique known as nanoparticle-tracking analysis (NTA). This technique is used by many researchers in the EV field. The ability to accurately replicate data between studies and laboratories is critical to advancing the knowledge surrounding EVs for use in liquid biopsy and cancer studies, in general. Thus, this chapter provides a step-by-step guide on isolating EVs using a variety of methods, for characterizing EVs using the NS300 NTA instrument, troubleshooting tips, and a robust guide to reporting key parameters that will help improve cross-lab reproducibility of NTA data.

Key words Extracellular vesicles, Exosomes, Microvesicles, NTA, Cancer, EV, Nanosight

1 Introduction

Molecular biology techniques have advanced tremendously throughout the twenty-first century, allowing for more high-throughput analysis of biological molecules and systems. With these advancements have come the capability to better explore the gamut of possibilities of biomolecules, particles, and systems. Overall, different types of microscopies (light and electron microscopy) have enabled the study of cells and biomolecules, which are in the microscale. However, in the past five decades, smaller-sized biological particles have emerged, which are in the nanoscale and thus require more specialized tools and techniques for their visualization and characterization.

Extracellular vesicles (EVs) are lipid-bilayer-encapsulated, bioactive-cell-derived nanoparticles; these were first discovered by Chargaff and West in 1946 [1]. Over the years, EVs have been classified based on size and mode of biogenesis into exosomes

Sherri L. Christian (ed.), *Cancer Cell Biology: Methods and Protocols*,
Methods in Molecular Biology, vol. 2508, https://doi.org/10.1007/978-1-0716-2376-3_23,

(30–150 nm sized and generated via the endocytic pathway), microvesicles (100–1000 nm sized and directly budded off cell membrane), and apoptotic bodies (>1000 nm sized, and released in the final stages of apoptosis) [2, 3]. Other types of EVs, which are newly emerging discoveries are <50 nm-sized EVs called exomeres [4] and cancer-cell-derived EVs that carry oncogenes, called oncosomes [5]. The size overlap between subtypes, varying cargo complexity, release from all tested living cells, and presence in all tested body fluids point toward the importance of extracellular vesicles as more than just waste and recycling machineries [6].

EVs, particularly exosomes and microvesicles, play roles in many aspects of normal physiology. Specifically, they facilitate intercellular communication, emerging recently as a mechanism for cell–cell communication [2]. Other cell- and organ-specific functions have been identified, which are beyond the scope of this chapter. However, in disease states, specifically cancers, EV release is emerging as a "hallmark of cancer onset, progression, and relapse." In comparison with release from normal cells, tumor cells have been shown to release up to 10-fold more EVs [7–9]. Based on their abundance and ubiquity, EVs are potentially a source of information about cell status at both normal and disease states and are considered by some to be the next generation of liquid biopsy biomarkers [10, 11]. Specifically, EVs could be used for monitoring patient response to therapy, and tracking relapse in hematological malignancies and solid tumours [12].

However, despite the promise of EVs in cancer, there have been major challenges in the field. Key among these are the lack of standardized isolation and characterization technologies [2, 13, 14]. The lack of suitable methods for detecting, analyzing, quantifying, and phenotyping EVs is a limitation that has slowed the pace of advancements in the field of EV research [15].

Nanoparticle-tracking analysis (NTA) is a direct and real-time technique for visualizing and analyzing nanoparticles in liquid suspension. NTA measures the size of particles in liquid media, by correlating the rate of Brownian motion (random motion of microscopic particles in a fluid due to constant bombardment by molecules in the surrounding fluid) to size of the particle, viscosity, and temperature of the liquid media. Based on a laser-illuminated microscope-associated technique, Brownian motion is tracked in real time by using a highly sensitive scientific CMOS camera [15]. Specifically, each particle in suspension is simultaneously and separately tracked by an internal particle-tracking image analysis program [15]. Because particles are tracked and visualized separately, the estimation of particle size and concentration is not limited by the intensity-weighted, z-average distribution associated with the particle sizing regimen in dynamic light scattering (DLS) and Photon Correlation Spectroscopy (PCS). Furthermore, because measuring of particle size and concentration is based on specific particle light scattering, it is possible to determine the size and concentration of a heterogenous mix of nanoparticles in suspension. Additionally, based on the limit of detection of nanoparticles, the NTA is more sensitive than nanoscale flow cytometers whose lower limit of measurement is 300 nm. Using high-intensity laser beams and low-background optical configuration, the NTA can measure the size of submicron particles of low particle refractive index as small as 30 nm such as exomeres and exosomes, and as large as 1–2 μm sized apoptotic bodies and oncosomes.

During an NTA run, particles in liquid suspension are introduced through a leur port onto a prism-edged optical flat. Next, a finely focused and optimally configured laser beam bombards the optical flat (and the introduced sample in suspension), with the optical flat's refractive index causing the laser beam to refract at the interface between the glass and the liquid suspension, in turn creating a region of intense light compressed to a low profile allowing easy microscopic visualization of nanoparticles [15]. To visualize and track the particles in liquid suspension, a scientific CMOS camera (sCMOS) captures the Brownian motion of particles with a 100 × 80 μm view field and records a video of this motion (*see* **Note 1**).

During the analysis phase of the NTA run, the recorded video is analyzed by a dedicated NTA computer software/pipeline that views and tracks the center of each particle one-frame at a time throughout the duration of the video, which is either 900 frames for 30 s or 1800 frames for 60 s [15]. Next, the software automatically calculates the value of the average distance traveled by each tracked particle, which is then used to calculate the diffusion coefficient of the particle (Dt). The Dt, known temperature of the sample (T), and solvent viscosity (η) is then put into Stoke Einstein's Equation to calculate the hydrodynamic diameter (d) of each particle (*see* **Note 2**):

$$\frac{\overline{(x,y)^2}}{4} = Dt = \frac{Tk}{3\pi\eta d}$$

To ensure that enough particles are tracked, and their size calculated in record time, the solution needs to contain a range of 10^6–10^9 particles/ml. In addition, a fluorescence mode is available, which can specifically detect, and measure particles labelled with compatible fluorophores using specific optical filters. These lasers, such as a 488 nm diode, can be used to excite conventional fluorophores and their emitted light on the nanoparticles detected and tracked accordingly.

Another application of the NTA is the application of electric fields across the liquid suspension. Upon application of the electric field, charged particles will electrophorese, the zeta potential (*see* **Note 3**) and polarity will then be used to calculate the electrophoretic velocity and directions of the particles respectively [15]. This is independent of the measurement of particle size or light scattering. Though this application has not been used as extensively as the others, this application allows for the identification of charged particles and discrimination of subpopulations (based on differing properties) within a heterogenous mixture.

Based on the emerging knowledge of EVs and mechanism of nanoparticle tracking, the NTA can measure EVs with high accuracy, in addition to being able to resolve heterogenous mix of EVs in a liquid suspension. Specifically, NTA can be used to characterize physical properties of EVs from cancer cell lines and patient's biofluids (e.g., plasma, cerebrospinal fluid (CSF), and bone marrow (BM) aspirates). Furthermore, differences in EV physical properties between different disease time points (at diagnosis versus post treatment versus after relapse) and between disease states can also be determined using NTA. Finally, EV subpopulations can be phenotyped by labeling EVs with fluorophore-labelled antibodies, which can then be detected using the NTA's fluorescence mode.

Despite the sensitivity of NTA for tracking nanoparticles and EVs, replicability of results between labs and research groups has been missing. Thus, this chapter highlights the need for reporting and maintaining NTA parameters such as camera levels, detection threshold, flow rate, dilution factor, and script settings, which affect data output.

2 Materials

2.1 Generating EV Conditioned Media from Suspension Cancer Cell Line

1. Disposable conical tubes.
2. Vesicle-free growth media (*see* **Note 4**).
3. 70% ethanol for disinfecting surfaces.
4. Tissue-culture treated dishes, plates, or flasks containing suspension cancer cells grown to up to 60% confluency.
5. Pipette tips and pipettes.
6. Pipette aids.
7. Low lint wipes.
8. 37 °C incubator with humidified atmosphere of 5% CO_2.
9. Biosafety cabinet, level IIA or greater.
10. Reagents and equipment to determine cell viability and total cell counts (Trypan Blue and hemacytometer, or Cell Counter).
11. Tabletop centrifuge.

2.2 Generating EV Conditioned Media from Adherent Cancer Cell Line

1. Disposable conical tubes.
2. Vesicle-free growth media (*see* **Note 4**).
3. 70% ethanol for disinfecting surfaces.
4. Tissue-culture treated dishes, plates, or flasks with adherent cancer cells grown to up to 70% confluency.
5. Pipette tips and pipettes.
6. Pipette aids.
7. Low-Lint wipes.
8. 37 °C incubator with humidified atmosphere of 5% CO_2.
9. Biosafety cabinet, level IIA or greater.
10. Reagents and equipment to determine cell viability and total cell counts (Trypan Blue and hemacytometer, or Cell Counter).
11. Tabletop centrifuge.
12. Vacuum or water aspirator.

2.3 Isolating EVs from Human Biofluid, Namely, Plasma, BM Aspirate, and CSF

1. Frozen plasma.
2. Disposable conical tubes.
3. 70% ethanol for disinfecting surfaces.
4. Pipette tips and pipettes.
5. Pipette aids.
6. Low-Lint wipes.
7. 37 °C incubator with humidified atmosphere of 5% CO_2.
8. Biosafety cabinet, level IIA or greater.
9. Tabletop centrifuge.

2.4 Isolation of EVs by Ultracentrifugation (UCF) (see Note 5)

1. Conditioned media or frozen biofluid.
2. Disposable conical tubes.
3. 50 ml propylene centrifuge tubes.
4. Refrigerated centrifuge.
5. Ultracentrifuge and type of rotor (fixed angle or swinging bucket).
6. Polyallomer tubes or polycarbonate bottles, suitable for the ultracentrifuge rotor used.
7. 70% ethanol for disinfecting surfaces.
8. Pipette tips and pipettes.
9. Pipette aids.
10. Low lint wipes.
11. Tabletop centrifuge.

2.5 Isolation of EVs by Size Exclusion Chromatography (SEC), Namely, qEV Izon (see Note 6)

1. Disposable conical tubes.
2. Conditioned media or frozen biofluid.
3. 70% ethanol for disinfecting surfaces.
4. Pipette tips and pipettes.
5. Pipette aids.
6. Low lint wipes.
7. SEC column, e.g., qEV (Izon).
8. 0.22 μm filter.
9. 1× Phosphate-buffered saline (PBS).
10. Tabletop centrifuge.
11. Optional: Centrifugation filter units (*see* **Note 7**).

2.6 Isolation of EVs Using Polyethylene Glycol (PEG), Namely, ExoQuick (Systems Biosciences) (see Note 8)

1. Disposable conical tubes.
2. Conditioned media or frozen biofluid.
3. 70% ethanol for disinfecting surfaces.
4. Pipette tips and pipettes.
5. Pipette aids.
6. Low-Lint wipes.
7. ExoQuick EV isolation solution.
8. ExoQuick resuspension buffer.
9. ExoQuick microsphere beads.
10. End-over-end tube rotator.
11. Tabletop centrifuge.

2.7 Isolation of EVs by Protein-Affinity-Based Isolation Namely, Vn96—New England peptide (see Note 9)

1. Disposable conical tubes.
2. Conditioned media or frozen biofluid.
3. 70% ethanol for disinfecting surfaces.
4. Pipette tips and pipettes.
5. Pipette aids.
6. Low lint wipes.
7. Tabletop centrifuge.
8. Vn96 peptide resuspended in resuspension buffer.
9. 0.1 μm filtered 1× PBS.
10. End-over-end tube rotator.

2.8 Preparation of Samples for Nanoparticle Tracking

1. Conditioned media, EVs in 1× PBS or resuspension buffer, biofluid.
2. Disposable centrifuge tubes.
3. 0.1 μm filtered 1× PBS (*see* **Note 10**).

4. Pipette tips and pipettes.
5. Microfuge tubes.

2.9 Running Samples on the Nanoparticle Tracker

1. 1 ml disposable syringes.
2. HPLC-grade water.
3. Microfuge tubes containing prepped samples.
4. Nanosight NS300 instrument.
5. Nanosight NS300 Low volume flow cell (LVFC).
6. Nanosight NS300 laser module.
7. Nanosight NS300 laser module protective wipe.
8. Medium-sized bucket of ice.

3 Methods

3.1 Generating EV Conditioned Media from Suspension Cancer Cell Line

1. Sterilize the hood by spraying the working surface with 70% ethanol and wiping down with low-lint wipes.
2. Transfer culture plate containing cancer cells from the incubator to the biosafety cabinet.
3. Transfer cell-suspension from culture dish into conical tube.
4. Centrifuge at 500 × g for 5 min, to pellet suspension cells.
5. Aspirate supernatant, resuspend cells in fresh media, and count cells (see Chapter 2 for more details).
6. Resuspend cells in appropriate amount of EV free media.
7. Transfer cells in EV free media into an appropriate cell culture plate, and culture for the desired amount of time, to generate conditioned media.

3.2 Generating EV Conditioned Media from Adherent Cancer Cell Line

1. Sterilize the hood by spraying the working surface with 70% ethanol and wiping down with low-lint wipes.
2. Transfer culture plate containing adherent cancer cells from the incubator to the biosafety cabinet.
3. Aspirate culture medium from culture plate.
4. Wash off residual culture medium once, using enough EV free media to cover the culture plate.
5. After wash, cover plate with required volume of EV free media.
6. Return cell culture plate to incubator to culture for the desired amount of time.

3.3 Preparation of EVs from Conditioned Media for NTA Characterization

1. Collect/pool media from culture plate.
2. Centrifuge at 500 × g for 5 min to pellet cells.
3. Collect supernatant (contains small and large EVs) and discard pellet (cells).

4. Centrifuge supernatant at 2000 × g for 5 min.
5. Pellet contains large EVs (apoptotic bodies).
6. For large vesicles, collect pellet and dissolve in appropriate buffer (*see* **Note 11**).
7. For small vesicles (exosomes and microvesicles), collect supernatant and dilute as appropriate/determined in 0.1 μm filtered 1× PBS (*see* **Notes 12–14**).

3.4 Preparation of EVs from Biofluid for NTA Characterization

1. If biofluid is frozen, thaw slowly on ice at room temperature.
2. Dilute as determined empirically (*see* **Note 12**), e.g., 1:1000 in 0.1 μm-filtered 1× PBS.
3. Transport on ice to NTA.
4. Run as below.

3.5 Isolation of EVs by Ultracentrifugation (UCF) (see Note 5)

1. Collect conditioned media into centrifuge tubes.
2. Centrifuge at 500 × g for 5 min to pellet cells.
3. Collect supernatant into fresh centrifuge tube and discard pellet.
4. Centrifuge at 2000 × g for 20 min at 4 °C to pellet debris and apoptotic bodies.
5. Collect supernatant into a polyallomer tube or polycarbonate bottle (appropriate for ultracentrifuge rotor).
6. Make a mark on the ultracentrifuge tube (using a waterproof marker), and place the tube in the rotor such that the mark faces upward.
7. Centrifuge at 10,000 × g for 30 min at 4 °C.
8. The pellet at this point contains large vesicles (e.g., microvesicles) (*see* **Note 15**).
9. Collect the supernatant into fresh ultracentrifuge tube and centrifuge at 100,000 × g for 70 min or longer as empirically determined at 4 °C.
10. Discard the supernatant, and collect the pellet (contains exosomes).
11. Wash the pellet by resuspending in at least 1 ml 1× PBS and pool all EV suspensions from all ultracentrifuge tubes used for the cell line.
12. Centrifuge at 100,000 × g for 1 h at 4 °C.
13. Discard the supernatant.

3.6 Preparation of UCF Isolated EVs for NTA Characterization

1. Wash EV pellet in 1× PBS.
2. Dilute pellet in 0.1 μm filtered PBS.
3. Dilute to an appropriate concentration range in 0.1 μm filtered 1× PBS and run diluted volume on the NTA.

3.7 Isolation of EVs by Size Exclusion Chromatography (SEC) (see Note 6)

1. Centrifuge conditioned media at 2000 × g for 10 min.
2. Collect supernatant. Filter using a 0.22 μm syringe filter for columns <20 ml or large filter systems for volumes >20 ml.
3. Add supernatant to the top of the SEC column after washing and prepping the column according to the manufacturer's guidelines.
4. Collect EV fractions by elution using 0.22 μm filtered 1× PBS (*see* **Note 7**).

3.8 Polyethylene Glycol (PEG)-Based EV Isolation, Namely, ExoQuick (see Note 8)

1. Centrifuge conditioned media at 3000 × g for 15 min.
2. Collect supernatant to fresh centrifuge tube.
3. Add ExoQuick reagent and invert the tube at least 5 times, to mix.
4. Store the ExoQuick-CM mix at 4 °C overnight (ON).
5. Pellet EVs by centrifugation at 1500 × g for 30 min.
6. Discard supernatant.
7. Centrifuge at 1500 × g for 5 min if residual supernatant remains.
8. Resuspend EV pellet in resuspension buffer.
9. Wash 1 unit of microsphere beads 3× with resuspension buffer.
10. Add EVs in resuspension buffer to washed beads.
11. Mix at room temperature on an end-over-end rotator for 15 min.
12. Centrifuge at 8000 × g for 5 min.
13. Collect supernatant (EVs in suspension).

3.9 Preparation of SEC and PEG Isolated EVs for NTA Characterization

Collect 300–600 ul of final EV suspension. Run straight on NTA, or dilute if outside concentration range (*see* **Note 13**).

3.10 Isolation of EVs by Protein-Affinity-Based Isolation Method, Namely, Vn96 (see Note 9)

1. Centrifuge at 500 × g for 5 min to pellet cells.
2. For biofluid, mix 1:1 with 0.1 μm filtered 1× PBS.
3. Collect supernatant or diluted plasma and add Vn96 solution (50–100 μg/ml of starting plasma volume).
4. Invert to mix.
5. Incubate at room temperature with rotation for 1 h.
6. Centrifuge at 17,000 × g for 15 min.
7. Discard supernatant.

8. Wash three times with at least 1 ml 0.1 μm 1× filtered PBS supplemented with protease inhibitors such as phenylmethylsulphonyl fluoride (PMSF), or protease inhibitor cocktail, and centrifuge at 17,000 × g for 10 min each.
9. After final wash, discard supernatant and collect EV pellet.

3.11 Preparation of Vn96 Isolated EVs for NTA Characterization

1. After washing the pellet, incubate pellet with Proteinase K or 2 M NaCl ON at 4 °C (*see* **Note 16**).
2. Collect mix and centrifuge at 17,000 × g for 15 min.
3. No pellet will be visible.
4. Dilute to bring concentration to acceptable range (*see* **Note 13**).

3.12 Nanosight System Setup and Sample Running (see Fig. 1)

1. Open the door of the NS300.
2. Lift the laser module and the unattached LVFC out onto the work bench.
3. Attach the LVFC to the surface of the module, ensuring that the sprung holes match up.
4. Insert the sprung screws (4), one at a time (*see* **Note 17**).
5. After fitting the LVFC onto the module top plate, slide the module back into the NS300.
6. Secure the module into place by pulling the red lever handle clockwise.
7. Close the door of the NS300.
8. Press the "on" button (left side of the NS300) and wait for the instrument to power on.
9. Open the NS 3.4 software on the computer.
10. Allow at least 60 s for setup and internal control.
11. Click on the "Start Camera" box.
12. Click auto setup, to allow the Nanosight to automatically optimize the camera level.
13. Next, adjust the camera level to 12–13, or as optimized (*see* **Note 18**).
14. Ensure the screen gain is set at 1, or as previously determined.
15. Prime the low volume flow cell (LVFC) following the subsequent steps.
16. Check that the outlet tubing (thick diameter tubing) is connected and inserted into the waste bottle.
17. Load a 1 ml syringe with HPLC water.
18. Open the door of the NS300 instrument, and remove laser module. Place the laser module vertically on the workbench.

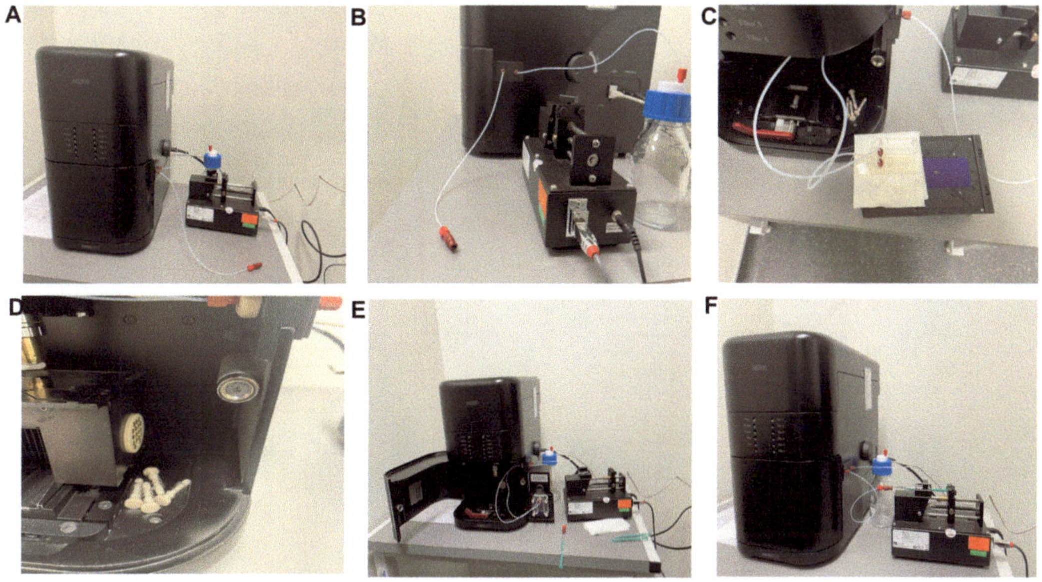

Fig. 1 The NS300 instrument shown at different positions before and during an NTA run. (**a**) The assembled and ready-to-use NS300, with the door closed showing the setup including the inlet tubing, syringe pump, wash bottle, and computer connecting cables (*see* **Note 34**). (**b**) Close-up view of the working side of the NS300, clearly showing all components shown in (**a**). (**c**) A clearer view of the LVFC placed with a lens cloth on the laser module. (**d**) A clearer view of the four sprung screws in their storage position. (**e**) The laser module placed on the workbench in a vertical position during the priming stage of an NTA run. (**f**) The inlet tubing of the NS300 with a 1 ml syringe loaded with sample and attached to the syringe pump in a ready-for-run position. See the green lights showing that the power, a single filter, and the laser are currently in use

19. Slowly introduce 100–300 ul of HPLC water into the inlet tubing, avoiding bubbles (*see* **Note 19**).
20. Watch to see the trajectory of flow, from the inlet tubing, through the tubings on the LVFC, onto the optical flat stage and out toward the outlet tubing.
21. Push all liquid through the inlet tubing one drop at a time at a rate of 50 μl per second, watching for the absence of bubbles and uninterrupted flow (*see* **Note 19**).
22. Place the laser module into the NS300, pressing the red lever hand up to secure it in place and close the door.
23. Open the NTA software on the computer, checking that all drivers are correctly engaged.
24. Click on "start" on the NTA computer screen, to engage the camera.
25. Click on "auto setup," to allow the camera setup a camera level for the sample.
26. Adjust camera level between 11 and 15 until all the particles on the screen can be seen clearly.

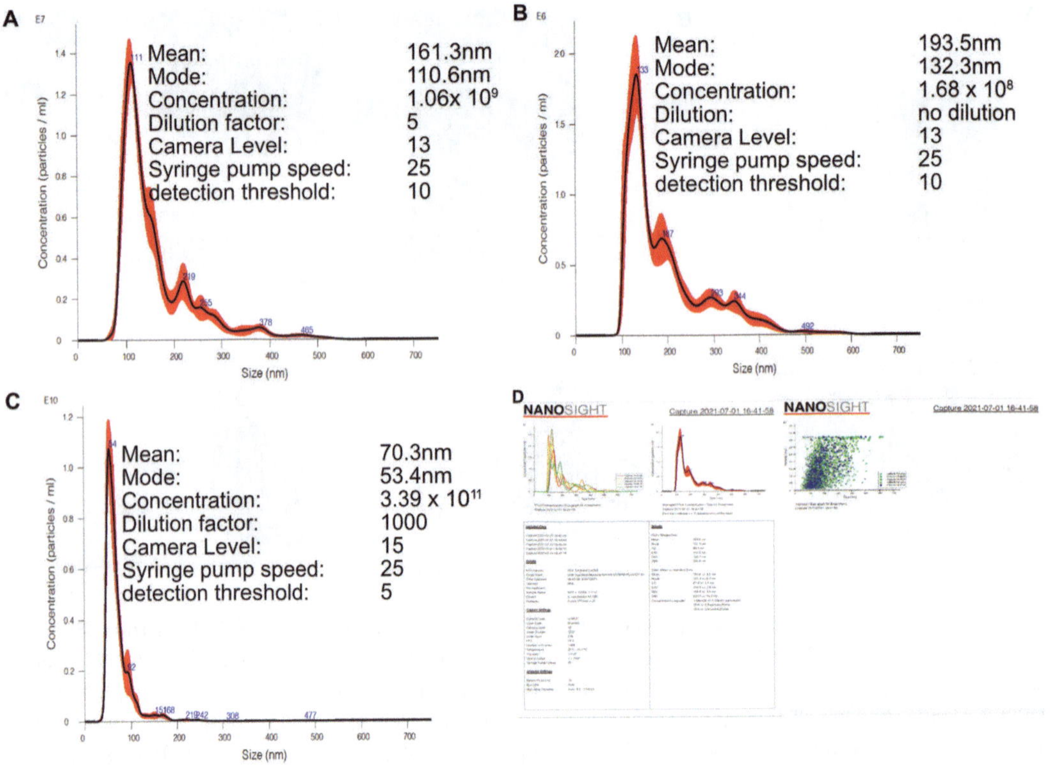

Fig. 2 Data output from characterization of extracellular vesicles using nanoparticle-tracking analysis (NS300 instrument and software). All EVs were diluted or collected in 0.1 μm filtered 1× PBS. (**a**) EVs in 24 h conditioned media from mouse murine lymphoma cell line WEHI 231. (**b**) SEC isolated EVs from WEHI 231 cell line. (**c**) EVs from frozen healthy human plasma. (**d**) Typical NS300 pdf file output

27. Observe the field of view on the camera, for particles flowing from right to left.
28. HPLC water should contain very few particles; thus, only few particles (<20 particles/s) should be seen on the screen. This shows that the NTA is ready for a run.
29. Observe the outflow of water into the waste bottle (*see* **Note 20**).
30. Turn off the camera, or it will time out on its own if left idle for a few minutes.
31. The first run should be characterization of background particles in the dilution buffer.
32. Open the door of the NTA, drop the red lever handles, and bring the module out onto the working bench.
33. Place the laser module in the vertical position in preparation for sample loading.
34. Load a 1 ml syringe with the dilution buffer, e.g., 0.1 μm filtered 1× PBS and inject into the inlet tubing (*see* **Note 21**).

35. Return the laser module into the NS300 machine and close the door. Attach the syringe loaded with the buffer to the leur port of the inlet tubing and place the attached syringe to the syringe pump.
36. To take a measurement, click on the standard measurement tab.
37. Set up script by selecting the number of captures and entering 5 (*see* **Note 22**), capture duration of 60 s, for syringe pump, enter flow rate of 25 (*see* **Note 23**), and check all boxes except the "advance sample prompt" box (*see* **Note 24**), For the dilution box, type the dilution factor (e.g., 1000 for 1:1000 dilution).
38. Next, click "create and run script." The script will appear on the left of the windowpane.
39. A prompt screen comes on and gives the space for you to save your sample with a recognizable name.
40. To edit the script, drag the script window to the center and modify the script if needed.
41. Load 500–800 uL of the diluted sample into a fresh 1 ml syringe, avoiding accumulation of bubbles in the syringe.
42. Attach the syringe to the leur port of the inlet tubing, twisting clockwise to secure the syringe tip in the leur port.
43. Inject 100–300 μl of sample, until a steady stream can be seen on the screen.
44. Attach syringe to the syringe pump, and secure the syringe on the pump by locking the clockwise lever (*see* **Note 25**).
45. Click start on the screen. For using the same settings as for the buffer, setup the script as above and click create and run script.
46. A prompt will appear on the screen, read, and click the prompt to indicate that all settings are accurate.
47. At the end of the run, a prompter screen appears to confirm parameters to process the run.
48. To process and export the data, set the detection threshold (based on previously optimized threshold and screen gain to your optimized parameters) (*see* **Notes 26–28**).
49. After setting the detection threshold, confirm the prompter and the processing of the run begins (*see* **Note 29**).
50. During the processing, there may be warnings such as "high noise." This can affect the reliability of the data (*see* **Note 30**).
51. At the end of the run, the software automatically opens the export settings.
52. Select all export options that apply to you such as exporting pdf.

53. Open the file location and check the pdf and video files of your data (*see* **Notes 31** and **32**). See Fig. 2, for sample image of acceptable NTA output.
54. Compare your data to the criteria for accepting or rejecting NTA data (*see* **Note 33**).
55. After each sample, flush the LVFC with at least 3–5 ml HPLC water (1 ml at a time), to remove the particles, and ensure <1% carryover (*see* **Note 34**).
56. After running all samples, flush the system with 5–8 ml of HPLC water.
57. Turn off the camera on the screen.
58. Go to the tab for temperature control and ensure that the temperature control is off (*see* **Notes 35** and **36**).
59. Disassemble the LVFC from the laser module carefully (*see* **Note 37**), and place disassembled components within the NS300.
60. Close the door of the Nanosight carefully and turn off the NS300 knob on the left side of the machine.
61. Go to the file position, copy the pdf, and excel files for analysis (*see* **Note 38**).
62. Close the software.
63. Set the computer to sleep mode.

4 Notes

1. Scientific CMOS camera (sCMOS) is an image sensor and fabrication technique mounted on a C mount that operates at a speed of 30 frames per second.
2. Stoke Einstein's equation components:

   T is the known temperature of the sample

   η is the solvent viscosity

   d is the hydrodynamic diameter

   k is the Boltzmann constant

   $(x, y)^2$ is the Mean square displacement

   Using the known temperature, solvent viscosity, Boltzmann constant, and mean square displacement, the NTA software calculates the hydrodynamic diameter of the tracked particles. The outputs of each NTA run are nanoparticle diameter (size), count-based particle concentration, and measurement of aggregation.

3. Zeta potential (ζ) is a physical property exhibited by all particles in suspension. It is the potential difference between the dispersion liquid and the surface of the particle immersed in the dispersant [16]. Factors that affect zeta potential include pH, conductivity of the liquid, and concentration of the formulation component.
4. A normal component of culture media is fetal bovine serum (FBS), or fetal calf serum (FCS), a source of nutrients, which help maintain cells in vitro. However, FBS/FCS is rich in EVs, lipoproteins, and protein complexes, which are basically undistinguishable from EVs derived from the cell line of interest, based on their size, density, and RNA composition [17]. This necessitates the development of culture media without these serum-derived EVs. However, not all cell lines can survive in the absence of FBS/FCS, and the degree of survival in the absence of FBS/FCS varies. There are different types of vesicle-free media for EV isolation, namely, EV-depleted FBS vesicle-free media [17, 18], FBS-free vesicle free media [19], Serum-free media [20], and FBS alternative supplemented media [21]. Because different cells react differently to EV depletion of FBS or complete serum depletion, for each cell line, the ability of cells to grow in vesicle-free media needs to be determined during the optimization stage of the experiment. If the cells can grow in vesicle-free media, an appropriate number of cells (determined based on the literature and optimized according to the experimental questions) should be grown for the optimized time (e.g., 1 h, 24 h). For cell lines that cannot survive in vesicle-free media, use of media supplemented with FBS alternative may be preferable.
5. Ultracentrifugation (UCF) is the gold standard for EV isolation: it is the most well-established technique so far. UCF uses a series of differential ultracentrifugation (centrifugation at centrifugal speeds of at least 100,000 × g) cycles. It is based on the principle that different sized particles sediment at different centrifugal forces and durations, based on their density and size difference from other particles in the sample of interest [22]. Hence, to isolate EVs of different sizes, conditioned media or plasma can be spun at different speeds, and the EVs collected as pellets. Typically, speeds of 300–500 × g pellet cells, 2000 × g pellet apoptotic bodies, and cell debris, while speeds of 10,000 × g and above (up to 100,000 × g) pellet microvesicles and exosomes. One main drawback of UCF for EV isolation is the cosedimentation of cellular organelles, lipoproteins, and protein aggregates, which are within the size range of EVs. Thus, UCF output tends to have low purity. Also, UCF requires very large starting volume with a resulting low yield. However, to mitigate some of these drawbacks, some

research groups have been known to combine UCF with other methods such as ultrafiltration and iodixanol density gradient.

6. Size exclusion chromatography (SEC) uses a size exclusion column, a porous stationary phase to sort EVs based on their size [22]. EVs, which have small hydrodynamic radii, can thus pass through the pores on the stationary phase to be eluted using buffers such as 1× PBS. The benefits of using SEC for EV isolation include high purity, quick turnaround time, and low cost per sample. However, SEC tends to require a high amount of starting material, to ensure high yield. To further improve its efficacy, SEC can be used in combination with other techniques such as UCF and ultrafiltration.
7. SEC isolated EVs eluted in 1× PBS can be concentrated down using 10kDa ultrafiltration EVs are eluted in fractions 7 through 12.
8. Polyethylene-glycol-based isolation (e.g., ExoQuick) is based on the use of polyethylene glycol (PEG), a nondenaturing water-soluble polymer capable of precipitating particles [23]. This method involves using an aqueous PEG solution (containing up to 2000 PEG monomers) to create a wrap around the EVs, which facilitates them forming aggregates that can be precipitated by low-speed centrifugation at 1500 × g [24, 25]. In recent years, there has been a rise in the number of publications that use PEG-based EV isolation. This may be because this technique requires a small starting volume, gives good yield, has a low cost per sample, and requires no large specialized lab apparatus. However, because PEG can also precipitate proteins, non-exosomal proteins tend to be coprecipitated with PEG EVs, which hampers the purity and specificity of this technique.
9. Protein-affinity-based isolation captures and pulls down EVs based on the presence of proteins such as heat shock proteins and tetraspanins on their surface. Specifically, Vn96, a 2.9 kDa 27-mer peptide that recognizes heat shock proteins on EVs, forms a complex with the EVs and allows for pull-down of these EV-Vn96 complex at 17,000 × g centrifugation [26]. The result of the 17,000 × g centrifugation is the EV-Vn96 pellet, a white translucent pellet, which sticks to the wall of the microfuge tube. The advantages of using Vn96 is the low starting material needed, high yield, and low sample turnaround time. However, the formation of EV-Vn96 pellet as well as potential of coprecipitation of cell-free DNA and non-exosomal proteins are drawbacks that may be important for decisions about downstream analyses [26, 27].
10. 1× Phosphate-buffered saline (PBS) is a buffer suitable for diluting extracellular vesicles prior to NTA analysis. Filtration

of the buffer through a 0.1 μm filter ensures that particles and aggregates in the buffer (whose size may overlap with EV size range) are removed. The filtered buffer can then be used to dilute the EV preparation, giving confidence that particles >100 nm within the prep are EVs from your sample, not from the buffer.

11. Large vesicles are vesicles above 1000 nm, which fall under the category of large microvesicles, apoptotic vesicles, and oncosomes.
12. For each new sample preparation, different dilutions should be tested, until a dilution within the range is identified. Typically, on the NTA computer screen, there should be 20–100 particles on the screen.
13. The optimal concentration range of the NTA is 10^6–10^9 per ml, outside which collected data will not be as reliable.
14. For resuspending EV pellets or diluting EVs, PBS is better than HPLC water due to osmolality differences between the EV lumen and the suspending buffer. Due to this osmolality difference, EVs in water may swell and burst, leading to inaccurate readings: swollen EVs are larger in size, while EVs that burst may result in particles that are smaller in size.
15. If you are interested in the larger microvesicles, this pellet should be collected, washed, and resuspended in PBS for use, or stored at −80 °C
16. After isolation of EVs using Vn96, the EVs are complexed with the Vn96 peptide. To use these EVs for some downstream analyses like NTA, the complex needs to be dislodged. This can be done using overnight (ON) incubation with Proteinase K or 2 M NaCL [26, 27]. After incubation, the mix can then be centrifuged at 17,000 × g for 15 min, with no visible pellet after the spin. The centrifuged supernatant can then be diluted for NTA. However, this dislodging may not be needed for downstream analyses that will not be impacted by the complex such as western blot and RNA isolation.
17. Fit the springs in the order of one screw on the left and another opposite screw on the right
18. For each sample, it is important to report the specific focus settings used, as this will serve as a guide when running other replicates for the same sample. For replicates of the same sample, it is important to maintain similar focus settings, as different focus settings change the overall concentration and skew the size average of the EVs.
19. When loading water or samples into 1 ml syringe, ensure you avoid the introduction of bubbles, which, if injected into the inlet tubing, could create additional pressure in the tubings.

This pressure, if allowed to build up, could slow down the free flow of liquid into and out of the NTA tubing system. Additionally, while it is possible to flush the tubings with air, that is not recommended for new users of the NS300 and should only be used when necessary. In cases when you must clear residual fluids in the tubings using air in the syringe, this should be done very carefully, as this could potentially rupture the gasket component if done with too much physical force applied to the syringe plunger.

20. If the setup of the tubings is accurate, water should be flowing out into the waste bottle in a dropwise fashion. If this does not happen, stop the camera, open the NTA door, dismantle the LVFC and the module and reassemble, and repeat the wash until water normally flows out of the outlet tubing into the waste bottle.
21. During NTA experiments, it is important to measure the particles in the suspension buffer, such as 0.1 μm filtered 1× PBS. This helps determine the level of background provided by the buffer.
22. Quintuplicate captures to ensure acceptable reproducibility, and for statistical analyses.
23. A slower flow rate (e.g., 25 is better than 100) is better for accuracy in estimating particle size.
24. The "advance sample capture" is used only for manual sample injection, without the pump.
25. Ensure that the syringe is not loaded more than half full, as this could cause the shaft to flex.
26. To determine the optimum detection threshold, reduce the threshold to the lowest and slowly increase. On the bottom right panel on the screen, there will be a count of red crosses (tracked) and blue crosses (not to be tracked). A red cross count of 10–100 denotes good conditions, ideally 20–50 (by eye, there should be <10 red crosses), and a blue cross count of <5.
27. Detection threshold is an important parameter, which MUST be kept uniform for all biological and technical replicates of an experiment. Samples run at different detection thresholds are not actually comparable, because detection threshold determines the number of particles tracked, which could in turn skew the averages (mean and mode diameter).
28. Screen gain is a number to represent how much light from a light source is reflected from the surface. Thus, a screen gain of 1 means the same amount of light projected from the source is reflected. Adjusting screen gain increases the contrast between the actual image and the feedback. It is better not to adjust screen gain during sample acquisition, as it does not improve data quality significantly.

29. The processing step in an NTA run takes at least 5 min and should be allowed to go on uninterrupted. Specifically, detection threshold should not be changed during the processing. If any errors were made, the processing should be allowed to run, and then another round of processing should be done with the adjusted parameters.
30. A high noise warning during processing would mean that there would be a "data not reliable because of high noise" warning on your data output. To solve this, reprocess with a few different detection thresholds (in line with the detection threshold rules above). If this does not fix the data, the only solution would be to make a fresh dilution and repeat the run, ensuring that focus and camera level are set to ensure all particles are seen clearly on the screen.
31. It is important to check your data immediately after the run, so you can decide if the run needs to be repeated, or if you need to reprocess with a different processing parameter (detection threshold and screen gain).
32. The video files of the NTA take up a large space on the computer drive. Thus, it would be better to save the files to an external hard drive, or to a cloud (if possible). This will free up space for the Nanosight. The space in the drive for saving files is important, as the NTA run will stop at any point if it runs out of space. It could be very disheartening to run a precious sample, and lose the data (without even processing), because of space. Hence, data storage space should be well managed.
33. Check for noise warnings, if there is a note at the end of your run saying, "data not reliable due to noise," large Newtonian rings around the particles or dim particles. If there is a noise warning, reprocess first with different detection thresholds. If the change in detection threshold does not fix the noise issue, then make a fresh dilution of the sample and repeat the run. If the warning is about vibration, check that instruments within the workspace that could increase the vibration of the workbench are silenced or turned off, and then repeat the run, with leftover sample, or with a fresh dilution. If you entered an incorrect dilution factor, you could manually recalculate, or repeat the run. For Newtonian rings around your particles, repeat the run, and adjust the focus settings until the particles are clear, single images without Newtonian rings. For dim particles, adjust the camera level and then the focus to ensure that even small particles can be seen clearer.
34. While there is no test to check for sample carryover, it is easy to determine this. During the prerun prep, there should be at least 20–50 particles on the screen at any given time: <1% carryover means that after enough washes, there should be

about 0–2 particles on the screen. Overall, you MUST flush the LVFC with HPLC water between samples to ensure adequate cleaning of one sample before loading another.

35. If the Nanosight is turned off without the temperature control being turned off, the next time the instrument is turned on, there is a possibility that the NTA software will be unable to detect the temperature module. This means that the instrument cannot be used. To fix this problem, dismantle the LVFC and laser module setup, turnoff the machine at the sockets, wait a few minutes, and then turn on the NTA instrument and the computer software.
36. It is also possible for the NTA software to be unable to detect the CMOS camera and other modules. This could happen when the room temperature is above 25 °C. To mitigate this problem, ensure that the NTA workspace is always kept cool. However, it is possible that for air conditioners that make sounds, the sound of the air conditioner (especially when placed near the workbench) may increase the vibrations. If this is the case, precool the room for at least an hour before the start of the run, turn off the air conditioner during the run (to minimize the noise and possible vibrations), and turn on the air conditioner while flushing the LVFC.
37. The LVFC should not be placed directly on the laser module surface. Instead, the LVFC should be placed on a lint free lens cloth on the laser module surface. This will prevent these components from scratching against each other when the Nanosight is not in use. The LVFC and the laser module are key components to the function of the Nanosight and should be handled carefully.
38. Once the NS300 is turned on, the power indicator should have a green light, which goes off when the NS300 is turned off. The module circle lights only when the laser module is inserted into the "powered on" NS300. The Comms light flashes briefly (when the instrument is turned on) and goes off shortly after. The Temp light comes on suddenly when the temperature of the room is too hot. This should not be ignored once this comes on. For the filter lights, usually, only one filter light is on at a time.

Acknowledgment

We acknowledge the team at Malvern (UK) for training on setup of the NTA and troubleshooting at different stages of the use of the NS300.

References

1. Chargaff E, West R (1946) The biological significance of the thromboplastic protein of blood. J Biol Chem 166:189–197
2. Raposo G, Stoorvogel W et al (2013) Extracellular vesicles: exosomes, microvesicles, and friends. J Cell Biol 200:373–383
3. Colombo M, Raposo G, Théry C (2014) Biogenesis, secretion, and intercellular interactions of exosomes and other extracellular vesicles. Annu Rev Cell Dev Biol 30:255–289
4. Shojaee S, Caeser R, Buchner M et al (2015) Erk negative feedback control enables pre-B cell transformation and represents a therapeutic target in acute lymphoblastic leukemia. Cancer Cell 28:114–128
5. Al-Nedawi K, Meehan B, Micallef J et al (2008) Intercellular transfer of the oncogenic receptor EGFRvIII by microvesicles derived from tumour cells. Nat Cell Biol 10:619–624
6. Yáñez-Mó M, Siljander PRM, Andreu Z et al (2015) Biological properties of extracellular vesicles and their physiological functions. J Extracell Vesicles 4:1–60
7. Balaj L, Lessard R, Dai L et al (2011) Tumour microvesicles contain retrotransposon elements and amplified oncogene sequences. Nat Commun 2:1–9
8. García-Romero N, Carrión-Navarro J, Esteban-Rubio S et al (2017) DNA sequences within glioma-derived extracellular vesicles can cross the intact blood-brain barrier and be detected in peripheral blood of patients. Oncotarget 8:1416–1428
9. Agarwal K, Saji M, Lazaroff SM et al (2015) Analysis of exosome release as a cellular response to MAPK pathway inhibition. Langmuir 31(19):5440–5448
10. Tamura T, Yoshioka Y, Sakamoto S et al (2021) Extracellular vesicles as a promising biomarker resource in liquid biopsy for cancer. Extracell Vesicles Circ Nucleic Acids 2:148–174
11. Zhou E, Li Y, Wu F et al (2021) Circulating extracellular vesicles are effective biomarkers for predicting response to cancer therapy. EBioMedicine 67:103365
12. Longjohn MN, Hudson J-ABJ, Smith NC et al (2020) Deciphering the messages carried by extracellular vesicles in hematological malignancies. Blood Rev 46:100734
13. Théry C, Aled C, Sebastian A et al (2006) Isolation and characterization of exosomes from cell culture supernatants and biological fluids. Curr Protoc Cell Biol 3(22):1–29
14. Margolis L, Sadovsky Y (2019) The biology of extracellular vesicles: the known unknowns. PLoS Biol 17:e3000363
15. Carr B, Wright M (2008) Nanoparticle tracking analysis: a review of applications and usage in the analysis of exosomes and microvesicles
16. Barba AA, Bochicchio S, Dalmoro A et al (2019) Polymeric and lipid-based systems for controlled drug release: an engineering point of view. Nanomater Drug Deliv Ther :267–304
17. Lehrich BM, Liang Y, Fiandaca MS (2021) Foetal bovine serum influence on in vitro extracellular vesicle analyses. J Extracell Vesicles 10: e12061
18. Guerreiro EM, Vestad B, Steffensen LA et al (2018) Efficient extracellular vesicle isolation by combining cell media modifications, ultrafiltration, and size-exclusion chromatography. PLoS One 13:e0204276
19. Lee YXF, Johansson H, Wood MJA et al (2019) Considerations and implications in the purification of extracellular vesicles – a cautionary tale. Front Neurosci 13:1067
20. Wang F, Cerione RA, Antonyak MA (2021) Isolation and characterization of extracellular vesicles produced by cell lines. STAR Protoc 2:100295
21. Auber M, Fröhlich D, Drechsel O et al (2019) Serum-free media supplements carry miRNAs that co-purify with extracellular vesicles. J Extracell Vesicles 8:1656042
22. Li P, Kaslan M, Lee SH et al (2017) Progress in exosome isolation techniques. Theranostics 7: 789
23. Ingham KC (1984) [20] Protein precipitation with polyethylene glycol. Methods Enzymol 104:351–356
24. Konoshenko MY, Lekchnov EA, Vlassov AV et al (2018) Isolation of extracellular vesicles: general methodologies and latest trends. Biomed Res Int 2018:8545347
25. Weng Y, Sui Z, Shan Y et al (2016) Effective isolation of exosomes with polyethylene glycol from cell culture supernatant for in-depth proteome profiling. Analyst 141:4640–4646
26. Ghosh A, Davey M, Chute IC et al (2014) Rapid isolation of extracellular vesicles from cell culture and biological fluids using a synthetic peptide with specific affinity for heat shock proteins. PLoS One 9:e110443
27. Roy JW, Taylor CA, Beauregard AP et al (2021) A multiparametric extraction method for Vn96-isolated plasma extracellular vesicles and cell-free DNA that enables multi-omic profiling. Sci Rep 11:1–15

INDEX

Sherri L Christian (ed.), *Cancer Cell Biology: Methods and Protocols*,
Methods in Molecular Biology, vol. 2508, https://doi.org/10.1007/978-1-0716-2376-3,

U

V

W

X

Z

MIX
Papier aus verantwortungsvollen Quellen
Paper from responsible sources
FSC® C105338

If you have any concerns about our products,
you can contact us on
ProductSafety@springernature.com

In case Publisher is established outside the EU,
the EU authorized representative is:
Springer Nature Customer Service Center GmbH
Europaplatz 3, 69115 Heidelberg, Germany

Printed by Libri Plureos GmbH
in Hamburg, Germany